METHODS IN MOLECULAR BIOLOGY

For further volumes:
http://www.springer.com/series/7651

For over 35 years, biological scientists have come to rely on the research protocols and methodologies in the critically acclaimed *Methods in Molecular Biology* series. The series was the first to introduce the step-by-step protocols approach that has become the standard in all biomedical protocol publishing. Each protocol is provided in readily-reproducible step-by-step fashion, opening with an introductory overview, a list of the materials and reagents needed to complete the experiment, and followed by a detailed procedure that is supported with a helpful notes section offering tips and tricks of the trade as well as troubleshooting advice. These hallmark features were introduced by series editor Dr. John Walker and constitute the key ingredient in each and every volume of the *Methods in Molecular Biology* series. Tested and trusted, comprehensive and reliable, all protocols from the series are indexed in PubMed.

Deadenylation

Methods and Protocols

Edited by

Eugene Valkov

RNA Biology Laboratory, Center for Cancer Research, National Cancer Institute, Frederick, MD, USA

Aaron C. Goldstrohm

Department of Biochemistry, Molecular Biology, and Biophysics, University of Minnesota, Minneapolis, MN, USA

Humana Press

Editors
Eugene Valkov
RNA Biology Laboratory
Center for Cancer Research
National Cancer Institute
Frederick, MD, USA

Aaron C. Goldstrohm
Department of Biochemistry
Molecular Biology, and Biophysics
University of Minnesota
Minneapolis, MN, USA

ISSN 1064-3745 ISSN 1940-6029 (electronic)
Methods in Molecular Biology
ISBN 978-1-0716-3480-6 ISBN 978-1-0716-3481-3 (eBook)
https://doi.org/10.1007/978-1-0716-3481-3

This Humana imprint is published by the registered company Springer Science+Business Media, LLC, part of Springer Nature.
The registered company address is: 1 New York Plaza, New York, NY 10004, U.S.A.

Preface

A stretch of polyadenosine, known as the poly(A) tail, at the 3′ of almost every mRNA is a critical determinant of its stability and expression. One essential process in mRNA regulation is deadenylation, the shortening of the poly(A) tail by enzymes called deadenylases, which can trigger mRNA degradation or translational repression. Over the years, there has been a rapidly accelerating effort to improve our understanding of the complex regulatory mechanisms governing deadenylation in a wide range of organisms and biological contexts. Most recently, the emergence of mRNA-based vaccines and therapeutics has only added to the impetus and urgency of this research focus. New methods, particularly the rapidly advancing improvements in long-read sequencing techniques and sophisticated computational tools for data analysis, are opening a new vista into the complex world of poly(A) tail metabolism on a scale ranging from a single molecule to a transcriptome. The biochemical methods and cellular analyses have also advanced the identification and characterization of critical factors, including deadenylases and their RNA-binding protein interactors. *Methods in Molecular Biology: Deadenylation*—a collection of the latest protocols by the leading experts in the field—was prompted by these developments and advances to facilitate and pave the way for future investigations of the complex regulatory networks that control mRNA stability and expression.

We thank all the authors for their contribution to this volume. In addition, we thank the series editor, Dr. John Walker, for his guidance and advice. Last but not least, we wish to thank the members of our laboratories and our families for their support throughout the preparation of this volume.

Frederick, MD, USA — *Eugene Valkov*
Minneapolis, MN, USA — *Aaron C. Goldstrohm*

Contents

Contributors

BLESSING AIRHIHEN • *School of Pharmacy, University of Nottingham, Nottingham, UK; School of Science and Technology, Nottingham Trent University, Nottingham, UK*

RENÉ M. ARVOLA • *Department of Molecular Genetics, The Ohio State University, Columbus, OH, USA*

MARINE BAPTISSART • *Reproductive and Developmental Biology Laboratory, National Institute of Environmental Health Sciences, National Institutes of Health, Durham, NC, USA*

CEDRIC BELAIR • *Laboratory of Genetics and Genomics, National Institute on Aging, Intramural Research Program, National Institutes of Health, Baltimore, MD, USA*

PERRY J. BLACKSHEAR • *Signal Transduction Laboratory, National Institute of Environmental Health Sciences, Research Triangle Park, NC, USA; Departments of Medicine and Biochemistry, Duke University Medical Center, Durham, NC, USA*

CHUNG-TE CHANG • *Institute of Biochemistry and Molecular Biology, National Yang Ming Chiao Tung University, Taipei, Taiwan*

QING CHEN • *Epigenetics and Stem Cell Biology Laboratory, National Institute of Environmental Health Sciences, Research Triangle Park, NC, USA*

JEFF COLLER • *Department of Molecular Biology and Genetics, School of Medicine, Johns Hopkins University, Baltimore, MD, USA; Department of Biology, Johns Hopkins University, Baltimore, MD, USA; Johns Hopkins Institute for NanoBioTechnology, Johns Hopkins University, Baltimore, MD, USA*

SHOWKAT A. DAR • *Laboratory of Genetics and Genomics, National Institute on Aging, Intramural Research Program, National Institutes of Health, Baltimore, MD, USA*

ANDRZEJ DZIEMBOWSKI • *International Institute of Molecular and Cell Biology, Warsaw, Poland; Faculty of Biology, University of Warsaw, Warsaw, Poland*

BENEDICT FABIA • *School of Computing, Korea Advanced Institute of Science and Technology (KAIST), Daejeon, Republic of Korea*

JULIE D. FORMAN-KAY • *Program in Molecular Medicine, Hospital for Sick Children, Toronto, ON, Canada; Department of Biochemistry, University of Toronto, Toronto, ON, Canada*

DOMINIQUE GAGLIARDI • *Institut de biologie moléculaire des plantes, CNRS, Université de Strasbourg, Strasbourg, France*

PIETRO GIRAUDO • *Institut de biologie moléculaire des plantes, CNRS, Université de Strasbourg, Strasbourg, France*

AARON C. GOLDSTROHM • *Department of Biochemistry, Molecular Biology, and Biophysics, University of Minnesota, Minneapolis, MN, USA*

ANKIT GUPTA • *Reproductive and Developmental Biology Laboratory, National Institute of Environmental Health Sciences, National Institutes of Health, Durham, NC, USA*

ROBERT W. HARKNESS • *Program in Molecular Medicine, Hospital for Sick Children, Toronto, ON, Canada; Department of Molecular Genetics, University of Toronto, Toronto, ON, Canada*

BREANNE M. HATFIELD • *Department of Chemistry, University of North Carolina, Chapel Hill, NC, USA*

STEPHANIE N. HICKS • *Signal Transduction Laboratory, National Institute of Environmental Health Sciences, Research Triangle Park, NC, USA*
GUANG HU • *Epigenetics and Stem Cell Biology Laboratory, National Institute of Environmental Health Sciences, Research Triangle Park, NC, USA*
JAMES R. IBEN • *Molecular Genetics Core, Eunice Kennedy Shriver National Institute of Child Health and Human Development (NICHD), National Institutes of Health (NIH), Bethesda, MD, USA*
ROSE IRWIN • *Program in Molecular Medicine, Hospital for Sick Children, Toronto, ON, Canada; Department of Biochemistry, University of Toronto, Toronto, ON, Canada*
MINJU KIM • *Department of Bio and Brain Engineering, Korea Advanced Institute of Science and Technology (KAIST), Daejeon, Republic of Korea*
PAWEL S. KRAWCZYK • *International Institute of Molecular and Cell Biology, Warsaw, Poland*
BRAD LACKFORD • *Epigenetics and Stem Cell Biology Laboratory, National Institute of Environmental Health Sciences, Research Triangle Park, NC, USA*
WI S. LAI • *Signal Transduction Laboratory, National Institute of Environmental Health Sciences, Research Triangle Park, NC, USA*
YOUNG-SUK LEE • *Department of Bio and Brain Engineering, Korea Advanced Institute of Science and Technology (KAIST), Daejeon, Republic of Korea*
YEVGEN LEVDANSKY • *RNA Biology Laboratory, Center for Cancer Research, National Cancer Institute, Frederick, MD, USA*
TIANWEI LI • *Molecular Genetics Core, Eunice Kennedy Shriver National Institute of Child Health and Human Development (NICHD), National Institutes of Health (NIH), Bethesda, MD, USA*
JONGMIN LIM • *Department of Bio and Brain Engineering, Korea Advanced Institute of Science and Technology (KAIST), Daejeon, Republic of Korea*
YUSHENG LIU • *College of Life Science, Northeast Forestry University, Harbin, China*
FALONG LU • *State Key Laboratory of Molecular Developmental Biology, Institute of Genetics and Developmental Biology, Chinese Academy of Sciences, Beijing, China; University of Chinese Academy of Sciences, Beijing, China*
SULOCHAN MALLA • *Laboratory of Genetics and Genomics, National Institute on Aging, Intramural Research Program, National Institutes of Health, Baltimore, MD, USA*
MANOLIS MARAGKAKIS • *Laboratory of Genetics and Genomics, National Institute on Aging, Intramural Research Program, National Institutes of Health, Baltimore, MD, USA*
RICHARD J. MARAIA • *Section on Molecular and Cell Biology, NICHD, NIH, Bethesda, MD, USA*
MARYATI MARYATI • *School of Pharmacy, University of Nottingham, Nottingham, UK; Faculty of Pharmacy, Universitas Muhammadiyah Surakarta, Surakarta, Indonesia*
SANDY MATTIJSSEN • *Section on Molecular and Cell Biology, NICHD, NIH, Bethesda, MD, USA*
FABIENNE MAUXION • *Institut de Génétique et de Biologie Moléculaire et Cellulaire (IGBMC), Centre National de Recherche Scientifique (CNRS) UMR 7104 – Institut National de Santé et de Recherche Médicale (Inserm) U1258 – Université de Strasbourg, Illkirch, France*
MARCOS MORGAN • *Reproductive and Developmental Biology Laboratory, National Institute of Environmental Health Sciences, National Institutes of Health, Durham, NC, USA*
SEWERYN MROCZEK • *International Institute of Molecular and Cell Biology, Warsaw, Poland; Faculty of Biology, University of Warsaw, Warsaw, Poland*

BRIAN N. PAPAS • *Integrative Bioinformatics, Biostatistics and Computational Biology Branch, National Institute of Environmental Health Sciences, National Institutes of Health, Durham, NC, USA*
LORENZO PAVANELLO • *School of Pharmacy, University of Nottingham, Nottingham, UK*
FILIP PEKOVIC • *Institute of Biochemistry and Biotechnology and Charles Tanford Protein Center, Martin Luther University Halle-Wittenberg, Halle, Germany; RNA Biology Laboratory, Center for Cancer Research, National Cancer Institute, Frederick, MD, USA*
JACKSON PETER • *Institut de biologie moléculaire des plantes, CNRS, Université de Strasbourg, Strasbourg, France*
DAVID PFLIEGER • *Institut de biologie moléculaire des plantes, CNRS, Université de Strasbourg, Strasbourg, France*
BERTRAND SÉRAPHIN • *Institut de Génétique et de Biologie Moléculaire et Cellulaire (IGBMC), Centre National de Recherche Scientifique (CNRS) UMR 7104 – Institut National de Santé et de Recherche Médicale (Inserm) U1258 – Université de Strasbourg, Illkirch, France*
QUENTIN SIMONNOT • *Institut de biologie moléculaire des plantes, CNRS, Université de Strasbourg, Strasbourg, France*
AUSTIN SPONAUGLE • *Department of Molecular Biology and Genetics, School of Medicine, Johns Hopkins University, Baltimore, MD, USA*
BAHAREH TORKZABAN • *Department of Molecular Biology and Genetics, School of Medicine, Johns Hopkins University, Baltimore, MD, USA*
AGNIESZKA TUDEK • *Institute of Biochemistry and Biophysics, Warsaw, Poland*
EUGENE VALKOV • *RNA Biology Laboratory, Center for Cancer Research, National Cancer Institute, Frederick, MD, USA*
ELMAR WAHLE • *Institute of Biochemistry and Biotechnology and Charles Tanford Protein Center, Martin Luther University Halle-Wittenberg, Halle, Germany*
JIAQIANG WANG • *College of Life Science, Northeast Agricultural University, Harbin, China*
XIUKUN WANG • *Epigenetics and Stem Cell Biology Laboratory, National Institute of Environmental Health Sciences, Research Triangle Park, NC, USA*
KEVIN M. WEEKS • *Department of Chemistry, University of North Carolina, Chapel Hill, NC, USA*
CHASE A. WEIDMANN • *Department of Biological Chemistry, Center for RNA Biomedicine, University of Michigan Medical School, Ann Arbor, MI, USA*
GERLOF SEBASTIAAN WINKLER • *School of Pharmacy, University of Nottingham, Nottingham, UK*
HÉLÈNE ZUBER • *Institut de biologie moléculaire des plantes, CNRS, Université de Strasbourg, Strasbourg, France*

Chapter 1

Reconstitution of Human CCR4-NOT Complex from Purified Proteins and an Assay of Its Deadenylation Activity

Yevgen Levdansky and Eugene Valkov

Abstract

We describe protocols to produce and reconstitute an active human CCR4-NOT complex. Individual recombinant subunits are expressed in *E. coli* or baculovirus-infected insect cells, purified using column chromatography, and reconstituted into a stable complex containing all eight core subunits. In addition, we describe the biochemical assay of deadenylation using the reconstituted complex.

Key words CCR4-NOT, Reconstitution, Deadenylation, mRNA decay, poly(A) tail

1 Introduction

1.1 CCR4-NOT Deadenylase Complex

In a eukaryotic cell, almost every mature messenger RNA (mRNA) has a polyadenosine stretch at its 3′ end, commonly called a "poly (A) tail." This molecular feature is necessary for translation and to protect mRNA from premature degradation [1]. The poly(A) tail's status determines mRNA stability and translatability. The enzymatic shortening regulates the length of the poly(A) tail in a process called "deadenylation," the first rate-limiting step in cytoplasmic mRNA decay [2].

CCR4-NOT complex is the principal factor responsible for the bulk of the deadenylation activity in a eukaryotic cell [3, 4]. The core of the human CCR4-NOT complex consists of eight subunits arranged into four structurally and functionally defined modules. NOT1 is a ~2400 amino acid-long scaffold subunit of the complex, which consists of α-helical repeat domains connected by flexible linkers. NOT6 (CCR4) and NOT7 (CAF1) deadenylase subunits organize into a nuclease module. NOT7 interacts with the central MIF4G domain of NOT1, and apart from its catalytic function, NOT7 serves as a bridge between NOT1 and CCR4 deadenylase, which, in turn, connects the latter to the rest of the complex [5, 6]. Toward the C-terminus of NOT1, neighboring to the

Eugene Valkov and Aaron C. Goldstrohm (eds.), *Deadenylation: Methods and Protocols*, Methods in Molecular Biology, vol. 2723, https://doi.org/10.1007/978-1-0716-3481-3_1,

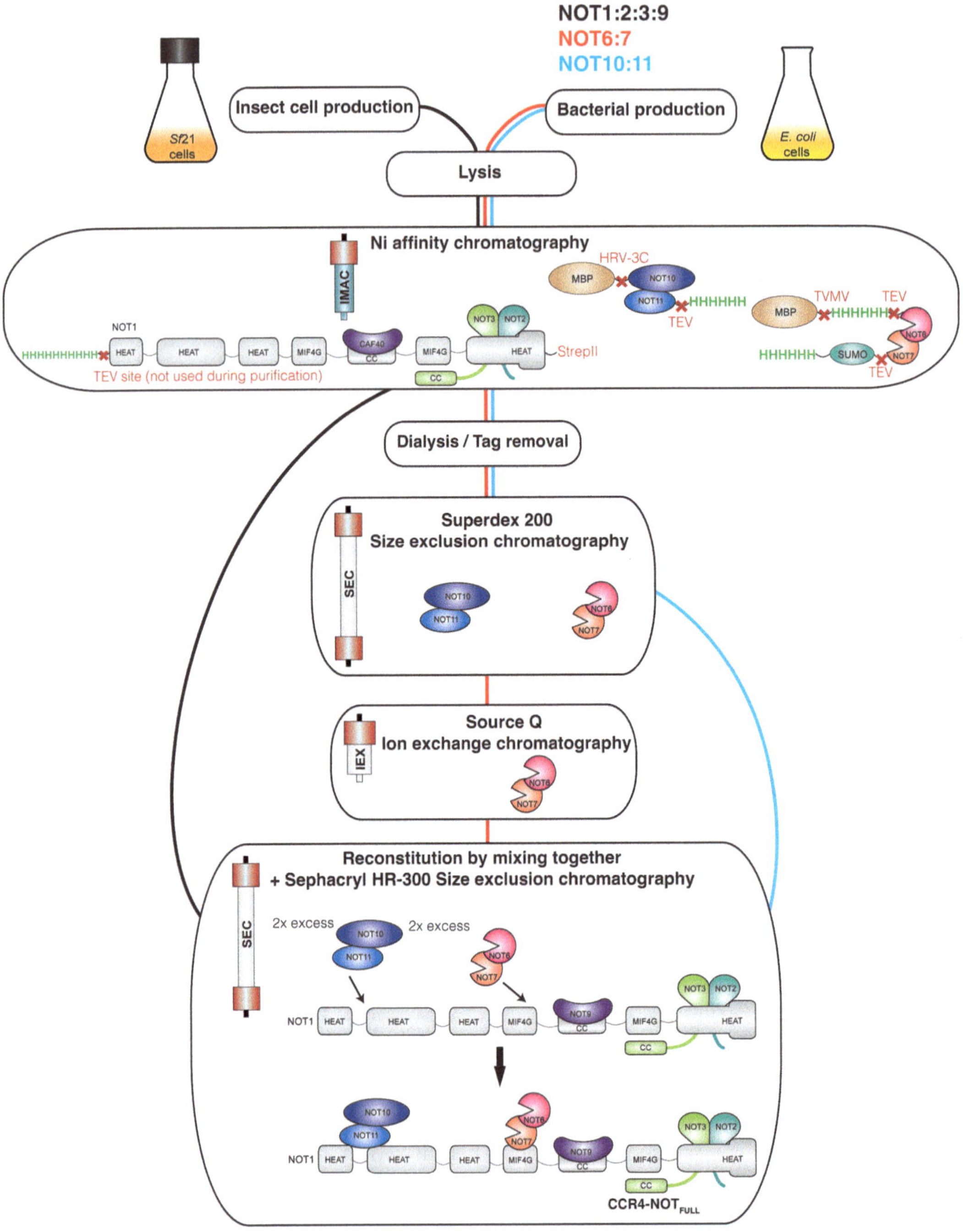

Fig. 1 Flowchart of the modular reconstitution of the human CCR4-NOT complex and purification of its components. The human CCR4-NOT complex was reconstituted from three subcomplexes: NOT1:2:3:9 backbone (produced in Sf21 insect cells), NOT6:7 catalytic module, and NOT10:11 module (both produced in *Escherichia coli*). NOT6:7 and NOT10:11 undergo several stages of purification (more detailed descriptions of purifications are in Figs. 2 and 3). After isolation from lysed bacterial cells using nickel-charged immobilized metal affinity chromatography (IMAC) column, tags are cleaved off during the dialysis. Both NOT6:7 and NOT10:11 heterodimers were separated from the cleaved-off tags by size-exclusion chromatography (SEC). To ensure the equimolar ratio of the subunits in the NOT6:7 subcomplex, it was subjected to a round of ion exchange (IEX) chromatography. To reconstitute the eight-subunit human CCR4-NOT complex, we added

central MIF4G domain, the coiled-coil domain of NOT1 interacts with the NOT9 (CAF40) subunit, and together, they form the NOT9 (CAF40) module. This module is necessary to stabilize NOT1, mediating protein–protein interactions with CCR4-NOT interactors and binding of nucleic acids [7–9]. The C-terminal part of NOT1 co-folds with NOT2 and NOT3 subunits into the NOT2:3 module [10]. This module is crucial for NOT1 stability and is another major protein–protein interaction site for RNA-binding proteins [11–13]. The N-terminal HEAT repeats of NOT1 and NOT10 subunits, together with the NOT11 subunit, form the NOT10:11 module, which interacts with other factors and supports the deadenylation activity of the whole complex [14, 15].

To study deadenylation by the human CCR4-NOT, obtaining milligram-level quantities of the highly purified complex is necessary. Purification of the CCR4-NOT complex from native sources such as human cells is hampered by the heterogeneity of the complex, multiple co-purifying proteins interacting with the CCR4-NOT, and overall low yields [16]. A recombinant approach enables us to overcome these challenges and establish a platform for studying the structural and biochemical properties of CCR4-NOT-mediated deadenylation (*see* Fig. 1).

1.2 In Vitro Deadenylation Assay

The in vitro deadenylation assay serves as the principal biochemical test of the enzymatic activity of purified deadenylase complexes such as the CCR4-NOT complex. The principles are described in Webster et al. [17].

Briefly, biochemically active deadenylases remove adenosines from the 3′ ends of RNA, which can be synthetic or prepared by in vitro transcription. In this assay, fluorescein-labeled synthetic RNA is used. A set of deadenylation reactions are prepared and followed in a time course. TBE-urea PAGE resolves end-point products of each reaction. Urea denatures secondary structures of RNA, and a high percentage of acrylamide enables resolution with single ribonucleotide precision. The set of time points should show a regular degradation pattern in the form of a “ladder.” To confirm that the result is enzyme-specific, a negative control is essential. A high-resolution scanning imager with fluorescence filters such as Typhoon is used for the readout of the assay.

This chapter describes the protocols for the production, purification, and reconstitution of the subunits that assemble the human

Fig. 1 (continued) purified NOT6:NOT7 and NOT10:11 modules in twofold excess each to the NOT1:2:3:9 isolated from insect cells using nickel-charged IMAC column. The reconstituted CCR4-NOT complex was separated from the excess heterodimers using SEC. (The figure was adapted from the manuscript by Raisch, Chang, Levdansky, and coworkers [18] under permission provided by a Creative Commons Attribution 4.0 International License (http:/creativecommons.org/licenses/by/4.0/))

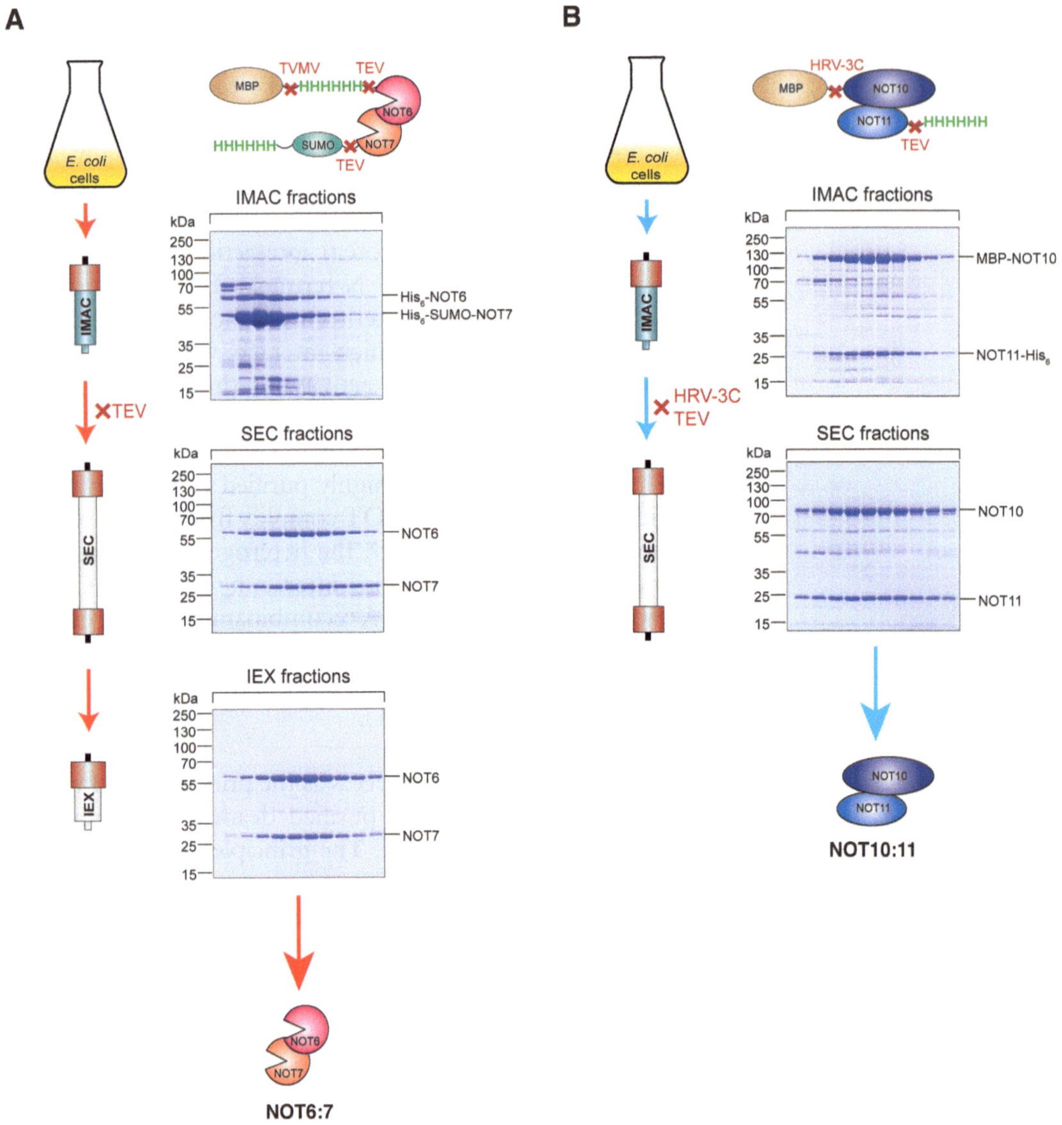

Fig. 2 Production and purification of NOT6:7 and NOT10:11 modules. (**a**) Coomassie-stained SDS-PAGE gels showing progressive improvement of purity and stoichiometry of the purified NOT6:7 subcomplex. NOT6, apart from the TEV-cleavable N-terminal His6 tag, has an upstream maltose-binding protein (MBP) tag. MBP tag is cleaved in vivo by the TVMV protease. NOT7 has an N-terminal His6 tag, fused in-frame with the TEV-cleavable SUMO tag. His6 tags on both proteins were utilized for the nickel affinity chromatography step. After the cleavage of tags by TEV during dialysis, NOT6:7 is separated from cleaved tags and the excess of NOT7 by size-exclusion chromatography (SEC). To ensure the equimolar ratio of NOT6 and NOT7, the heterodimer is subjected to a round of ion exchange (IEX) chromatography. (**b**) Coomassie-stained SDS-PAGE gels showing progressive improvement of purity and stoichiometry of the purified NOT10:11 subcomplex. NOT10 has an N-terminal HRV-3C-cleavable MBP tag, and NOT11 has a C-terminal TEV-cleavable His6 tag; the His6 tag was utilized for the nickel affinity chromatography step. After the cleavage of tags by TEV and HRV-3C proteases during dialysis, NOT10:11 is separated by SEC. (The figure was adapted from the manuscript by Raisch, Chang, Levdansky, and coworkers [18] under permission provided by a Creative Commons Attribution 4.0 International License (http:/creativecommons.org/ licenses/by/4.0/))

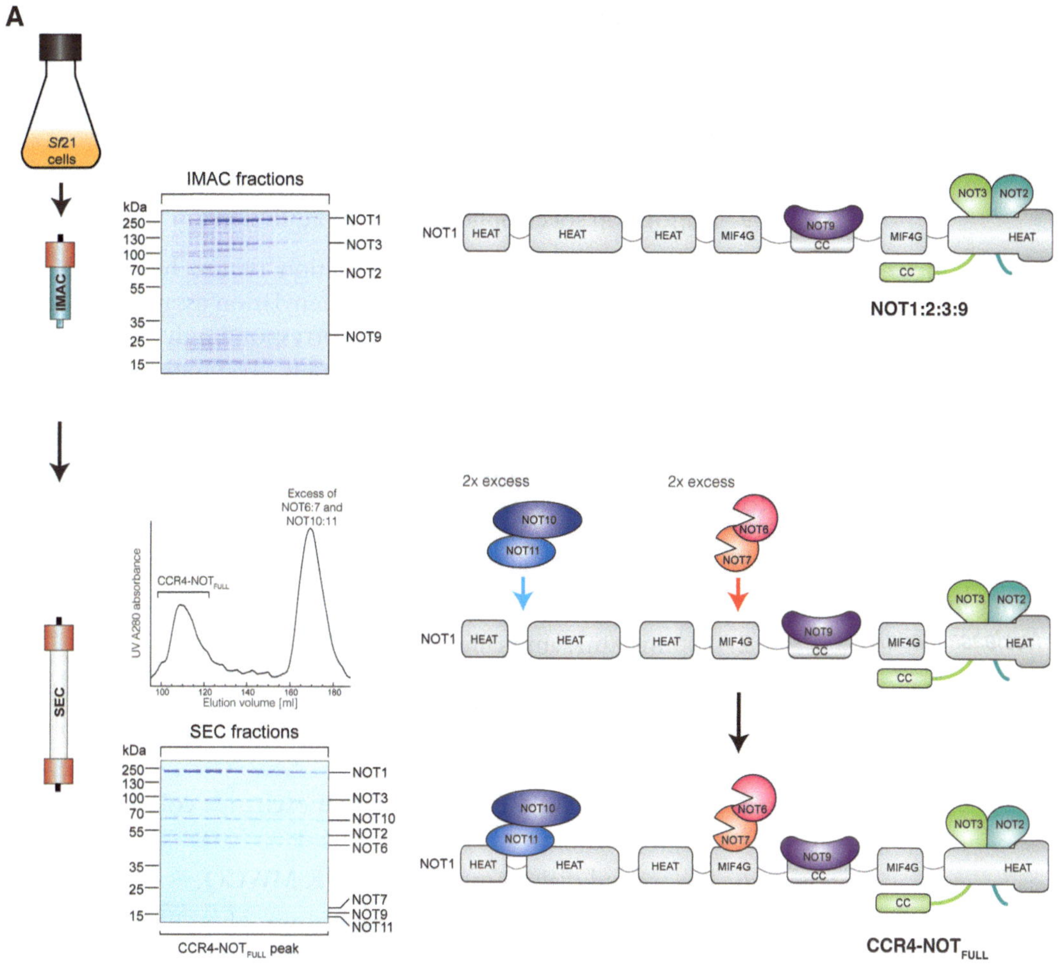

Fig. 3 Production and purification of NOT1:2:3:9 and reconstitution of the full human CCR4-NOT complex. (**a**) Coomassie-stained SDS-PAGE gel of the NOT1:2:3:9 subcomplex purified by nickel affinity chromatography. (**b**) Coomassie-stained SDS-PAGE gel of the reconstituted full human CCR4-NOT complex after the size-exclusion chromatography step. (The figure was adapted from the manuscript by Raisch, Chang, Levdansky, and coworkers [18] under permission provided by a Creative Commons Attribution 4.0 International License (http:/creativecommons.org/ licenses/by/4.0/))

CCR4-NOT complex and a biochemical assay to monitor its poly (A) degradation activity at single-nucleotide resolution.

2 Materials

1. LB Lennox growth medium.
2. *Escherichia coli* BL21 Star (DE3) cell strain.
3. Peristaltic pump.
4. ÄKTA pure chromatography system (Cytiva) or equivalent.

5. 5 mL immobilized metal affinity chromatography (IMAC) columns charged with Ni^{2+} (*see* **Note 1**).
6. Amicon 15K centrifugal concentrators 10K MWCO.
7. Sonifier.
8. EmulsiFlex-C5 (Avestin).
9. NanoDrop spectrophotometer or equivalent.

Materials specific for the production of the human CCR4-NOT complex and for setting up deadenylation assays are listed in Subheadings 2.1, 2.2, 2.3, and 2.4, correspondingly. All chemicals should be at least molecular biology grade. To prepare the buffers, use Milli-Q grade water (18 MΩ-cm at 25 °C). All the buffers must be filtered through a 0.22 μm nitrocellulose membrane using a vacuum source or syringe-driven filters. Buffers used for chromatography must be degassed.

2.1 Production and Purification of the Nuclease Module

1. Plasmids pMCSG19c-NOT6 (AmpR) and pET28b-SUMO-NOT7 (KanR).
2. 100 mg/mL ampicillin.
3. 50 mg/mL kanamycin.
4. HiLoad 26/600 Superdex 200 pg size-exclusion column (Cytiva).
5. SOURCE 15Q 4.6/100 PE anion exchange column (Cytiva); *see* **Note 2**.
6. Pre-wetted dialysis membrane 3.5K MWCO.
7. Ni-A_6:7 Buffer: 50 mM potassium phosphate, 300 mM NaCl, and 25 mM imidazole (pH 7.5) [500 mL total].
8. Ni-B_6:7 Buffer: 50 mM potassium phosphate, 300 mM NaCl, and 500 mM imidazole (pH 7.5) [100 mL total].
9. Dialysis_6:7 Buffer: 50 mM HEPES/NaOH, 300 mM NaCl, and 2 mM DTT (pH 7.5) [2 L total].
10. GF_6:7 Buffer: 10 mM HEPES/NaOH, 300 mM NaCl, and 2 mM DTT (pH 7.5) [1 L total].
11. Dilution_6:7 Buffer: 50 mM HEPES/NaOH and 2 mM DTT (pH 7.5) [50 mL total].
12. Q-A_6:7 Buffer: 10 mM HEPES/NaOH, 75 mM NaCl, and 2 mM DTT (pH 7.5) [250 mL total].
13. Q-B_6:7 Buffer: 10 mM HEPES/NaOH, 1000 mM NaCl, and 2 mM DTT (pH 7.5) [100 mL total].

2.2 Production and Purification of the NOT10:11 Module

1. Plasmids pnYC-NpM-NOT10 (CamR) and pnEA-NOT11-CvH (AmpR).
2. *Escherichia coli* BL21 Star (DE3) cell strain.

3. 100 mg/mL ampicillin.
4. 34 mg/mL chloramphenicol in 100% (v/v) ethanol.
5. Slide-A-Lyzer MINI devices 2 mL 3.5K MWCO.
6. 3.0 M $(NH_4)_2SO_4$.
7. Tobacco etch virus (TEV) protease [home-made].
8. Human rhinovirus type 14 3C (HRV-3C) protease [home-made].
9. Pre-wetted dialysis membrane 3.5K MWCO.
10. Ni-A_10:11 Buffer: 50 mM HEPES/NaOH, 300 mM NaCl, and 25 mM imidazole (pH 7.5) [500 mL total].
11. Ni-B_10:11 Buffer: 50 mM HEPES/NaOH, 300 mM NaCl, and 500 mM imidazole (pH 7.5) [100 mL total].
12. Dialysis_10:11 Buffer: 50 mM HEPES/NaOH, (pH 7.5), 300 mM NaCl, 10% (v/v) glycerol, and 2 mM DTT (pH 7.5) [2 L total].
13. GF_10:11 Buffer: 10 mM HEPES/NaOH, 300 mM NaCl, 10% (v/v) glycerol, and 2 mM DTT (pH 7.5) [2 L total].

2.3 Production, Purification of the NOT1:2:3:9 Ternary Subcomplex, and Reconstitution of the CCR4-NOT Complex

1. Recombinant baculovirus generated from the bacmid pACE-Bac-NOT1-NOT9-NOT2-NOT3.
2. *Spodoptera frugiperda* Sf21 insect cell line.
3. Sf-900 II serum-free medium (Gibco).
4. 0.45 μM PVDF syringe-driven filters.
5. HiPrep 26/60 Sephacryl S-300 HR size-exclusion column (Cytiva).
6. Ni-A_1:2:3:9 Buffer: 50 mM HEPES/NaOH, 10 mM potassium phosphate, 500 mM NaCl, and 40 mM imidazole (pH 7.5) [500 mL total].
7. Ni-B_1:2:3:9 Buffer: 50 mM HEPES/NaOH, 10 mM potassium phosphate, 300 mM NaCl, and 500 mM imidazole (pH 7.5) [100 mL total].
8. GF_1:2:3:9 Buffer: 10 mM HEPES/NaOH, 200 mM NaCl, and 2 mM DTT (pH 7.5) [1 L total].

2.4 In Vitro Deadenylation Assay

1. Serological pipettes and filter tips.
2. 50 mL glass beaker and a magnetic stirrer.
3. 10× Nuclease Buffer (NB): 200 mM PIPES/NaOH, 400 mM NaCl, 100 mM KCl, and 20 mM $Mg(OAc)_2$ (pH 6.8) [10 mL total].
4. 10× "No-Salt" Nuclease Buffer (NS-NB): 200 mM PIPES/NaOH, 100 mM KCl, and 20 mM $Mg(OAc)_2$ (pH 6.8) [10 mL total].

5. RNA Loading Buffer: 95% (v/v) deionized formamide; 17.5 mM EDTA (pH 8.0), and 0.01% (w/v) bromophenol blue [10 mL total].
6. High-voltage power supply unit.
7. Glass plates, spacers, combs, clamps, and a vertical electrophoresis tank.
8. 10× Tris–borate–EDTA (TBE) buffer.
9. Urea.
10. 10% (w/v) ammonium persulfate (APS).
11. Tetramethylethylenediamine (TEMED).

3 Methods

The described methods demonstrate how to purify recombinant subcomplexes of the human CCR4-NOT complexes NOT6:7, NOT10:11, and NOT1:2:3:9 (described in Subheadings 3.1, 3.2, and 3.3, respectively) with subsequent in vitro reconstitution of the full eight-subunit human CCR4-NOT complex (Subheading 3.3) using heterologous expression systems and protein chromatography. The key advantage of this in vitro reconstitution approach is that it enables to obtain highly reproducible homogeneous preps of the human CCR4-NOT with milligram-level quantities that can be used for various downstream structural and biochemical applications. Validation of the CCR4-NOT deadenylation activity using in vitro deadenylation assay is described in Subheading 3.4.

3.1 Production and Purification of the NOT6:7 Heterodimer

Plasmid maps and their construction were previously described [18]. Unless otherwise indicated, all purifications should be performed at 4 °C.

1. Co-transform *E. coli* BL21 Star (DE3) cells with pMCSG19c-NOT6 (AmpR) and pET28b-SUMO-NOT7 (KanR) plasmids.
2. Pick three to four colonies with a wooden toothpick or pipette tip and inoculate a 200 mL LB pre-culture. Add ampicillin (100 μg/mL) and kanamycin (50 μg/mL). Incubate at 37 °C with shaking at 220 rpm.
3. Inoculate 6 L LB Lennox media with the pre-culture (30 mL/L). Grow at 37 °C with shaking until OD_{600} of 0.3–0.4 is reached.
4. Reduce the temperature to 20 °C. After 1 h, induce expression by adding 1 mM IPTG.
5. After 16–18 h, harvest the cells by centrifugation at 4000 g for 15 min.

6. Add ~25 mL of Ni-A_6:7 buffer per ~10 g of pellet. Resuspend the pellet well by homogenizing or aspirating.
7. Lyse the cells using EmulsiFlex. Two to three passes are usually sufficient for full lysis. Alternatively, by sonication with 5 s on/off cycle for 2 min 30 s of total on time at 20% amplitude.
8. Clarify the lysate by centrifugation at 40,000 g for 45 min at 4 ° C.
9. Transfer the lysate carefully without disturbing the pellet to one or several 50 mL conical tubes.
10. Apply the lysate at 5 mL/min using a peristaltic pump on the HiTrap IMAC column charged with nickel and pre-equilibrated in Ni-A_6:7 buffer.
11. Transfer the column to the ÄKTA pure system. Perform a 100 mL wash in Ni-A_6:7 buffer, and then elute the bound fraction with a linear gradient over 80 mL to 100% Ni-B_6:7 buffer at 5 mL/min, and hold at 100% Ni-B_6:7 buffer for 10 mL (*see* Fig. 2a).
12. Combine the peak fractions and add TEV protease at the ratio of 1:50 (w/w) of protease to total protein.
13. Transfer to the pre-wetted dialysis membrane. Leave to dialyze in Dialysis_6:7 buffer with mild stirring overnight at 4 °C.
14. Equilibrate HiLoad 26/600 Superdex 200 pg column with GF_6:7 buffer on an ÄKTA pure system.
15. Transfer the protein from the dialysis membrane into a 50 mL conical tube. Concentrate to 10 mL using Amicon 15K concentrators (10K MWCO) and then load on HiLoad 26/600 Superdex 200 pg via a capillary or glass super loop. Elute at 2.5 mL/min.
16. The peak of interest elutes at around 180–190 mL. A larger, later-eluting peak is the excess of NOT7.
17. Combine the peak fractions and concentrate to 1.0–1.2 mL using an Amicon 15K concentrator (10K MWCO).
18. Transfer the concentrated protein into a 5 mL syringe. Carefully and slowly aspirate Dilution_6:7 buffer equivalent to three volumes of the protein sample [3.0–3.6 mL].
19. Apply diluted NOT6:7 on a SOURCE 15Q 4.6/100 PE column pre-equilibrated in Q-A_6:7 buffer via a capillary loop.
20. Wash with 10 column volumes of Q-A_6:7 buffer.
21. Elute the complex with a shallow gradient of Q-B_6:7 buffer (to reach 15% B over approximately 50 mL) at 1 mL/min. The NOT6:7 heterodimer elutes in a single sharp peak when the conductivity is 15–16 mS/cm.

22. Combine the peak fractions, measure the concentration with NanoDrop, and subsequently concentrate to 5–7 mg/mL using Amicon 15K concentrator (10K MWCO).
23. Aliquot in 10–50 μL in thin-walled PCR tubes. Flash-freeze in liquid nitrogen and store at −80 °C.

3.2 Production and Purification of the NOT10:11 Heterodimer

Plasmid maps and their construction were previously described [18]. Unless otherwise indicated, all purifications should be performed at 4 °C.

1. Co-transform *E. coli* BL21 Star (DE3) cells with pnYC-NpM-NOT10 (CamR) and pnEA-NOT11-CvH (AmpR) plasmids.
2. Pick three to four colonies with a wooden toothpick or pipette tip and inoculate a 200 mL LB pre-culture. Add ampicillin (100 μg/mL) and chloramphenicol (34 μg/mL). Incubate at 37 °C with shaking at 220 rpm.
3. Inoculate 6 L LB Lennox media with the pre-culture (30 mL/L). Grow at 37 °C with shaking until OD_{600} of 0.3–0.4 is reached.
4. Reduce the temperature to 20 °C. After 1 h, induce expression by adding 1 mM IPTG.
5. After 16–18 h, harvest the cells by centrifugation at 4000 g for 15 min.
6. Add ~25 mL of Ni-A_10:11 buffer per ~10 g of pellet. Resuspend the pellet well by homogenizing.
7. Lyse cells by using EmulsiFlex. Two to three passes are usually sufficient for full lysis and alternatively by sonication with 5 s on/off cycle for 2 min 30 s of total on time at 20% amplitude.
8. Clarify the lysate by centrifugation at 40,000 g for 45 min at 4 °C.
9. Transfer the lysate carefully without disturbing the pellet to one or several 50 mL conical tubes.
10. Apply the lysate at 5 mL/min using a peristaltic pump on the HiTrap IMAC column charged with nickel and pre-equilibrated in Ni-A_10:11 buffer.
11. Transfer the column to the ÄKTA pure system. Perform a 100 mL wash in Ni-A_10:11 buffer, and then elute the bound fraction with a linear gradient over 80 mL to 100% Ni-B_10:11 buffer at 5 mL/min, and hold at 100% Ni-B_6:7 buffer for 10 mL. The complex elutes in a single peak with the maximum at around 35% of Ni-B_10:11 buffer (*see* Fig. 2b).
12. The N-terminal maltose-binding protein (MBP) tag of NOT10 and the C-terminal His6 tag of NOT11 are cleaved overnight by adding HRV-3C and TEV proteases, respectively, during dialysis. First, set up dialysis in a 2 L Dialysis_10:11

buffer using the pre-wetted dialysis membrane. Add HRV-3C and TEV proteases to the protein sample, each at a ratio of 1: 50–100 (w/w) protease to protein. Some protein precipitation may be observed. Centrifugation should remove this precipitate, and the supernatant is transferred to the new tube.

13. Add ice-cold 3.0 M $(NH_4)_2SO_4$ at a 1:1 ratio to the cleared protein solution. Incubate on ice for 10–15 min.
14. Centrifuge the precipitated protein at maximum speed for 15 min using a refrigerated tabletop centrifuge; 20–25 mL per 50 mL conical tube is best for forming a compact pellet.
15. Remove the supernatant by aspirating it with a 25 mL pipette without disturbing the pellet.
16. Gently resuspend the pellet using 2 mL of GF_10:11 buffer. The resulting solution may be slightly cloudy initially but should be clear eventually.
17. Transfer the resuspended protein to a fresh 50 mL conical tube. Add GF_10:11 buffer to 10 mL.
18. Equilibrate HiLoad 26/600 Superdex 200 pg column with GF_10:11 buffer on an ÄKTA pure system.
19. Load on HiLoad 26/600 Superdex 200 pg via a capillary loop or a glass superloop. Elute at 2.5 mL/min. The peak of interest is between 150 and 190 mL.
20. Concentrate the protein by precipitation as in **steps 13–16**. Before proceeding, measure the concentration with NanoDrop. Resuspend the pellet in the volume, which would yield the final concentration of 5–7 mg/mL.
21. Transfer the resuspended protein to a Slide-A-Lyzer MINI device(s) and leave on mild stirring overnight in 1 L of GF_10:11 buffer.
22. Aliquot in 50–100 μL in thin-walled PCR tubes. Flash-freeze in liquid nitrogen and store at −80 °C.

3.3 NOT1:2:3:9 Purification and Reconstitution of the CCR4-NOT Complex

Baculovirus generation was previously described [18]. Unless otherwise indicated, all purifications should be performed at 4 °C.

1. Grow Sf21 insect cells in 50 mL cultures using Sf-900 II media at ~2.0×10^6 cells/mL. Add the V1 amplified virus of NOT1: 2:3:9 (generated from pACEBac-NOT1-NOT9-NOT2-NOT3 bacmid) at a 1:100 (v/v) ratio. Harvest the infected cells 48 h after the cells stop dividing. Flash-freeze in liquid nitrogen and store at −80 °C.
2. Thaw tubes with cell pellets, corresponding to two to four 50 mL cultures, by placing them into a beaker with tepid water for 2 min. Then place on ice.

3. Add 16 mL of Ni-A_1:2:3:9 buffer per pellet into each tube and resuspend by aspirating.
4. Lyse by sonication with 0.2 s on/0.8 s off cycle for 10 s of total on time at 10% amplitude. Repeat once.
5. Clarify the lysate by centrifugation at 40,000 g for 45 min at 4 °C.
6. Transfer the cleared lysate into fresh 50 mL conical tubes and then filter using a 0.45 μm syringe-driven filter.
7. Apply the lysate at 5 mL/min using a peristaltic pump on the HiTrap IMAC column charged with nickel and pre-equilibrated in Ni-A_1:2:3:9 buffer.
8. Transfer the column to the ÄKTA pure system. Perform a 30 mL wash in Ni-A_1:2:3:9 buffer and then elute the bound fraction with a linear gradient to 100% Ni-B_1:2:3:9 buffer at 5 mL/min (*see* Fig. 3).
9. Combine the peak fractions. Add ~0.4 mg of each purified NOT6:7 and NOT10:11 heterodimers per 1 mg of NOT1: 2:3:9.
10. Incubate for 1.5 h on ice. Then load onto HiPrep 26/60 Sephacryl S-300 column equilibrated in GF_1:2:3:9 buffer on an ÄKTA pure system.
11. The peak of the reconstituted full human CCR4-NOT complex elutes at 110–112 mL. The first 90–105 mL peak contains predominantly nucleic acids and some aggregated protein.
12. Combine the peak fractions and concentrate to 1.0–1.5 mg/mL using Amicon 15K concentrator (10K MWCO).
13. Aliquot in 10–50 μL in thin-walled PCR tubes. Flash-freeze in liquid nitrogen and store at −80 °C.

3.4 In Vitro Deadenylation Assay

The deadenylation reaction is set up as a master mix from which set aliquots corresponding to individual time points are taken out. To ensure the highest reproducibility and optimal visualization of the products (*see* **Note 3**), the reaction volume of every time point is set at 60 μL. For the in vitro deadenylation assays, proteins and RNA should be kept on ice until immediate use. Use only filtered tips and serological pipettes to transfer the solutions.

1. Prepare 1× NB and 2× NB from 10× NB stock and 1× NS-NB from 10× NS-NB stock.
2. Prepare labeled microcentrifuge tubes with 180 μL of RNA Loading Buffer to quench the corresponding time point of the assay.
3. Prepare RNA, CCR4-NOT, and other components (if applicable) that all of them are at 40 mM salt concentration before mixing (*see* **Note 4**). The order of the added

components to the master mix depends on the exact assay. Pipette RNA first and the CCR4-NOT complex last if the setup permits.

4. Pipette the RNA control sample as a 1× volume reaction without deadenylase (adjust to 60 μL with 1× NB). Prepare control samples with the other components if the other components are tested.
5. Add the last component to the master mix, transfer the tube to the heating block setup at 37 °C, and start the timer.
6. At the corresponding time point, remove 60 μL (one reaction volume) from the master mix reaction and quench in the corresponding tube with the RNA loading buffer. Add 180 μL of RNA loading buffer to quench control samples after finishing the assay. Quenched reactions can be stored at 4 °C short term (a few days) and at −20 °C for a longer term (a few months).
7. Prepare the TBE–urea–acrylamide solution. In a 50 mL beaker, add 16 g of urea and then add 4 mL of 10× TBE and 20 mL of 40% (w/v) 19:1 acrylamide solution. Scale up accordingly if necessary. Leave for ~20–30 min to mix on a magnetic stirrer at 250–300 rpm.
8. Meanwhile, cast the clean glass plates using clear office tape. Seal the sides and bottom first and then wrap around for additional sealing of the corners. Assemble carefully and try to squeeze out any air between the tape and the glass; otherwise, bubbles may enter the gel, compromising the final scan image (*see* **Note 5**).
9. Add 400 μL of 10% (w/v) APS to the TBE–urea–acrylamide solution. Then, add 40 μL of TEMED and wait 7–10 s. Pour the gel gently but relatively quickly and place the comb (*see* **Note 6**). The optimal time for a gel to polymerize and cool to room temperature is 35–50 min (*see* **Note** 7).
10. Carefully remove the tape, place the gel on the longer side facing outside, and clamp it with the clips. Once the tape and the comb are removed, proceed quickly. Pour 1× TBE to fill the running chamber.
11. Rinse the wells by aspirating the TBE with a filtered tip.
12. First, load 10 μL of the RNA loading buffer to unused wells. Rinse the wells again by aspirating the TBE with a filtered tip before loading the samples.
13. Run the gel at 300 V for 2–2.5 h (*see* **Notes 8** and **9**).
14. Image the gel using the Typhoon scanner or equivalent (*see* Fig. 4b).

A

A_{20}	5′-6-FAM-UCUAAAUAAAAAAAAAAAAAAAAAAAA-3′
A_1	5′-6-FAM-UCUAAAUA-3′
A_0	5′-6-FAM-UCUAAAU-3′
UCU	5′-6-FAM-UCU-3′

B

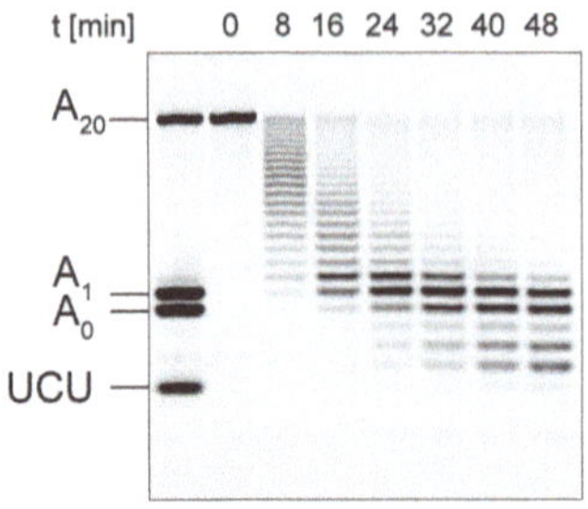

Fig. 4 In vitro deadenylation assay of the full human CCR4-NOT complex. (**a**) Synthetic 6-FAM-labeled (green) RNA substrate, which consists of 7-mer body (red) and stretch of 20 adenosine residues and a set of markers (A_1, A_0, UCU) used in a deadenylation assay time-course experiment. (**b**) Scanned TBE–urea PAGE gel of the deadenylation assay with the full human CCR4-NOT complex (50 nM) and an equimolar amount of synthetic 7-mer-A_{20} RNA. (The figure was adapted from the manuscript by Raisch, Chang, Levdansky, and coworkers [18] under permission provided by a Creative Commons Attribution 4.0 International License (http:/ creativecommons.org/ licenses/by/4.0/))

4 Notes

1. Alternatively, HisTrap FF or similar nickel-charged resins packed into columns from Cytiva or other suppliers can be used.
2. Similar high-resolution anion exchange columns, such as Mono Q, can be used.
3. It is much easier and safer to work with synthetic fluorescently labeled RNAs. However, in some scenarios, radioactively P^{32}-labeled RNAs generated by in vitro transcription (IVT) will be of more benefit, especially if those RNAs are over 100 nucleotides long. Synthetic fluorescently labeled RNAs over 100 nucleotides long may have a rather high cost, and their synthesis may result in low yields if ordered from third parties. Production of the RNA for the deadenylation assays using IVT with subsequent radioactive labeling may help to resolve both issues.
4. Although every assay is unique in its exact setup, there are several considerations to follow: (1) For *n* time points, prepare *n*+1 reaction volumes—e.g., for four time points, calculate five reaction volumes (*see* Table 1)—such approach eliminates variability in the pipetted volume of the last time point; (2) any single component does not exceed 25% of total volume; (3) the minimal volume to be pipetted for any component at any stage is 5 μL. Such an approach eliminates most pipetting errors and contributions of the stock buffers and their components.

Table 1
Typical layout of the deadenylation assay

	One reaction [60 μL]	Four plus one reactions [300 μL]
RNA (0.2 μM)	15 μL	75 μL
CCR4-NOT (0.25 μM)	12 μL	60 μL
Other components	x (up to 15 μL per component)	x (up to 75 μL per component)
1× NB	[60 – 15 – 12 – x] μL	[300 – 75 – 60 – x] μL

Typical layout of the assay:

The deadenylase complexes are rather sensitive to the salt concentration, which must be strictly controlled. All of the components of the assay are aimed to be in a 1× NB concentration regime before being combined.

RNA

50 nM concentration of the FAM-labeled synthetic RNA gives optimal results on a gel. Synthetic RNAs are typically delivered lyophilized and are then resuspended in Milli-Q grade water to 100 μM storage stocks. The working stock of the RNA is then prepared in a series of dilutions.

RNA is diluted to 4 μM to give a primary working stock, which is then diluted with one volume of 2× NB and 18 volumes of 1× NB (e.g., 5 μL [4 μM stock] + 5 μL [2× NB] + 90 μL [1× NB] = 100 μL [0.2 μM in 1× NB]).

CCR4-NOT

Salt from the storage buffer must be reduced to 40 mM as in 1× NB by diluting with 1× NS-NB.

CCR4-NOT is in the GF_1:2:3:9 buffer, which contains 200 mM NaCl. To one volume of CCR4-NOT, add four volumes of 1× NS-NB. To dilute to the final targeted concentration, use 1× NB.

Other Components

The impact of other components (proteins, peptides, small molecules, etc.) interacting with RNA or CCR4-NOT can be tested in the assay. Salt from the storage buffer of all components must be reduced to 40 mM as in 1× NB by diluting with 1× NS-NB. If a component is in Milli-Q water, add 2× NB at 1:1 ratio. To adjust to the final targeted concentration, use 1× NB.

5. Cleanliness of the glass plates is crucial since it affects the gel's general quality and how easily it can be removed from the glass plates for scanning. It is easy to damage the gel during this step, and cleaning minimizes this risk. Glass plates are rinsed with water and then with 1% (w/v) SDS solution. Subsequently, rinse with 70% (v/v) ethanol and wash extensively with water. They must dry without additional wiping (even with lint-free wipes) because they may leave traces and fibers, affecting the final image quality.
6. Before pouring the gel, prepare the combs and clamps to have them immediately available. The gel will polymerize very quickly, within 30–60 s. APS must be added first, followed by TEMED; otherwise, the gel will polymerize too quickly. If the gel is poured immediately, it will be too wet, and trapped bubbles between the glass plates and tape are more likely to get into the gel. Moreover, the gel may start leaking. In ~10 s, the TBE–urea–acrylamide solution will be more "jelly-like" and far less likely to leak. Immediately insert the combs, adjust their positions, and set the clamps to squeeze unpolymerized acrylamide between the comb and the glass plates.
7. Acrylamide polymerization is an exothermic reaction, and the glass plates will get hot. Still, carrying out the procedure in the cold room is not advised as polymerization will be too rapid, affecting the gel quality. If a TBE–urea–polyacrylamide gel is not used within 1.5–2 h from the polymerization reaction's start, it will dry out. If used, it is hard to wash off the urea from the wells, and urea in sample lanes may significantly distort the run, position of bands, and subsequent image analysis.
8. If the used RNA substrates are relatively short (<35 nt), terminate at one-quarter to one-third of the length from the bottom of the glass plate. For longer substrates (>35 nt), proceed until the bromophenol blue dye reaches the end of the glass plate. For substrates over 60 nt, the run can be tracked using 0.01% (w/v) xylene cyanol dye.
9. One-hour pre-run is optional; however, we have not observed a substantial improvement in the quality of the run and the final scanned image. Running the gel at 300 V enables the separation of the products quickly yet without loss of quality of the final image. Running the gel at higher voltages distorts the bands (the bands become "smiley" and smeary). It creates much additional heat, which may damage the gel upon disassembly of the glass plates.

Acknowledgments

We wish to thank members of the Valkov Lab for their comments on this protocol and fruitful discussions. E.V. and Y.L. are supported by the Intramural Research Program, Center for Cancer Research, National Cancer Institute, part of the National Institutes of Health.

References

1. Passmore LA, Coller J (2021) Roles of mRNA poly(A) tails in regulation of eukaryotic gene expression. Nat Rev Mol Cell Biol 23:1–14
2. Raisch T, Valkov E (2022) Regulation of the multisubunit CCR4-NOT deadenylase in the initiation of mRNA degradation. Curr Opin Struct Biol 77:102460
3. Yi H, Park J, Ha M et al (2018) PABP cooperates with the CCR4-NOT complex to promote mRNA Deadenylation and block precocious decay. Mol Cell 70:1081–1088.e5
4. Eisen TJ, Eichhorn SW, Subtelny AO et al (2020) The dynamics of cytoplasmic mRNA metabolism. Mol Cell 77:786–799.e10
5. Petit A-P, Wohlbold L, Bawankar P et al (2012) The structural basis for the interaction between the CAF1 nuclease and the NOT1 scaffold of the human CCR4-NOT deadenylase complex. Nucleic Acids Res 40:11058–11072
6. Chen Y, Khazina E, Izaurralde E et al (2021) Crystal structure and functional properties of the human CCR4-CAF1 deadenylase complex. Nucleic Acids Res 49:6489–6510
7. Garces RG, Gillon W, Pai EF (2007) Atomic model of human Rcd-1 reveals an armadillo-like-repeat protein with in vitro nucleic acid binding properties. Protein Sci 16:176–188
8. Sgromo A, Raisch T, Bawankar P et al (2017) A CAF40-binding motif facilitates recruitment of the CCR4-NOT complex to mRNAs targeted by Drosophila Roquin. Nat Commun 8:14307
9. Chen Y, Boland A, Kuzuoğlu-Öztürk D et al (2014) A DDX6-CNOT1 complex and W-binding pockets in CNOT9 reveal direct links between miRNA target recognition and silencing. Mol Cell 54:737–750
10. Boland A, Chen Y, Raisch T et al (2013) Structure and assembly of the NOT module of the human CCR4-NOT complex. Nat Struct Mol Biol 20:1289–1297
11. Bhandari D, Raisch T, Weichenrieder O et al (2014) Structural basis for the Nanos-mediated recruitment of the CCR4-NOT complex and translational repression. Genes Dev 28:888–901
12. Enwerem III, Elrod ND, Chang C-T et al (2021) Human Pumilio proteins directly bind the CCR4-NOT deadenylase complex to regulate the transcriptome. RNA 27:445–464
13. Bawankar P, Loh B, Wohlbold L et al (2013) NOT10 and C2orf29/NOT11 form a conserved module of the CCR4-NOT complex that docks onto the NOT1 N-terminal domain. RNA Biol 10:228–244
14. Mauxion F, Basquin J, Ozgur S et al (2022) The human CNOT1-CNOT10-CNOT11 complex forms a structural platform for protein-protein interactions. Cell Rep 42: 111902
15. Höpfler M, Absmeier E, Peak-Chew S-Y, et al (2023) Mechanism of ribosome-associated mRNA degradation during tubulin autoregulation, https://www.biorxiv.org/content/10.1101/2023.01.12.523749v1
16. Albert TK, Lemaire M, van Berkum NL et al (2000) Isolation and characterization of human orthologs of yeast CCR4–NOT complex subunits. Nucleic Acids Res 28:809–817
17. Webster MW, Stowell JAW, Tang TTL et al (2017) Analysis of mRNA deadenylation by multi-protein complexes. Methods 126:95–104
18. Raisch T, Chang C-T, Levdansky Y et al (2019) Reconstitution of recombinant human CCR4-NOT reveals molecular insights into regulated deadenylation. Nat Commun 10:3173

Chapter 2

In Vitro Reconstitution of the *Drosophila melanogaster* CCR4-NOT Complex to Assay Deadenylation

Filip Pekovic and Elmar Wahle

Abstract

The CCR4-NOT complex is a multi-subunit poly(A)-specific 3′ exoribonuclease that catalyzes the deadenylation of mRNA. In this chapter, we describe procedures to express and purify recombinant *Drosophila melanogaster* CCR4-NOT. Furthermore, we provide protocols for preparing radioactively labeled RNA substrates and conducting in vitro deadenylation assays.

Key words Deadenylation, Poly(A) tail, mRNA decay, CCR4-NOT, Deadenylases

1 Introduction

Deadenylases are poly(A)-specific 3′-exoribonucleases that degrade the poly(A) tail, present at the 3′-end of almost all eukaryotic mRNAs, in a regulated manner [1–3]. The poly(A) tail is important for the efficient translation and stability of its mRNA [3, 4]. The shortening of the poly(A) tail by deadenylases is the first step in the mRNA decay pathway and can lead to translational repression [2, 5]. There are two major deadenylases in eukaryotes: the PAN2-PAN3 complex and the CCR4-NOT complex [6–9]. The PAN2-PAN3 deadenylation complex is responsible for shortening the poly(A) tails above 90 nucleotides in length [7]. Below 90 nucleotides, the CCR4-NOT complex removes the poly (A) tail, priming the mRNA for decay [3, 7]. The CCR4-NOT complex consists of up to eight conserved subunits (Fig. 1), of which CCR4 and CAF1 exhibit exoribonuclease activity and form a heterodimer [6, 10, 11]. NOT1 is the large central scaffold to which all other subunits bind. CAF40 and the NOT2/NOT3 heterodimer are universally conserved subunits providing interaction surfaces for RNA-binding proteins that recruit the CCR4-NOT complex to specific substrates. NOT10 and NOT11 constitute a heterodimer limited to higher eukaryotes and associated with

Eugene Valkov and Aaron C. Goldstrohm (eds.), *Deadenylation: Methods and Protocols*, Methods in Molecular Biology, vol. 2723,
https://doi.org/10.1007/978-1-0716-3481-3_2,

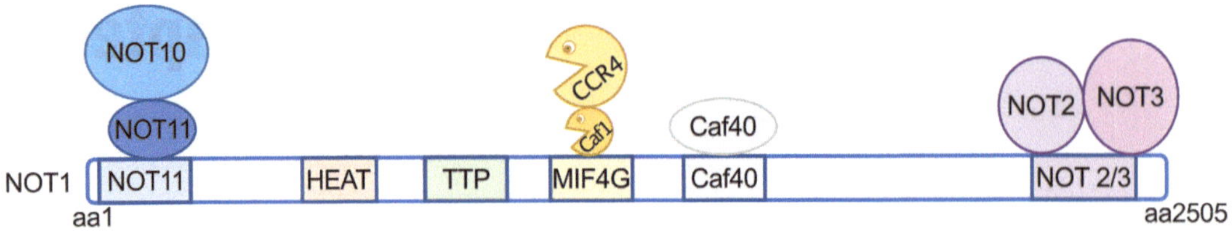

Fig. 1 Schematic structure of *Drosophila* CCR4-NOT. The full CCR4-NOT complex consists of eight subunits assembled in a modular composition. The scaffold protein NOT1 recruits all subunits. Domains and protein interaction sites of NOT1 (e.g., HEAT, MIF4G, and TTP) are indicated. (Figure is adapted from [13])

the N-terminus of NOT1 [12]. The understanding of how the non-catalytic subunits contribute to the deadenylation activity of the exoribonuclease heterodimer has been facilitated by the reconstitution of a seven-subunit complex from *S. pombe* and the eight-subunit CCR4-NOT complexes from *H. sapiens* and *D. melanogaster* [10, 11, 13]. It was shown in all cases that the deadenylation activity is strongly enhanced in the full complex compared to the CCR4-CAF1 heterodimer alone. This is most likely achieved by additional RNA-binding surfaces on CAF40 and, in the case of higher eukaryotes, the NOT10-NOT11 heterodimer [11, 14]. In addition, in vitro deadenylation systems employing the reconstituted complexes allow the detailed study of targeted deadenylation mediated through interacting RNA-binding proteins (RBPs) (e.g., Mmi1 [10], Puf3 [15], Zfs1/TTP [15, 16], and Smaug [13]). Such systems can give insights into critical interactions between participating proteins and their functional consequences.

In the first part of this chapter, we describe a procedure to reconstitute the *Drosophila melanogaster* CCR4-NOT complex and smaller subcomplexes in vitro (Fig. 2). All subunits are full length. NOT1, NOT2, NOT3, CAF40, and the nuclease heterodimer are all co-expressed in insect cells utilizing the baculovirus expression system [17], which has also been used for the human and *S. pombe* CCR4-NOT complexes [10, 11]. Only the NOT10-NOT11 heterodimer is co-expressed in *E. coli* and combined with the other subunits after purification. In our experience, the expression of truncated versions of the proteins in *E. coli*, following the same strategy as for the human MINI subcomplex [11], has not been successful for the fly complex due to the insolubility of most subunits. Therefore, it is strongly advised to use the baculoviral system even if truncated versions of the proteins are to be studied. We have used the FLAG-tag for initial affinity capture of the complexes, but other tags (e.g., Strep-tag) on CCR4 or CAF40 will also work and may be more cost-efficient. In our hands, the strategy applied for the purification of the human complex, namely the addition of an N-terminal tag to NOT1 [11], was unsuccessful for the *Drosophila* complex.

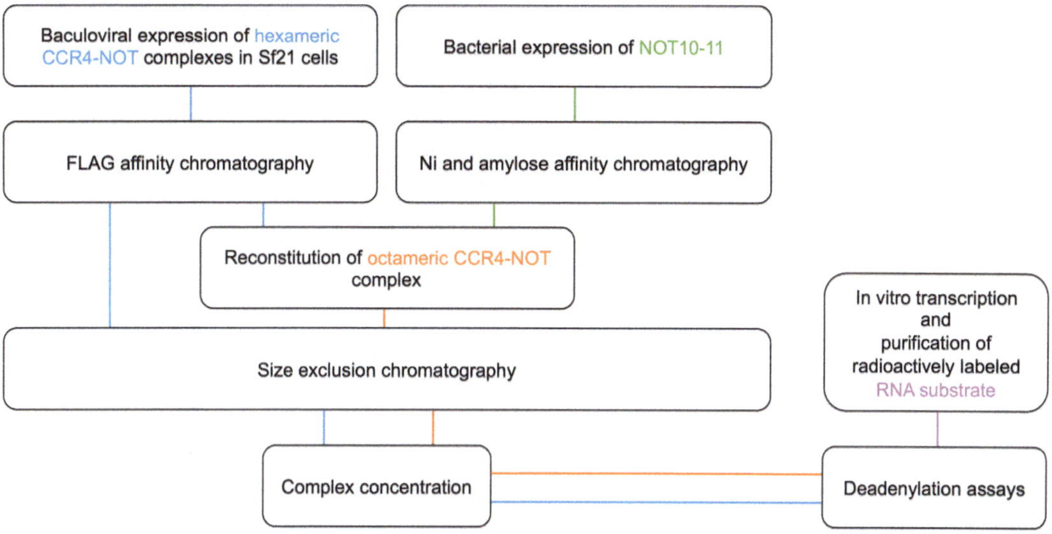

Fig. 2 Flowchart of the reconstitution of *Drosophila* CCR4-NOT to in vitro assays of deadenylation. Both bacterial and insect cell expression systems are utilized to produce components of the CCR4-NOT complex. Purified NOT10 and NOT11 are added to the hexameric CCR4-NOT complex to reconstitute the full CCR4-NOT complex. Complexes of the right size and composition are isolated by size-exclusion chromatography. The concentrated CCR4-NOT complexes are tested in their deadenylation activity on in vitro synthesized radioactively labeled RNA

The second part of this chapter focuses on producing radioactively labeled substrate RNA and the in vitro deadenylation assay (Fig. 2), similar to what has been described in other studies [10, 11, 18, 19]. For radioactive labeling of the substrate RNA, we use [α-^{32}P]-UTP in the in vitro transcription reaction. Thus, the poly (A) tail remains unlabeled, and deadenylation is assayed by increased mobility of the RNA in denaturing polyacrylamide gels. In principle, it should be possible to label the poly(A) tail using [α-^{32}P]-ATP and directly measure the release of AMP or a decrease in the acid-precipitable polymer, but we have not tested this method. To append a poly(A) tail of defined length, it is best to have the poly(A) tail encoded downstream of the desired RNA sequence in a plasmid and perform a run-off transcription [20, 21]. DNA templates with up to 100 encoded A residues have worked reasonably well in our hands, but the sequence encoding the tail can be unstable when propagated in *E. coli*. Alternatively, one can use a purified poly(A) polymerase to add any desired poly(A) length to a previously synthesized RNA. While this provides more flexibility, the procedure is more cumbersome, particularly if tails of a defined length are desired. A relatively short RNA substrate (below 500 nucleotides) will make separating the deadenylated product from the polyadenylated substrate by denaturing gel electrophoresis easier, particularly when partially degraded poly (A) tails remain. It is recommended to perform an oligo(dT)-

RNase H digestion of the purified polyadenylated RNA substrate as a size marker for the fully deadenylated product. The same procedure can also ascertain that RNA shortening is due to deadenylation. Depending on the sensitivity of your phosphorimager and under the labeling conditions described here, no less than 5 fmol of radioactively labeled RNA should be loaded per gel lane; otherwise, small amounts of deadenylated product or intermediates may be below the detection threshold. We obtained the best results with around 25 fmol of radioactively labeled RNA. As an alternative to radiolabeled RNA, we have also used synthetic RNA oligonucleotides bearing a 5′ fluorescence tag. No less than 100 fmol of fluorescently labeled RNA should be loaded per gel lane, with best results around 250 fmol.

2 Materials

Follow the general precautions for working with RNA, avoiding RNase contaminations. All solutions should be prepared with analytical-grade reagents and distilled, deionized water. Aqueous solutions that come into contact with RNA should be made RNase-free by diethylpyrocarbonate (DEPC) treatment: Add 0.1% (v/v) DEPC to the aqueous solution, incubate at 37 °C with agitation overnight, and autoclave to remove any traces of non-reacted DEPC. Many solutions can be treated similarly, but amine-containing solutions cannot, as DEPC reacts with amines. In this case, use DEPC-treated water for preparing the solution. Even if not treated with DEPC, solutions are autoclaved or sterilized by filtration if sensitive to heat. Use disposable plastic ware for solutions containing RNA or destined to come into contact with it whenever possible. In our experience, pipet tips and Eppendorf tubes can be used without any particular pretreatment, but handle them with gloves and keep their containers closed. Wear gloves for all procedures involving the handling of RNA, insect cells, and toxic substances like acrylamide and radioisotopes. The use of radioisotopes requires training, institutional approval/licensing, and protective equipment. While being handled, proteins and RNA should be kept on ice or in a cold room if not stated otherwise.

For all buffers listed below, we adjust the pH in a stock solution (e.g., 1 M Tris-HCl) at room temperature and then dilute it into the final solution.

2.1 Materials to Produce Recombinant D. melanogaster CCR4-NOT

1. *E. coli* cells carrying bacmid DNA (e.g., DH10 MultiBac [22]).
2. *E. coli* cells for heterologous protein expression (e.g., BL21 Star (DE3)).
3. Transfer plasmids for introducing components of CCR4-NOT by homologous recombination into bacmid DNA: pFBDM-

NOT1_NOT2 (NOT1 under control of the polyhedrin promoter, NOT2 under control of the p10 promoter), pFBDM-CAF40-FLAG_NOT3 (CAF40 under control of the polyhedrin promoter, NOT3 under control of the p10 promoter), and pFastBac Dual-CAF1_CCR4-FLAG (CAF1 under control of the polyhedrin promoter, CCR4 under control of the p10 promoter). (A list of vectors can be found in [23]) (*see* **Note 1**).

4. Bicistronic vector for co-expressing NOT10 and NOT11 in *E. coli*: pETDuet-1-His_8-MBP-NOT10_His_6-NOT11 [13] (*see* **Note 1**).
5. mRNA from *Drosophila*. We have used RNA from 15 min to 135 min-old embryos (*see* **Note 2**).
6. 50 mg/mL kanamycin in water.
7. 7 mg/mL gentamycin in water.
8. 10 mg/mL tetracycline in water.
9. 100 mg/mL ampicillin in water.
10. 20 mg/mL 5-bromo-4-chloro-3-indolyl-beta-D-galactopyranoside (X-Gal) in dimethylsulfoxide (DMSO).
11. 1 M isopropyl-β-D-thiogalactoside (IPTG) in water.
12. P1 buffer: 50 mM Tris–HCl (pH 8.0), 10 mM EDTA (pH 8.0), and 100 μg/mL RNase A.
13. P2 buffer: 200 mM NaOH, 1% SDS.
14. P3 buffer: 3 M potassium acetate (pH 5.5).
15. Sf21 insect cells for virus production and protein expression.
16. Sf-900 II serum-free medium (SFM) for insect cells.
17. FuGENE HD transfection reagent.
18. 0.4% (w/v) trypan blue solution.
19. Cell disruptor (e.g., Sonicator).
20. Autoinduction solution 50 × 5052: 25% glycerol (v/v), 2.5% glucose (w/v), and 10% α-lactose monohydrate (w/v). Dissolving the lactose can be sped up by heating the solution.
21. Autoinduction solution 20xM: 0.5 M Na_2HPO_4, 0.5 M KH_2PO_4, 1 M NH_4Cl, and 0.1 M Na_2SO_4.
22. 1 M $MgSO_4$.
23. Lysis buffer: 500 mM NaCl, 50 mM HEPES-NaOH (pH 7.4), 7 mM K_2HPO_4, 3 mM KH_2PO_4, 10% w/v sucrose, 1 μM pepstatin A, and 1 mM phenylmethylsulfonyl fluoride (PMSF) (*see* **Note 3**). Keep ice-cold.
24. Wash buffer: 500 mM NaCl, 50 mM HEPES-NaOH (pH 7.4), 7 mM K_2HPO_4, 3 mM KH_2PO_4, and 10% w/v sucrose. Keep ice-cold.

25. FLAG elution buffer: Supplement the wash buffer with 200 ng/μL FLAG peptide for elution of protein immediately before use. Keep ice-cold.
26. Anti-FLAG M2 affinity matrix.
27. Ni-NTA agarose.
28. Amylose resin.
29. 2 M Imidazole (adjusted to pH 7.5 with HCl).
30. GST-tagged HRV 3C protease.
31. Glutathione agarose.
32. ÄKTA chromatography system for protein purification.
33. Superose 6 chromatography column (separation range of 5 kDa to 5 MDa) for size-exclusion chromatography (*see* **Note 4**).
34. PES-based concentrators, 10 kDa molecular weight cut-off (*see* **Note 5**).

2.2 Materials for In Vitro Synthesis of Radioactively Labeled Substrate RNA

1. DNA template to be linearized for run-off transcription (for details, *see* Subheading 3.2).
2. 20 U/μL BsaI restriction enzyme.
3. Bacteriophage T3 RNA polymerase (50 U/μL) (*see* **Note 6**).
4. 10× T3 transcription buffer (typically supplied by the vendor of the enzyme).
5. 40 U/μL RNase inhibitor, murine.
6. 1 M dithiothreitol (DTT) solution in DEPC-treated water. Always add to buffers immediately before use, as DTT is sensitive to oxidation.
7. 10 mM nucleotide solutions (ATP, CTP, GTP, and UTP, each diluted in DEPC-treated water from a 100 mM buffered stock solution at neutral pH) (*see* **Note 7**).
8. [α-^{32}P]-labeled nucleotide to be incorporated. We use [α-^{32}P]-UTP at 3000 Ci/mmol.
9. 3 M sodium acetate (pH 5.2).
10. Ethanol, abs.
11. Ethanol, 70%.
12. Acrylamide/bisacrylamide (19:1), 40% w/v solution.
13. Urea.
14. Ammonium persulfate, 10% w/v in water.
15. *N*,*N*,*N*,*N*′-Tetramethyl-ethylenediamine.
16. Formamide RNA loading buffer: 95% formamide (deionized), 17.5 mM EDTA (pH 8), 0.01% w/v bromophenol blue, and 0.01% w/v xylene cyanol.

17. 10× TBE buffer: 1 M Tris base, 1 M boric acid, 20 mM EDTA (disodium salt).
18. DNA size markers: Digest plasmid DNA with a restriction enzyme resulting in a suitable range of fragments and label them with polynucleotide kinase and [γ-^{32}P]-ATP.
19. Plastic wrap.
20. Scalpel.
21. Thermostable glass plates for electrophoresis.
22. 0.2 mm thick spacers and combs. The comb should result in gel wells that can accommodate 30 μL.
23. Gel elution buffer: 20 mM Tris–HCl (pH 7.5), 0.25 M sodium acetate (pH not adjusted), 1 mM EDTA (pH 8.0), 0.25% w/v SDS.
24. Storage phosphor screen.
25. Phosphor imager.
26. X-ray cassette.
27. Scintillation cocktail.
28. Scintillation counter.

2.3 Materials for In Vitro Deadenylation Assays

1. Radioactively labeled polyadenylated RNA substrate (*see* **Note 8**).
2. 10× deadenylation buffer (*see* **Note 9**): 500 mM potassium acetate, 300 mM HEPES-KOH (pH 7.4), and 20 mM magnesium acetate. Store at –20 °C.
3. BSA, nuclease-free: 1 g/l in DEPC-treated water.
4. Stop solution: 95% formamide (deionized), 17.5 mM EDTA (pH 8.0), and 0.01% bromophenol blue. Optional: addition of 0.01% w/v xylene cyanol (*see* **Note 10**).
5. Gel materials as described above (Subheading 2.2, **steps 12–22**). Combs may be used that generate smaller wells, for example, 20 μL or less.
6. Materials for phosphor imaging as described above (Subheading 2.2, **steps 24–26**).

3 Methods

This section focuses on the steps essential to prepare baculoviruses expressing *D. melanogaster* CCR4-NOT. The background for the baculovirus expression system is extensively described in Ref. [17, 23–25].

3.1 Expression and Purification of D. melanogaster CCR4-NOT

3.1.1 Preparation of Baculoviruses

1. Transform pFBDM-NOT1-NOT2, pFBDM-CAF40-FLAG_-NOT3, and pFastBac Dual-CAF1_CCR4-FLAG separately into 50 μL of electro- or chemocompetent *E. coli* DH10 MultiBac cells.
2. Add 950 μL of LB medium.
3. Incubate cells at 37 °C under constant shaking at 800 rpm for 4–6 h to allow the transformed DNA to be integrated into the bacmid DNA.
4. Plate 20 μL of cells on LB agar plates containing 50 μg/mL kanamycin, 7 μg/mL gentamicin, 10 μg/mL tetracycline, 100 μg/mL X-gal, and 40 μg/mL IPTG to select for recombinant DH10 MultiBac transformants.
5. Incubate plates for at least 24 h at 37 °C before picking colonies (*see* **Notes 11** and **12**). If recombination into the bacmid DNA was successful, colonies will remain white, whereas negative clones will turn blue.
6. For each plasmid, inoculate 6 mL of LB medium (+50 μg/mL kanamycin, +7 μg/mL gentamicin, and +10 μg/mL tetracycline) with a clearly white colony.
7. Incubate at 37 °C under constant shaking at 220 rpm for at least 18 h.
8. Harvest the cells by centrifugation at 3750 g for 15 min.
9. Resuspend the cell pellet in 250 μL buffer P1.
10. Add 250 μL of buffer P2 for lysis and mix by gently inverting the tube 4–6 times.
11. Add 350 μL of buffer P3 to neutralize the solution and mix by gently inverting the tube 4–6 times.
12. Centrifuge at 20,000 g for 20 min to separate cell debris and genomic DNA from the bacmid DNA (*see* **Note 13**).
13. Take 650 μL of the supernatant and mix gently with 800 μL of ice-cold 2-propanol by inverting the tube multiple times.
14. Incubate for 20 min on ice.
15. Pellet precipitated bacmid DNA by centrifugation at 20,000 g for 15 min.
16. Remove supernatant.
17. Add 500 μL ethanol (70% v/v) to the pellet.
18. Centrifuge at 20,000 g for 5 min.
19. Remove supernatant.
20. Repeat **steps 17–19**.
21. Air-dry the pellet in the cell culture hood.
22. Resuspend in 40 μL ultrapure water by gently tapping the tube. (Do not vortex!)

23. Store the bacmid DNA protected from light at 4 °C (*see* **Note 14**). The final bacmid DNA concentration ranges between 5 and 10 mg/mL.
24. Use the purified bacmid DNA to transfect Sf21 cells: In a cell culture hood, seed 1.5 × 10^6 Sf21 cells in a total volume of 2 mL medium per well of a 6-well plate.
25. Incubate for at least 30 min at 27 °C until cells become semi-adherent.
26. In parallel, prepare the transfection mixture by mixing 5 μL of FuGENE HD with 10 μL bacmid DNA and 200 μL of culture medium (e.g., Sf-900 II SFM).
27. Incubate the transfection mixture for at least 15 min at 27 °C.
28. Add the transfection mixture dropwise to the semi-adherent cells.
29. Seal the 6-well plate with parafilm.
30. Incubate for 3–5 days at 27 °C until signs of infection appear (*see* **Note 15**).
31. Collect the supernatant, which is your V0 baculovirus.
32. Store the baculovirus protected from light at 4 °C. The V0 generation can be stable for up to 1 year.
33. Due to the low virus titer of the V0 generation, further amplification in suspension culture is necessary: Infect 0.8 × 10^6/mL Sf21 cells in 50 mL medium in a 250 mL Erlenmeyer flask with 0.5 mL of your V0 virus.
34. Incubate at 27 °C and 120 rpm on a rotary shaker until cell viability is below 70% (measured by trypan blue staining). Typically, this will be the case after 4–5 days.
35. Harvest virus by spinning down the cells at 500 g for 5 min and collecting the supernatant, which contains your V1 baculovirus.
36. Store the baculovirus protected from light at 4 °C.

3.1.2 Expression and Purification of a Hexameric CCR4-NOT Complex

Here, we describe the procedure to reconstitute a stable hexameric CCR4-NOT complex consisting of NOT1, NOT2, NOT3, CAF40, CCR4, and CAF1. NOT10 and NOT11 can be co-expressed and purified from bacteria to be added to the hexameric CCR4-NOT complex to form an octameric complex [13]. In this case, full-length NOT1 has to be used since NOT10/NOT11 associates with the N-terminus of NOT1. In the absence of NOT10/NOT11, the N-terminus of NOT1 is dispensable [13]. Note that there are two naturally occurring isoforms of NOT1; isoform PC has the NOT10/NOT11 binding site, whereas isoform PE lacks it (https://flybase.org/). The procedure to generate the NOT10/NOT11 heterodimer and reconstitute the octameric complex will be described following the protocol for the reconstitution of the hexameric CCR4-NOT.

Perform all protein purification steps after cell lysis in a cold room. Keep samples on ice whenever possible.

1. For the reconstitution of the hexameric CCR4-NOT, co-infect Sf21 cells with three baculoviruses that express the following combination of proteins: (1) NOT3 and CAF40-FLAG, (2) CCR4-FLAG and CAF1, and (3) NOT2 and NOT1. These three viruses are used in a 2:1:2 ratio. A good starting point is to add the virus expressing the catalytic heterodimer in a 1:400 (v/v) ratio to 2×10^6/mL Sf21 cells and the other two viruses in a 1:200 (v/v) ratio.
2. Incubate cells on a rotary shaker as described above (Subheading 3.1.1, **step 34**).
3. The following procedure is for 200 mL of Sf21 cells at 2×10^6 cells/mL in a 1-l Erlenmeyer flask, which usually yields 250–500 μg of purified CCR4-NOT.
4. The following day measure the cell density, dilute to 2×10^6/mL cells if necessary, and continue the incubation.
5. Harvest the cells 72 h post-infection by centrifuging the cell suspension at 250 g for 7 min.
6. Remove the supernatant.
7. Resuspend the pellet in ice-cold lysis buffer corresponding to one-tenth of the culture volume (*see* **Note 16**).
8. All the following steps should be carried out on the ice or at 4 °C.
9. Lyse cells by mild sonication on ice. The goal is to lyse the cells and reduce the viscosity of the solution. In our experience, 8–10 cycles of a Branson 250 sonifier each consisting of 5 s on and 45 s off, with an output of 4 using a 1/4″ tip is sufficient (*see* **Note 17**).
10. The lysate is cleared by centrifugation in a pre-cooled high-speed centrifuge at 20,000 g for 30 min.
11. Equilibrate the FLAG affinity matrix. For this purpose, transfer 500 μL of FLAG matrix (packed volume) into a microcentrifuge tube (*see* **Note 18**) and centrifuge at 500 g for 3 min to separate the matrix from the storage buffer.
12. Remove the storage buffer and resuspend the matrix in 1 mL lysis buffer.
13. Transfer the matrix to a 50 mL tube and add 20 mL of lysis buffer.

14. Incubate under constant gentle rotation on a tube roller for at least 10 min.
15. Centrifuge at 500 g for 3 min to separate and remove the buffer from the matrix.
16. Add cleared lysate to the matrix.
17. Incubate for at least 1 ½ h under constant rotation.
18. Centrifuge at 500 g for 3 min and collect supernatant (your flow-through).
19. Add 20 mL of wash buffer.
20. Incubate under constant rotation for 10 min.
21. Centrifuge at 500 g for 3 min and collect supernatant (this is your first wash fraction).
22. Repeat the wash step three more times.
23. Resuspend the matrix in 1 mL of wash buffer and transfer to a fresh microcentrifuge tube.
24. Centrifuge at 500 g for 3 min and remove the wash buffer.
25. Add 500 μL of elution buffer.
26. Incubate under constant rotation for 30 min.
27. Centrifuge at 500 g for 3 min.
28. Collect supernatant (i.e., the eluate) in a clean microcentrifuge tube.
29. Equilibrate the size-exclusion chromatography (SEC) column (Superose 6 10/300 GL (24 mL) or 3.2/300 (2.4 mL)) with at least two column volumes of wash buffer.
30. Load appropriate volume of FLAG eluate (max. 500 μL for Superose 6 10/300 GL or 100 μL for Superose 6 3.2/300) and perform chromatography at a flow rate of 0.5 mL/min (Superose 6 10/300 GL) or 0.05 mL/min (Superose 6 3.2/300).
31. Collect 0.5 mL fractions (0.05 mL for Superose 6 3.2/300) after the void volume and store on ice until further use. The hexameric or octameric complexes elute in single monodisperse peaks after the void volume and, due to their size, can be easily separated from excess CCR4-CAF1 and CAF40, which elute far later (Fig. 3a, b).
32. Analyze aliquots of the collected fractions via SDS-PAGE and Coomassie staining (*see* **Note 19**).
33. Concentrate fractions containing the hexameric complex with a PES-based concentrator (MWCO 10 kDa). If desired, do the same for fractions containing the other proteins.

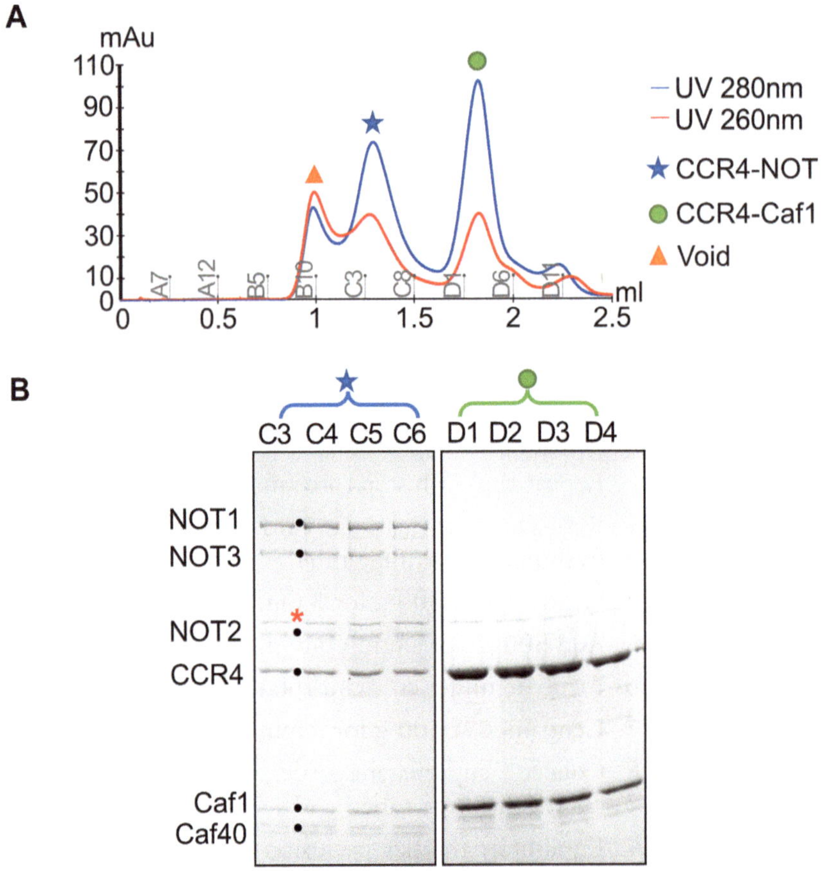

Fig. 3 Purification of CCR4-NOT by size-exclusion chromatography. The figure shows the purification of a hexameric CCR4-NOT_{MINI} complex [13] by SEC (Superose 6 3.2/300). NOT1 is N-terminally truncated in this complex, starting at the MIF4G domain, which mediates binding to CAF1 (*see* Fig. 1). The interaction surface for NOT10 and NOT11 is missing. (**a**) UV profile of the SEC. (**b**) Analysis of fractions by SDS-PAGE and Coomassie staining. Only fractions with stoichiometric protein ratios should be pooled (e.g., fraction C3-C6). A contaminating band above NOT2 (indicated by a red asterisk) is most likely a heat shock protein. The CAF1/CCR4 heterodimer is well separated from the six-subunit complex. (Figures (**a**) and (**b**) are adapted from [13])

34. Determine protein concentration either by in-gel quantification compared to a series of protein standards of known concentration (e.g., BSA) or measurement of absorbance at 280 nm, using the extinction coefficients of all subunits (*see* **Note 20**). Alternatively, a NanoDrop device can be used.
35. Divide preparations into suitable aliquots (*see* **Note 21**).
36. Freeze in liquid nitrogen and store at −80 °C.

3.1.3 Expression and Purification of an Octameric CCR4-NOT Complex

1. Transform pETDuet-1-His_8-MBP-NOT10_His_6-NOT11 into competent *E. coli* Bl21 Star (DE3) cells.
2. Add 950 μL of LB medium.
3. Incubate cells at 37 °C under constant shaking at 800 rpm for 1 h.

4. Plate 20 μL on LB agar plates containing 100 μg/mL ampicillin.
5. Incubate plates at 37 °C until visible colonies have grown (at least 14 h).
6. Pick one colony and inoculate 25 mL of LB medium with 100 μg/mL ampicillin in a 125 mL Erlenmeyer flask.
7. Grow the culture at 37 °C under constant shaking at 220 rpm for at least 14 h.
8. Add 10 mL of this culture to 0.5 L LB medium (plus antibiotic) in a 2 L Erlenmeyer flask.
9. Grow at 37 °C until the $OD_{595\ nm}$ reaches 0.5.
10. Shift the temperature to 20 °C and incubate for another hour.
11. Add the autoinduction solutions (final concentrations: 1× 5052, 1× M, 2 mM $MgSO_4$).
12. Incubate for 24 h.
13. Harvest the cells by centrifugation at 5000 g for 10 min.
14. Resuspend the pellet in 50 mL lysis buffer supplemented with 10 mM imidazole.
15. Lyse the cells by sonication (*see* Subheading 3.1.2, **step 9**; on time extended to 10 s).
16. Centrifuge lysate for 20 min at 15,000 g.
17. Apply the cleared lysate to 500 μL Ni-NTA agarose equilibrated in lysis buffer (*see* Subheading 3.1.2, **steps 11–16**).
18. Incubate for 2 h under constant rotation at 4–8 °C.
19. Wash the matrix four times with at least five column volumes of wash buffer plus 20 mM imidazole (*see* Subheading 3.1.2, **steps 19–21**).
20. Elute the proteins by incubating the matrix with 10 mL of wash buffer plus 160 mM imidazole for 30 min.
21. Centrifuge at 500 g for 3 min.
22. Collect the supernatant. This is your eluate.
23. Add the eluate to 1 mL amylose resin equilibrated in the wash buffer.
24. Incubate for 2 h under constant rotation at 4–8 °C.
25. Wash the matrix four times (*see* Subheading 3.1.2, **steps 19–21**) with at least five column volumes of wash buffer.
26. Add 2 mL of wash buffer containing HRV 3C protease to the matrix to cleave off the NOT10/NOT11 heterodimer. Use an amount of protease approximately equimolar with NOT10/NOT11 (concentration of NOT10/NOT11 estimated from the absorbance of the Ni-NTA eluate at 280 nm).
27. Pellet the resin by centrifugation at 500 g for 3 min.

28. Transfer the supernatant to a new tube containing 100 μL of Glutathione Sepharose 4B equilibrated in wash buffer.
29. Incubate for 90 min, allowing the protease to bind to the glutathione matrix.
30. Separate the matrix by centrifugation at 500 g for 3 min.
31. Concentrate the supernatant containing NOT10/NOT11 with a PES concentrator (10 kDa MWCO).
32. Determine the protein concentration by measuring the absorbance at 280 nm. Assuming a 1 : 1 ratio of NOT10 and NOT11, use the extinction coefficient of the two proteins to calculate their molar concentrations according to the Beer–Lambert law.
33. Divide into suitable aliquots.
34. Freeze in liquid nitrogen and store at −80 °C.
35. To reconstitute the octameric CCR4-NOT complex, add a twofold excess of NOT10/11 heterodimer to hexameric CCR4-NOT and incubate for 2 h at 8 °C.
36. Purify the octameric complex by SEC as described for the hexamer (*see* Subheading 3.1.2, **steps 29–36**).

3.2 Preparation of Radiolabeled Substrate RNA

Here, we describe the method to generate a substrate RNA for deadenylation reactions. The RNA is synthesized in a run-off transcription using linearized plasmid DNA as a template. The plasmid must contain a bacteriophage promoter (T3 in our case) upstream of the sequence to be transcribed. Synthesis of a polyadenylated RNA with a defined tail length is most easily performed by insertion of a corresponding poly(dT) template at the 3′ end of the transcription unit. In this case, the plasmid should be linearized with a restriction endonuclease generating a 3′-end that will be transcribed to poly(A) with no additional nucleotides (*see* **Note 22**) [26]. For this purpose, we use BsaI, a type IIS restriction enzyme. The recognition site is placed five nucleotides downstream of the poly(dT) sequence so that the enzyme will cleave outside its recognition site in the upstream poly(dT) sequence. The synthesized RNA is gel purified to remove any side products of transcription or fragments generated by nuclease contamination (*see* **Note 23**).

1. Prepare the linearized plasmid DNA by digesting at least 20 μg of plasmid DNA with the BsaI restriction endonuclease.
2. Check for completeness of digestion by agarose gel electrophoresis.
3. Purify the linearized DNA using standard procedures. Aim for a final DNA concentration of 0.5–1 μg/μL.
4. Set up the run-off transcription on ice in a 25 μL reaction with the following components: 1× T3 transcription buffer, 1 μg

linearized DNA template, 1 mM of ATP, CTP, and GTP each, 0.2 mM UTP, 40 μCi [α-^{32}P]-UTP, 1 mM DTT, 800 units/mL murine RNase inhibitor, and 125 units of T3 RNA polymerase. Add RNA polymerase last after mixing all other components.

5. Mix gently and incubate for 2 h at 37 °C.
6. In the meantime, prepare a denaturing polyacrylamide gel. Coat the side touching the gel with a siliconizing surfactant for one glass plate. This will make it easier to detach the glass plate from the gel.
7. Prepare the gel solution fresh by dissolving solid urea (8.3 M) in 5% w/v acrylamide/bisacrylamide (19:1) in 1× TBE buffer.
8. Add 1% v/v APS and 0.1% v/v TEMED to initiate gel polymerization.
9. Cast the gel and let it polymerize for at least 20 min.
10. Pull the comb out and assemble the gel into the gel box.
11. Run the gel in 1× TBE with a constant power of 1 W/cm for 30 min to preheat the gel to ca. 60 °C. This will prevent the formation of secondary RNA structures.
12. At any time during the transcription reaction, take 1 μL of the reaction mixture and dilute 1:50 in water. This is your input sample for quantifying total radioactivity by scintillation counting.
13. At the end of the transcription reaction, add 76 μL of ultrapure water (DEPC-treated), 10 μL of sodium acetate (3 M, pH 5.2), and 275 μL of absolute ethanol to the transcription reaction.
14. Mix, incubate briefly, and centrifuge for at least 20 min at 20,000 g.
15. Collect supernatant containing non-incorporated nucleotides in a microcentrifuge tube.
16. To ascertain that you have not lost the RNA pellet, which will not be visible, hold the tubes containing the pellet and supernatant separately in front of a Geiger counter. Both fractions should contain roughly comparable amounts of radioactivity. Do not discard the supernatant until you are sure that you have recovered the pellet.
17. Add 350 μL of 70% ethanol to the RNA pellet.
18. Centrifuge for 2 min at 20,000 g.
19. Collect supernatant in a microcentrifuge tube.
20. Proceed as in **step 16** to ascertain that you have recovered the pellet. The pellet should now contain significantly more radioactivity than the supernatant.
21. Air-dry pellet.

22. Dissolve the pellet in 10 μL water.
23. Add 20 μL of RNA loading buffer only after the pellet is completely dissolved.
24. Heat the RNA at 95 °C for 3 min to denature.
25. Rapidly cool down on ice.
26. Load your sample with a radiolabeled size marker in a neighboring lane.
27. Run the gel until the bromophenol blue band reaches the end of the gel (*see* **Note 24**).
28. Disassemble the gel sandwich so that the gel remains stuck to one of the glass plates.
29. Wrap the glass plate with the gel in plastic wrap.
30. Put two pieces of filter paper at two different positions close to the lanes containing your sample(s) and the size marker.
31. Mix 10 μL of the radioactive supernatant from **step 6** with 10 μL of formamide loading buffer containing bromophenol blue dye and place one or two drops on each filter paper. The radioactive filter paper with the colored dots will be used as position markers to locate your RNA sample in the gel so that it can be cut out. Therefore, carefully put sticky tape on the radioactive filter paper to fix it in a specific position.
32. Place the wrapped gel in an X-ray cassette.
33. Put a phosphorimaging screen on top.
34. Close the cassette and wait at least 5 min before scanning the screen.
35. Print out the image in actual size, either on transparent film or on white paper. Use it to locate your RNA in the gel by arranging the paper so the radioactive signal overlays with the dye-stained dots on the filter paper.
36. Check if your RNA band has the expected size compared to the size markers (*see* **Note 25**).
37. Cut out the gel piece containing your RNA with a scalpel and put the gel piece into a microcentrifuge tube.
38. Add 400 μL of gel elution buffer.
39. Mix and centrifuge the gel piece for a few seconds so that it is completely submerged in the buffer.
40. Freeze the mixture once at −20 °C to loosen the gel structure.
41. Leave overnight at room temperature for the RNA to diffuse out of the gel.
42. The next day vortex the mixture and carefully transfer the supernatant into a new microcentrifuge tube containing 800 μL of 96% ethanol.

43. Centrifuge for 20 min at 20,000 g.
44. Remove supernatant.
45. Here and in the subsequent washing steps, proceed as in **step 16** to ensure you have not lost the RNA pellet. The supernatant should now be almost devoid of radioactivity.
46. Add 350 μL 70% v/v ethanol to the pellet and agitate briefly.
47. Centrifuge for 5 min at 20,000 g.
48. Remove supernatant.
49. Repeat **steps 44–46**.
50. Air-dry the pellet.
51. Dissolve purified RNA in 20 μL ultrapure water (DEPC-treated).
52. Take 1 μL of the RNA solution and put into a vial with scintillation cocktail. This is your elution sample.
53. Add 10 μL of your diluted input sample (**step 12**) to another vial with scintillation cocktail.
54. Measure the radioactivity for both input and elution samples with a scintillation counter. From the results, you can calculate the concentration of your purified RNA.
55. First, calculate the molar amount of UTP present in the reaction; in our example, 0.2×10^{-3} mol/l $\times$ 25 $\times$ 10–6 L = 5×10^{-9} mol = 5000 pmol.
56. Then, calculate the specific activity (cpm/pmol) of the UTP pool in your transcription reaction from the radioactivity in your input sample (*see* **Note 26**):

$$\text{Specific activity} = \frac{\text{counts(input)} \times \text{dilution factor} \times V(\text{transcription})}{V(\text{measured}) \times n(\text{UTP})}$$

$$\text{In our example : specific activity} = \frac{\text{counts(input)} \times 50 \times 25\ \mu\text{L}}{10\ \mu\text{L} \times 5000\ \text{pmol}}$$

Next, use the specific activity of UTP and the radioactivity in your elution sample to calculate the amount of purified RNA based on the *U* content of your RNA:

$$n(\text{RNA}) = \frac{\text{counts(elution)} \times V(\text{RNA solution})}{V(\text{measured}) \times \text{specific activity} \times \text{number of } U \text{ residues in RNA}}$$

For example, for an RNA with 19 uridine residues:

$$n(\text{RNA}) = \frac{\text{counts(elution)} \times 20\ \mu\text{L}}{1\ \mu\text{L} \times \text{specific activity} \times 19}$$

Finally, divide the amount of purified RNA by the volume of the solution (in our example 20 μL) to determine the RNA concentration:

$$c(\text{RNA}) = \frac{n(\text{RNA})}{V(\text{RNA solution})}$$

Expect concentrations in the range of 0.5–3 μM depending on the size of the RNA.

57. Dilute the RNA to an appropriate working concentration (e.g., 0.5 μM) with water (DEPC-treated). Store at –20 °C.

3.3 *In Vitro* Deadenylation Assay with Recombinant *D. melanogaster* CCR4-NOT

1. The general deadenylation reaction contains deadenylation buffer, radioactively labeled RNA substrate, and CCR4-NOT complex.
2. For radioactive substrate RNA, use 5–20 nM per 20 μL reaction (*see* **Notes 27** and **28**).
3. Prepare deadenylation reactions in a volume of at least 20 μL for easier handling. Use multiples of your standard reaction volume for extended kinetics, etc. (*see* **Note 29**).
4. First, add 1/10 of the desired final volume of 10× deadenylation buffer (in this example 2 μL per reaction) to the appropriate amount of water, taking into account the volumes of any additional reagents that you are going to add (*see* **Note 30**).
5. Add 2 mM DTT and nuclease-free BSA (to 150 μg/mL) and mix well.
6. Add RNase inhibitor (to 800 U/mL), any additional reagents, and mix.
7. Add substrate RNA, mix again, and preincubate for 5 min at 25 °C.
8. Start the deadenylation reaction by adding the CCR4-NOT complex, mix, and continue to incubate at 25 °C (*see* **Notes 31** and **32**).
9. Take 5 μL aliquots of the reaction at different time points, quench with 15 μL ice-cold stop solution, and store on ice (*see* **Note 33**).
10. Denature the RNA for 5 min at 95 °C and then chill on ice.
11. Prepare a denaturing polyacrylamide gel as described in Subheading 3.2, **steps 6–11** (*see* **Note 34**). Acrylamide concentration will vary depending on the sizes of the substrate RNA and the fully deadenylated product. For small substrates below 60 nt, use 20% acrylamide/bisacrylamide (19:1) gels, use 12% gels for intermediate substrates (100–300 nt), and 5% gels for long substrates (above 500 nt).

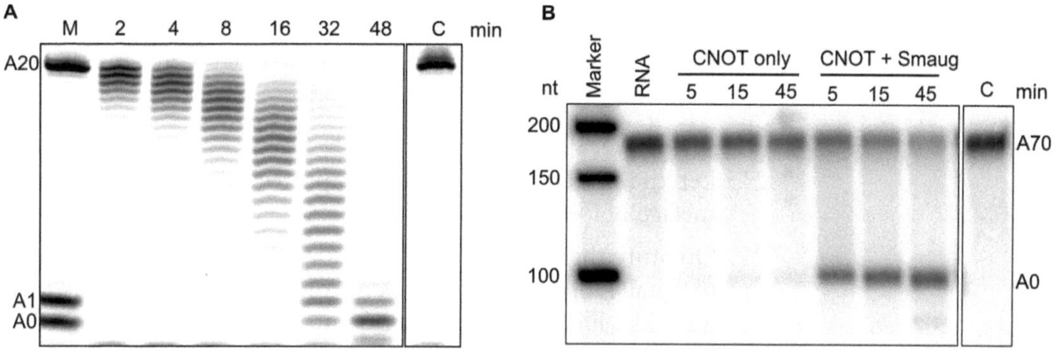

Fig. 4 Time course of basal and targeted deadenylation. (**a**) CCR4-NOT$_{MINI}$ complex (*see* Fig. 3) was incubated with an eightfold excess of fluorescently labeled substrate RNA carrying a poly(A) tail of 20 nucleotides, and aliquots were withdrawn at the indicated time points. Size markers (M) and control reaction without the addition of CCR4-NOT (C) were loaded for comparison. The fact that the entire RNA population was deadenylated synchronously, even though CCR4-NOT was limiting, indicates a distributive reaction mechanism. This means the enzyme dissociates from its RNA substrate after cleaving off one or perhaps a few nucleotides. By random re-association with substrates, CCR4-NOT deadenylates the entire population. (**b**) Radioactively labeled substrate RNA carrying a poly(A) tail of 70 nucleotides and a Smaug recognition element was preincubated with the RBP Smaug or with buffer. Afterward, the CCR4-NOT$_{MINI}$ complex was added. RNA was present at a tenfold excess over enzyme. Aliquots were withdrawn at the indicated time points. Size markers (M) and control reactions with only RNA or RNA with Smaug (**c**) were loaded for comparison. Smaug-independent deadenylation activity was low due to the presence of yeast tRNA. In the Smaug-containing reaction, a fraction of the substrate RNA was deadenylated completely, while the remaining fraction remained full length. This indicates a processive reaction in which the enzyme remains associated with one substrate molecule through many cycles of catalysis. Only after the deadenylation of this RNA molecule is complete does CCR4-NOT dissociate and reassociate with another RNA molecule

12. Load the samples on the denaturing polyacrylamide gel. Alternatively, samples can be stored at −20 °C for at least a month. The sensitivity of detection will drop with storage due to radioactive decay. If you have stored the samples, perform the heat denaturation immediately before electrophoresis.
13. Always run suitable size markers on the gel. Denature DNA size markers for 5 min at 95 °C before loading. Also, load an aliquot of substrate RNA that has not been incubated. This will provide a marker for the unreacted substrate and a control for its quality. Finally, include a reaction that has been incubated in the absence of CCR4-NOT. This will reveal nonspecific RNA degradation during incubation under reaction conditions (Fig. 4a, b).
14. You can estimate the migration of your substrate and product by following the dye migration (*see* [27]).
15. At the end of the run, remove one of the glass plates so that the gel is still sticking to the other plate (*see* **Note 35**).
16. Wrap the gel in plastic wrap.
17. Put the gel in an X-ray cassette.

18. Place a phosphorimaging screen on the gel.
19. Put the closed cassette at −20 °C to freeze the gel and prevent diffusion of the RNA.
20. Expose overnight or longer if necessary.
21. Let the screen pre-warm for 5 min on a room temperature surface before scanning (Fig. 3).
22. Quantify the amount of fully deadenylated product compared to total radioactivity with densitometric image analysis software like Fiji [28] or ImageQuant.

3.4 Assaying Targeted Deadenylation

The deadenylation of specific RNAs can be accelerated by RBPs that bind to specific sites in the RNA and recruit the CCR4-NOT complex (Fig. 4b). The overall procedure for the in vitro deadenylation assay employing such deadenylation specificity factors follows the methods described above (*see* Subheading 3.3). To analyze the specific effect of RBPs on the deadenylation of their target RNAs, a comparison should be made between the deadenylation of bound target RNA and non-bound control RNA. A point mutation in the RBP binding site that prevents protein binding provides the ideal negative control. With this control, one can tell if the RBP specifically recruits CCR4-NOT to its target RNA to enhance deadenylation or if the RBP acts as a stimulator of basal CCR4-NOT activity independently of target RNA binding. Deadenylation assays are sensitive to contaminating RNases. A control reaction with just the RBP present in the absence of CCR4-NOT will reveal nuclease contamination in the RBP preparation. Another negative control without any protein will reveal nuclease contamination in the reaction buffer.

1. Mix wild-type and mutant control RNA separately with your RBP in the deadenylation buffer (*see* **Notes 37** and **38**).
2. Include at least one reaction per RNA and RBP combination to which you do not add CCR4-NOT and another to which you add no protein.
3. In parallel, prepare reactions without the addition of RBP, which will be used to determine the basal level of deadenylation by CCR4-NOT alone.
4. Preincubate all reactions at the appropriate temperature (in case of *Drosophila* proteins, 25 °C) to allow complex formation between the substrate RNA and the RBP (*see* **Note 37**).
5. Start the deadenylation reaction with the addition of CCR4-NOT.
6. Mix gently and perform a time course of deadenylation as described above (*see* Subheading 3.3).

4 Notes

1. A detailed cloning strategy and list of DNA primers used for the amplification of the ORFs are included in [13]. We have added N-terminal His tags to NOT1, NOT2, NOT3, and CAF40. We have not used the His tags for purification or tested whether they affect expression. For the production of the eight-subunit CNOT complex, including the NOT10/NOT11 heterodimer, the NOT1 variant PC must be used, which contains the NOT10/NOT11 binding site. NOT1 variant PE lacks the binding site.
2. Note that the expression of some proteins involved in deadenylation is limited to early embryos. Depending on the cDNA to be amplified, RNA from older stages may or may not work.
3. PMSF decays rapidly in aqueous solutions. Prepare a 100 mM stock solution by dissolving 17.4 mg PMSF per mL of isopropanol and add the necessary volume to the protein purification buffer immediately before use with vigorous stirring.
4. A significant portion of CCR4-NOT sticks to the column filters. Therefore, it is strongly advised to remove the filters to improve yields.
5. No cellulose-based concentrator units should be used because CCR4-NOT tends to stick to the cellulose membrane.
6. Depending on the promoter present in the plasmid, other RNA polymerases like T7 or SP6 may have to be used.
7. NTPs must be stored in a buffered solution at neutral pH; otherwise, the phosphoanhydride bonds will be hydrolyzed. Either purchase NTPs in a buffered solution or dissolve them in Tris buffer or similar. Check the pH by applying a drop of the solution to pH indicator paper and add more base if necessary. Unless you rely on commercial nucleotide solutions, determine the exact NTP concentrations based on their molar extinction coefficients and absorbance measurements.
8. Alternatively, 5′-fluorescently labeled RNA substrates with desired poly(A) tail lengths can be commercially synthesized and HPLC-purified. We have tested only A20 tails, but longer tails (e.g., A60) can also be synthesized.
9. We have only tested the reconstituted CCR4-NOT complexes at pH 7.4. It has been shown for the human CCR4-CAF1 heterodimer that CAF1 activity is strongly influenced by pH and magnesium ion concentration [29]. In the context of the CCR4-NOT complex, we have observed no deadenylation activity in the absence of magnesium ions and optimal activity above 1 mM magnesium acetate. Furthermore, the deadenylation activity of the complex is strongly inhibited by high salt

concentrations (>100 mM potassium acetate). Therefore, total salt concentration, including salts introduced with protein solutions, should be kept below 100 mM.

10. Depending on the size of the analyzed RNA species, the xylene cyanol dye may comigrate and interfere with fluorescence scans. The dyes will not interfere with the detection of radioactive substrate.
11. It is recommended to re-streak the initial white colony on another LB agar blue/white selection plate to confirm that the colony is white and has the desired construct integrated into the bacmid DNA. The blue/white selection plate can be incubated for another 24 h at 6 °C to intensify the blue staining, making the selection easier.
12. The integration of the correct sequence into the bacmid DNA should be checked by colony PCR using primers hybridizing to sites flanking the mini-*attn*T7 site, where the integration occurs.
13. Be very gentle with isolated bacmid DNA in all steps because it is quite fragile due to its large size. Do not vortex the DNA solution; mix by gently flicking the tube or inverting it multiple times.
14. Bacmid DNA is best used in the first month after preparation but can be stable for up to 1 year.
15. Signs of infection include larger cells, floating cells, and holes in the cell layer.
16. At this stage, cells can be frozen in liquid nitrogen and stored at −80 °C until further processing. In this case, thaw cells in a room temperature water bath and then proceed with sonication.
17. Lysis status can be checked by examining cells under a light microscope.
18. Cut off the end of the pipette tip to minimize the damage to the matrix due to pipetting.
19. Analyze the fractions of the SEC chromatography by SDS–PAGE and pool only fractions that show few or no contaminating bands and approximately stoichiometric ratios of subunits as judged by Coomassie staining. In a good preparation, the subunits of the CCR4-NOT complex are the dominant bands in the gel except for one contamination running a little above NOT2 (Fig. 3b). The pooled and concentrated fractions should be tested in a deadenylation assay. To estimate the deadenylation activity, we used a 5′-fluorescently labeled RNA substrate with an A20 tail downstream of a seven nucleotide sequence (Fig. 4a). An eightfold excess of substrate RNA over CCR4-NOT usually results in a deadenylation rate of

around 0.5 adenosines removed per minute. This is similar for all CCR4-NOT subcomplexes tested except the CCR4-CAF1 heterodimer, which is about 50-fold less active.

20. Absorbance measurements have the disadvantage that any contaminating protein contributes, so they are accurate only with highly purified proteins. However, due to the very high absorbance of RNA at 260 nm, even small RNA contaminations become visible. In-gel quantification is not affected by protein contaminations (unless bands coincide), but the results are not very accurate since the extent to which proteins are stained with Coomassie varies. RNA contaminations will not be detected by in-gel quantification.
21. For use in deadenylation reactions, thaw a tube and mix gently before removing the aliquots. The remaining protein can be frozen in liquid nitrogen and stored at −80 °C for further use. In our hands, an aliquot can be frozen and thawed at least ten times without noticeable activity loss.
22. RNA polymerases will add a small number of random non-templated nucleotides to the 3′ ends of a fraction of their transcripts [26]. Since CCR4-NOT is adenosine-specific, these random nucleotides may retard deadenylation.
23. Note that the transcript is always slightly heterogeneous in length. This may be due to some degree of genetic instability of the homopolymeric coding sequence during propagation of the plasmid and/or slippage of RNA polymerase during transcription.
24. Even though the RNA has been ethanol-precipitated and the pellet washed, some non-incorporated nucleotides are usually carried over. These are likely to have run out of the gel. At the end of electrophoresis, treat the anode buffer as radioactive waste. Do not pour it down the drain unless you have counted an aliquot in the scintillation counter and demonstrated the absence of radioactivity. (A Geiger counter is not very useful for this purpose, as water absorbs the radiation quite efficiently.)
25. Cut out the main RNA band if your size marker did not contain enough radioactivity to become visible during the short exposure time. After you have purified the RNA, run an aliquot on an analytical gel next to a size marker. Expose long enough that you can see both and make sure that your RNA has the expected size.
26. Specific activity is defined as the amount of radioactivity (counts per minute, cpm) per compound (mol) amount. The UTP pool is composed of the radioactive and non-radioactive UTP. Since the RNA polymerase cannot distinguish between labeled UTP and unlabeled UTP, the specific activity is calculated for the sum of both sources of UTP. In this calculation,

the total amount of UTP is assumed to be equal to the amount contributed by the non-radioactive nucleotide; the contribution of ^{32}P-labeled UTP to the total amount is minor due to the very high specific activity of the labeled nucleotide and can be ignored. As the specific activity of UTP does not change upon its incorporation into RNA, the specific activity of RNA can be calculated from the specific activity of UTP and the number of U residues present in the RNA. During subsequent manipulations of the RNA, the chemical amount present can always be calculated from the radioactivity. Remember, though, that the specific activity decreases with time. The specific activity can be calculated from the initial value and the $t_{1/2}$ of ^{32}P (14.3 days).

27. Lower concentrations also work, but longer exposure times may be necessary for the detection of radioactivity.
28. For fluorescent substrate, RNA optimal signal is detected with 50 nM per reaction (Cy2 fluorescent label).
29. Typically, two or more reactions are run in parallel that differ in the nature or concentration of one component. To perform *n* reactions in parallel, assemble a total volume corresponding to *n*+1 reactions to accommodate pipetting inaccuracies. This reaction mixture should contain all reagents that are common to the reactions to be performed in parallel. Mix well, dispense 1× reaction volumes into *n* tubes, and then add the reagent(s) in which the reactions differ.
30. The addition of BSA greatly increases the stability of the CCR4-NOT complex under reaction conditions. Other reagents to be added may include, for example, RNA-binding proteins targeting specific RNAs, yeast, or bacterial tRNA to improve the specificity of targeted deadenylation and/or reduce RNase activity by serving as a nonspecific carrier, or PEG to simulate molecular crowding. If they are part of your standard reaction mix, one or more reagents can be incorporated into the 10× buffer.
31. Usually, CCR4-NOT corresponding to half the substrate RNA concentration shows high deadenylation activity (ca. 1 adenosine removed/min [13]).
32. Some additional hints for assembling reactions: Thaw frozen reagents completely and mix before you remove an aliquot to put into the reaction mix. Pay attention to the order in which you assemble reagents. Never dilute a protein into pure water. Instead, mix water and 10× reaction buffer first. Only add an RNase inhibitor when DTT is present; otherwise, the inhibitor may lose activity. Once the protein is present, mix gently, as foaming can denature proteins. Either vortex very gently or hold the upper end of the tube between two fingers and flick it

gently with a finger of the other hand. Add RNA late, certainly after RNase inhibitor and, if included, carrier tRNA. This decreases the likelihood of nonspecific degradation. We usually ignore small changes in buffer composition introduced by the addition of minor volumes of reagents in "normal" buffers. The larger the volume added and the more unusual the buffer composition (typically high salt), the more important it becomes to perform controls to exclude buffer effects. Water and simple salt solutions can be pipetted repeatedly with the same tip without losing accuracy. However, protein solutions wet the surface of the tip, so a fresh tip has to be used every time.

33. If your samples contain higher protein concentrations, purification of the RNA by proteinase K digestion and/or phenol extraction, followed by ethanol precipitation, may be necessary. Ethanol precipitation may require a carrier (e.g., glycogen or tRNA).
34. Run radioactive samples on thin gels (0.2 mm) to obtain the best resolution, whereas fluorescent samples are best run on thicker gels (1 mm).
35. For fluorescent RNA: Remove the gel from both glass plates and submerge it in water. Directly put the gel on the imaging surface and scan for fluorescence corresponding to the fluorescent label.
36. Deadenylation by CCR4-NOT is inhibited by salt concentrations above 100 mM. Ensure that adding your RBP does not introduce excessive salt into the reaction. If necessary, concentrate and/or desalt your RBP.
37. Normally, the RNA should be saturated with the RBP in the deadenylation assay. For this purpose, the RBP should be present at a concentration at least equimolar with its binding site or in slight stoichiometric excess. Also, at least the protein, if not the RNA, should be present at a concentration sufficiently high above the K_D. The extent of complex formation can be checked by native gel electrophoresis: Set up a series of binding reactions containing radiolabeled RNA at the same fixed concentration as in your deadenylation assay and add concentrations of the RBP varying over a suitable range. Load all samples on a native gel. Typically, a low-percentage polyacrylamide gel in 0.5× TBE can be used. The RNA-RBP complex will migrate slower than the free RNA so that the dependence of RNA binding on protein concentration can be determined. Under the appropriate conditions and with quantification of the data, an apparent K_D can also be estimated. Use a comparable RNA with a mutant binding site to control

for binding specificity. Most importantly, include a no-protein control, as retarded bands can also be due to conformational heterogeneity of the RNA.

38. We have observed that deadenylation promoted by *Drosophila* RBPs Smaug and Cup is strongly stimulated by a macromolecular crowding reagent (PEG 20,000 at 3%) [13]. We do not know if this is generally true for all RBP-dependent deadenylation reactions. The basal deadenylation activity of CCR4-NOT is not affected by PEG. CAF1 is highly sensitive to pH and Mg^{2+} and Zn^{2+} ions [29], so these are variables that one might wish to test also in the context of RBP-dependent deadenylation.

39. At the concentrations typically used, RNA–protein complex formation will usually take seconds. However, there are exceptions, for example, if the RNA is present in different conformations, of which only one will bind the RBP. In this case, a slow conformational transition of the RNA can limit the rate of complex formation. One can investigate the time dependence of complex formation by varying the pre-incubation time before the addition of CCR4-NOT and checking if this results in changes in the deadenylation kinetics.

References

1. Goldstrohm AC, Wickens M (2008) Multifunctional deadenylase complexes diversify mRNA control. Nat Rev Mol Cell Biol 9: 337–344
2. Eisen TJ, Eichhorn SW, Subtelny AO et al (2020) The dynamics of cytoplasmic mRNA metabolism. Mol Cell 77:786–799
3. Passmore LA, Coller J (2021) Roles of mRNA poly (A) tails in regulation of eukaryotic gene expression. Nat Rev Mol Cell Biol 23:93–106
4. Webster MW, Chen Y-H, Stowell JAW et al (2018) mRNA Deadenylation is coupled to translation rates by the differential activities of Ccr4-Not Nucleases. Mol Cell 70:1089–1100
5. Muhlrad D, Decker CJ, Parker R (1994) Deadenylation of the unstable mRNA encoded by the yeast MFA2 gene leads to decapping followed by 5′-->3′ digestion of the transcript. Genes Dev 8:855–866
6. Temme C, Zhang L, Kremmer E et al (2010) Subunits of the Drosophila CCR4-NOT complex and their roles in mRNA deadenylation. RNA 16:1356–1370
7. Schäfer IB, Yamashita M, Schuller JM et al (2019) Molecular basis for poly (A) RNP architecture and recognition by the Pan2-Pan3 Deadenylase. Cell 177:1619–1631
8. Harnisch C, Moritz B, Rammelt C et al (2012) Activity and function of Deadenylases. Enzymes (Essen) 31:181–211
9. Wahle E, Winkler GS (2013) RNA decay machines: Deadenylation by the Ccr4–Not and Pan2–Pan3 complexes. Biochim Biophys Acta 1829:561–570
10. Stowell JAW, Webster MW, Kögel A et al (2016) Reconstitution of targeted Deadenylation by the Ccr4-Not complex and the YTH domain protein Mmi1. Cell Rep 17:1978–1989
11. Raisch T, Chang C-T, Levdansky Y et al (2019) Reconstitution of recombinant human CCR4-NOT reveals molecular insights into regulated deadenylation. Nat Commun 10:3173
12. Mauxion F, Basquin J, Ozgur S et al (2023) The human CNOT1-CNOT10-CNOT11 complex forms a structural platform for protein-protein interactions. Cell Rep 42: 111902
13. Pekovic F, Rammelt C, Kubíková J et al (2023) RNA binding proteins Smaug and Cup induce CCR4-NOT-dependent deadenylation of the nanos mRNA in a reconstituted system. Nucleic Acids Res 51:3950

14. Sgromo A, Raisch T, Bawankar P et al (2017) A CAF40-binding motif facilitates recruitment of the CCR4-NOT complex to mRNAs targeted by Drosophila Roquin. Nat Commun 8:14307
15. Webster MW, Stowell JA, Passmore LA (2019) RNA-binding proteins distinguish between similar sequence motifs to promote targeted deadenylation by Ccr4-Not. elife 8:e40670
16. Lai WS, Stumpo DJ, Wells ML et al (2019) Importance of the conserved carboxyl-terminal CNOT1 binding domain to Tristetraprolin activity in vivo. Mol Cell Biol 39:e00029–e00019
17. Berger I, Fitzgerald DJ, Richmond TJ (2004) Baculovirus expression system for heterologous multiprotein complexes. Nat Biotechnol 22:1583–1587
18. Webster MW, Stowell JAW, Tang TTL et al (2017) Analysis of mRNA deadenylation by multi-protein complexes. Methods 126:95–104
19. Jeske M, Temme C, Wahle E (2014) Assaying mRNA deadenylation in vitro. Methods Mol Biol 1125:297–311
20. Jeske M, Meyer S, Temme C et al (2006) Rapid ATP-dependent Deadenylation of nanos mRNA in a cell-free system from Drosophila embryos. J Biol Chem 281:25124–25133
21. Jeske M, Wahle E (2008) Chapter 6 Cell-free Deadenylation assays with Drosophila embryo extracts. Methods Enzymol 448:107–118
22. Pelosse M, Crocker H, Gorda B et al (2017) MultiBac: from protein complex structures to synthetic viral nanosystems. BMC Biol 15:99
23. Fitzgerald DJ, Berger P, Schaffitzel C et al (2006) Protein complex expression by using multigene baculoviral vectors. Nat Methods 3:1021–1032
24. Felberbaum RS (2015) The baculovirus expression vector system: a commercial manufacturing platform for viral vaccines and gene therapy vectors. Biotechnol J 10:702–714
25. Chambers AC, Aksular M, Graves LP et al (2018) Overview of the Baculovirus expression system. Curr Protoc Protein Sci 91:5.4.1–5.4.6
26. Sarcar SN, Miller DL (2018) A specific, promoter-independent activity of T7 RNA polymerase suggests a general model for DNA/RNA editing in single subunit RNA polymerases. Sci Rep 8:13885
27. Sambrook J, Russell DW (2001) Molecular cloning: a laboratory manual. Cold Spring Harbor Laboratory Press, New York
28. Schindelin J, Arganda-Carreras I, Frise E et al (2012) Fiji - an open source platform for biological image analysis. Nat Methods 9:676–682
29. Chen Y, Khazina E, Izaurralde E et al (2021) Crystal structure and functional properties of the human CCR4-CAF1 deadenylase complex. Nucleic Acids Res 49:6489–6510

Chapter 3

Analysis of Human Endogenous mRNA Deadenylation Complexes by High-Resolution Gel Electrophoresis

Chung-Te Chang

Abstract

In eukaryotic cells, poly(A) tails stabilize mRNA molecules and play a pivotal role in enhancing translational efficiency. Consequently, the enzymatic shortening of these poly(A) tails by deadenylase enzymes has a critical role in post-transcriptional gene regulation. However, deadenylases are usually large, multisubunit, and multifunctional complexes, which complicates their biochemical analysis. This chapter presents a methodology for isolating human deadenylation complexes from endogenous sources and conducting an in vitro deadenylation assay to examine their enzymatic activity. The reactions involving fluorescently labeled synthetic polyadenylated RNAs are subsequently analyzed using high-resolution denaturing polyacrylamide gel electrophoresis.

Key words Poly(A) tail, Deadenylase, CCR4-NOT, Post-transcriptional regulation

1 Introduction

Messenger RNA turnover is a critical step of gene expression, and it depends on the accurate control by both transcriptional regulation and post-transcriptional regulation. The initial and rate-limiting step in mRNA decay is usually shortening the poly(A) tail by deadenylase enzymes [1–4]. These enzymes are important for mRNA stability and for regulating translational efficiency [5–7]. Therefore, dissecting deadenylation complexes using purified factors and performing in vitro deadenylation assays to avoid the effect of multiple interferences is preferable.

Most studies have reported methods for purifying deadenylation complex components from a recombinant system [8–10]. Since deadenylation complexes are formed by multiple proteins in the cell, it is unclear whether the activities of isolated nucleases are representative of their endogenous activities. Furthermore, acquiring deadenylation components from a recombinant

Eugene Valkov and Aaron C. Goldstrohm (eds.), *Deadenylation: Methods and Protocols*, Methods in Molecular Biology, vol. 2723, https://doi.org/10.1007/978-1-0716-3481-3_3,

system is laborious and time-consuming, and the enzymatic activity needs to be assessed from batch to batch.

To gain a deeper insight into the deadenylase molecular mechanism and cellular roles, I used the endogenous CNOT7 complex in human cells as an example. I describe detailed experimental steps and key considerations necessary for reliable and accurate results. Nevertheless, the technique described here could also be applied to analyze the deadenylase activities of other endogenous deadenylation factors or in vitro purified reconstituted proteins. Overall, this highly sensitive and accurate method for visualizing and quantifying RNA fragments in samples provides a valuable tool for analyzing deadenylation activity patterns.

2 Materials

Prepare all solutions using ribonuclease-free ultrapure water and analytical-grade reagents. Prepare and store all reagents at 4 °C (unless otherwise indicated). Follow all waste disposal regulations.

2.1 Overexpressing the Deadenylation Factor in HEK-293T Cells

1. HEK-293T cell line (ATCC CRL-3216).
2. pT7-EGFP-C1-CNOT7 plasmid (Addgene #37325) (*see* **Note 1**).
3. Complete growth medium: Dulbecco's modified Eagle's medium (DMEM) with 10% (v/v) fetal bovine serum (FBS).
4. Opti-MEM (*see* **Note 2**).
5. Lipofectamine 2000 Transfection Reagent.

2.2 Immunoprecipitation of the GFP-Tagged Deadenylation Factor

1. 1× Phosphate-buffered saline (PBS).
2. NET buffer: 50 mM Tris–HCl, (pH 7.5), 150 mM NaCl, 1 mM ethylenediaminetetraacetic acid (EDTA), (pH 8.0), 0.1% (v/v) Triton X-100, and 10% (v/v) glycerol.
3. NET-500 buffer: 50 mM Tris–HCl, (pH 7.5), 500 mM NaCl, 1 mM EDTA, (pH 8.0), 0.1% (v/v) Triton X-100, and 10% (v/v) glycerol.
4. Protein G Sepharose beads.
5. Anti-GFP antibody.

2.3 In Vitro Deadenylation Reaction

1. RNA-polyA substrate: 5′-6-FAM-UCUAAAU$(A)_{20}$ (*see* **Note 3**).
2. Nuclease buffer: 50 mM Tris–HCl (pH 7.5), 150 mM NaCl, 2 mM $MgCl_2$, 1 mM dithiothreitol (DTT), 10% (v/v) glycerol. Store at −20 °C.
3. 2× RNA loading buffer: 95% (v/v) formamide, 0.1% (w/v) sodium dodecyl sulfate (SDS), 0.01% (w/v) bromophenol blue, and 0.35 mM EDTA (pH 8). Store at −20 °C.

2.4 Denaturing Polyacrylamide Gel

1. Urea.
2. 5× Tris–borate–EDTA (TBE) buffer.
3. 40% (w/v) acrylamide/bis solution (19:1).
4. 10% (w/v) ammonium persulfate (APS): Make fresh daily.
5. N, N, N, N'-Tetramethyl-ethylenediamine (TEMED).

3 Methods

3.1 Overexpression of the mRNA Decay Factor in Human Cells

1. Cultivate HEK-293T cells at a density of 3×10^6 cells in a 10 cm tissue culture containing 10 mL of complete growth medium and allow them to incubate for 24 h before transfection until they reach approximately 80% confluency.
2. Dilute 20 μg of pT7-EGFP-CNOT7 plasmid DNA in 500 μL of Opti-MEM.
3. In a separate tube, dilute 50 μL of Lipofectamine 2000 transfection reagent in 500 μL of Opti-MEM.
4. Combine the diluted DNA with the diluted Lipofectamine 2000, mix them gently, and then incubate the mixture for 20 min at room temperature to allow complex formation.
5. Once the transfection complex is prepared, add it dropwise to the cells in the dish. It is important to gently agitate the dish to ensure uniform distribution of cells and then incubate them at 37 °C in a CO_2 incubator for 48 h.
6. After 48 h incubation, the cells are ready for downstream applications.

3.2 Purification of the mRNA Decay Complex from Human Cells

1. Remove the medium from the 10 cm tissue culture and place the dish on ice to start the experiment. Next, rinse the cells with 1 mL of ice-cold 1× PBS to remove any remaining medium.
2. To extract the cellular components, add 1 mL of NET buffer to the 10 cm tissue culture dish and use a cell scraper to detach the cells from the dish carefully. Then, transfer the lysate to a 1.5 mL tube and incubate on ice for 5 min to ensure complete lysis.
3. Following incubation, centrifuge the lysate at 13,000 g for 5 min at 4 °C to pellet any remaining cell debris.
4. Collect the supernatant in a new 1.5 mL tube and add 2.5 μg of anti-GFP antibody to the tube. Incubate the tube with rotation for 1 h at 4 °C to allow the antibody to specifically bind to the GFP-tagged protein of interest (*see* **Note 4**).
5. Resuspend the Protein G beads stock solution and add 50 μL of bead suspension to a new 1.5 mL tube containing 1 mL of NET buffer to wash the beads (*see* **Note 5**).

6. Centrifuge the tube at 1000 g for 1 min to pellet the immunoprecipitated complex and carefully aspirate the supernatant.
7. Transfer the sample to the washed Protein G beads and incubate the lysate–bead mixture while rotating for 1 h at 4 °C to allow the Protein G beads to bind to the antibody-bound GFP-tagged protein complex.
8. Centrifuge the mixture at 1000 g for 1 min at 4 °C and discard the supernatant.
9. Wash the beads twice with 1 mL of NET buffer to eliminate nonspecific binding. Gently mix the beads with the wash buffer and ensure the beads are well-suspended in the buffer for each wash. After mixing, centrifuge the tube at 1000 × *g* for 1 min at 4 °C and then discard the supernatant.
10. To minimize the presence of contaminants on the beads that could potentially affect the accuracy of assay results, wash the beads twice with 1 mL of NET-500 buffer as described in **Step 9** (*see* **Note 6**).
11. To remove residual wash buffer and minimize any potential interference in the downstream assay, wash the beads with 1 mL of nuclease buffer and centrifuge them at 1000 g for 1 min at 4 °C.
12. Use a gel-loading tip to remove as much wash buffer as possible from the beads to reduce protein loss and minimize the sample volume, ultimately improving assay results.
13. Add 50 μL of nuclease buffer to the CNOT7 complex-bound beads. This step prepares the beads for use in the in vitro deadenylation assay by ensuring the appropriate buffer conditions are present for optimal enzymatic activity.

3.3 In Vitro Deadenylation Assay

1. In a 1.5 mL tube, combine 30 μL of immunoprecipitated CNOT7 complex beads with 5 μL of 25 μM 5′-6-FAM-labeled RNA-PolyA (final concentration of 0.1 μM). Gently mix the components and adjust the final volume to 50 μL with nuclease buffer (*see* **Note 7**).
2. Incubate the reaction mixture at 37 °C for the indicated time (*see* **Note 8**).
3. To stop the reaction, add 150 μL of 2× RNA loading buffer to the reaction mixture and thoroughly mix it (*see* **Note 9**).
4. Once the deadenylation reaction is complete, the sample should be incubated at 95 °C for 10 min to inactivate the deadenylase enzyme and stop the reaction. The sample should be immediately transferred to the ice to prevent further degradation of the RNA fragments.

5. After incubation, the sample should be centrifuged at 13,000 g for 1 min to pellet the beads. The supernatant can then be carefully collected for downstream analysis.

3.4 Analysis of RNA Fragments by Denaturing Polyacrylamide Gel Electrophoresis

1. To prepare the denaturing polyacrylamide gel (*see* **Note 10**), dissolve 15 g of urea, 4 mL of 5× TBE buffer, and 20 mL of 40% (w/v) acrylamide/bis solution (19:1) in a 100 mL beaker to achieve a final concentration of approximately 6 M urea. Stir vigorously on a magnetic stirrer until all components dissolve completely (*see* **Note 11**).
2. Before pouring, add 40 μl of TEMED and 0.4 mL of 10% (w/v) APS to the acrylamide solution and mix thoroughly (*see* **Note 12**). When pouring the acrylamide solution into a gel sandwich, tilt the sandwich at a 45° angle and pour the solution slowly and carefully. Be cautious when inserting the comb to avoid introducing bubbles into the solution.
3. Once the gel has polymerized, carefully remove the comb and pre-run the gel at 300 V for 30 min in 1× TBE buffer to eliminate ions that increase conductivity and may interfere with the result.
4. To ensure proper loading of the deadenylation assay sample onto the gel, it is crucial to first rinse the wells with 1× TBE buffer to remove the remaining urea, and then, carefully load 10 μL of the sample into well using a gel-loading tip.
5. To prevent sample heating and distortion, run the gel at 200 V for 1 h. Then, increase the voltage to 300 V until the bromophenol blue dye front reaches two-thirds of the gel to ensure complete separation of the RNA fragments.
6. After the electrophoresis is complete, remove the gel sandwich from the apparatus. To avoid damaging the gel, carefully separate the glass plates while laying the sandwich flat on paper towels (*see* **Note 13**).
7. Place the gel in a container with distilled water and agitate gently for 10 min to remove the urea, which can interfere with downstream applications. Change the water several times during this process to ensure thorough removal of the urea.
8. Once the gel is ready, the RNA fragments can be accurately quantified and compared using a Typhoon RGB Biomolecular Imager (GE Healthcare) or equivalent equipment (Fig. 1). This allows for precise analysis of the sample bands, and the resulting images and data can be processed using image analysis software to extract information about the size, abundance, and distribution of the RNA fragments.

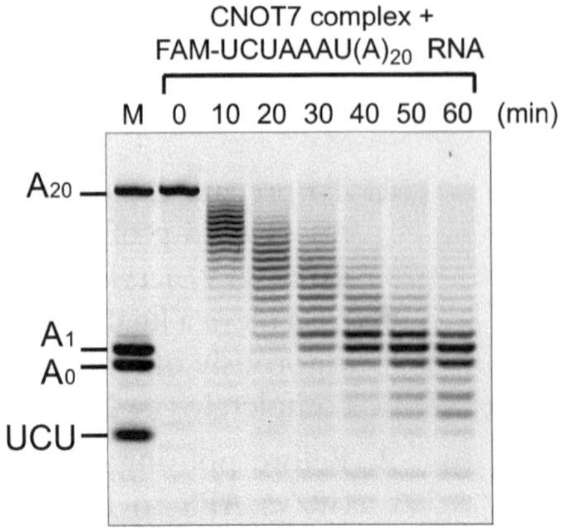

Fig. 1 Visualization of deadenylation kinetics by denaturing polyacrylamide gel electrophoresis

The immunoprecipitated CNOT7 complex was mixed with a 5′-6-FAM-labeled RNA body consisting of a 7-mer (UCUAAAU) and a tail containing 20 adenosines (A_{20}). The reaction was then analyzed at indicated time points, and the resulting RNA fragments were separated using a 20% denaturing polyacrylamide gel. Finally, the RNA fragments were visualized by scanning the gel with a Typhoon RGB Biomolecular Imager.

4 Notes

1. This step refers to using the human CNOT7 protein as an example in the assay. However, you can use any other decay factors of your interest to analyze their activity.
2. Serum-free or serum-reduced media can be used for transfection instead of Opti-MEM to avoid potential interference from serum components that may affect either the transfection efficiency or the assay outcome.
3. This step involves using a specific RNA sequence to analyze deadenylation activities. For this purpose, the RNA sequence used in previous studies is employed. However, it is important to note that alternative RNA substrates can be used in the assay to investigate different patterns of deadenylation activities.
4. Using anti-GFP antibodies from different manufacturers may require different amounts to achieve optimal immunoprecipitation results.
5. For preparing a small volume of Protein G Sepharose beads, it is recommended to create a master mix of the required buffers and then distribute the beads into 1.5 mL tubes. This method ensures consistent bead quality and minimizes batch-to-batch variability.

6. For optimal results, it may be necessary to test and adjust the salt concentration to find the most effective concentration for the specific experiment.
7. Using recombinant purified decay factors in the assay can be a useful alternative to purifying the protein of interest from the cells. This approach allows for better control of potential interfering factors that may be present in the sample, thus reducing variability and increasing the reliability of the assay results.
8. The incubation time may vary depending on the experiment. Testing different time points can optimize the assay, and assuming that the optimum incubation time is between 30 and 60 min is a good starting point.
9. I have noticed that the quality of the assay results can be improved by increasing the concentration of the RNA loading buffer. As such, using a higher concentration of RNA loading buffer is recommended to obtain better results.
10. To achieve better separation of RNA fragments, it is recommended to use a larger-size gel, such as a midi gel system. This will allow the RNA fragments to travel a greater distance, resulting in improved separation.
11. To dissolve 6 M urea, it takes about 0.5~1 h at room temperature. Heating the gel mixture can accelerate the dissolution of the urea, but it can also cause acrylamide degradation, leading to the lower resolution of RNA fragments during electrophoresis.
12. Adding TEMED and then ammonium persulfate, respectively, can achieve slower gel polymerization, leading to better resolution of RNA fragments during electrophoresis.
13. The urea gel is extremely fragile, and to avoid causing tearing, you need to disassemble the plate with a spatula slowly and cut the edge of the gel.

Acknowledgments

This work was supported by the National Science and Technology Council of Taiwan and the Yen Tjing Ling Medical Foundation.

References

1. Decker CJ, Parker R (1993) A turnover pathway for both stable and unstable mRNAs in yeast: evidence for a requirement for deadenylation. Genes Dev 7:1632–1643
2. Yamashita A, Chang TC, Yamashita Y et al (2005) Concerted action of poly(A) nucleases and decapping enzyme in mammalian mRNA turnover. Nat Struct Mol Biol 12:1054–1063
3. Chen CYA, Shyu AB (2011) Mechanisms of deadenylation-dependent decay. Wiley Interdiscip Rev RNA 2:167–183

4. Temme C, Simonelig M, Wahle E (2014) Deadenylation of mRNA by the CCR4-NOT complex in Drosophila: molecular and developmental aspects. Front Genet 5:143
5. Braun JE, Huntzinger E, Fauser M et al (2011) GW182 proteins directly recruit cytoplasmic deadenylase complexes to miRNA targets. Mol Cell 44:120–133
6. Subtelny AO, Eichhorn SW, Chen GR et al (2014) Poly(A)-tail profiling reveals an embryonic switch in translational control. Nature 508:66–71
7. Lima SA, Chipman LB, Nicholson AL et al (2017) Short poly(A) tails are a conserved feature of highly expressed genes. Nat Struct Mol Biol 24:1057–1063
8. Ito K, Inoue T, Yokoyama K et al (2011) CNOT2 depletion disrupts and inhibits the CCR4-NOT deadenylase complex and induces apoptotic cell death. Genes Cells 16:368–379
9. Webster MW, Stowell JAW, Tang TTL et al (2017) Analysis of mRNA deadenylation by multi-protein complexes. Methods 126:95–104
10. Raisch T, Chang CT, Levdansky Y et al (2019) Reconstitution of recombinant human CCR4-NOT reveals molecular insights into regulated deadenylation. Nat Commun 10:3173

Chapter 4

Quantitative Biochemical Analysis of Deadenylase Enzymes Using Fluorescence and Chemiluminescence-Based Assays

Blessing Airhihen, Lorenzo Pavanello, Maryati Maryati, and Gerlof Sebastiaan Winkler

Abstract

Deadenylase enzymes play a key role in mRNA degradation and RNA processing. In this chapter, we describe two activity assays for the quantitative biochemical analysis of deadenylase enzymes, which can easily be adapted for other nuclease enzymes. The assays use distinct principles of detection, which are based on differential annealing of a probe complementary to the substrate RNA or detection of adenosine monophosphate (AMP). The assays are sensitive, flexible, and can be used in low-throughput tube-based formats and 96-well or 384-well plate formats. The assays rely on plate reader detection and can be carried out using manual pipetting or robotic liquid handling equipment. In addition to two activity assays, we describe differential scanning fluorimetry (thermal shift assay) as a complementary assay that allows the direct characterization of ligand binding to deadenylase enzymes. The assays can be useful for the characterization of deadenylase variants and are particularly suitable for the discovery and development of small-molecule inhibitors of deadenylase enzymes.

Key words Deadenylase, Nuclease, Enzyme, Screening, Inhibitor, Assay

1 Introduction

Deadenylation of the poly (A) tail of cytoplasmic mRNA is the first step in the 5′-3′ mRNA degradation pathway in eukaryotic cells [1, 2]. Several deadenylase enzymes have been identified, including the Caf1 and Ccr4 subunits of the Ccr4-Not complex and the Pan2 subunit of the Pan2-Pan3 complex [3, 4]. Deadenylase enzymes are also important in other pathways. For example, PARN and TOE1 are important for processing diverse noncoding RNAs, including small nucleolar RNA, small nuclear RNA, and human TR, the telomerase RNA component [5–7].

Various assays are available to determine the biochemical activity of deadenylase enzymes in vitro. Most commonly, a gel-based assay is used, where degradation of the oligo(A) or poly

Eugene Valkov and Aaron C. Goldstrohm (eds.), *Deadenylation: Methods and Protocols*, Methods in Molecular Biology, vol. 2723, https://doi.org/10.1007/978-1-0716-3481-3_4,

(A) substrate is monitored using denaturing polyacrylamide gel electrophoresis [8]. In this method, RNA can be detected using fluorescent or radionuclide labels, or RNA is stained. Another well-established method is based on the solubility of nucleotides in trichloroacetic acid (TCA). In this assay, poly(A) uniformly labeled using radionuclides is used, followed by precipitation of the intact substrate by TCA and quantification of soluble nucleotides by scintillation counting [8]. Further alternatives are also available, including a colorimetric assay based on the association of RNA and methylene blue or size-exclusion chromatography [9–11]. While these assays are robust and have proven to be useful, they require laborious and time-consuming steps, the use of radioactive isotopes, or are not easy to quantify.

The assays described here have several advantages [12–14]. They are sensitive, can be carried out in small volumes, are suitable for automated liquid handling, and are based on quantitative readout of chemiluminescence or fluorescence. As such, these assays are suitable for the analysis of enzyme variants containing amino acid substitutions and the determination of K_M and V_{max} values and are particularly suitable for the discovery of drug-like inhibitors by the screening of compound libraries and the pharmacological characterization of inhibitors, for example, by comparison of IC_{50} values [12, 14].

Several small-molecule inhibitors for Caf1/CNOT7 and PARN have been reported [13–16]. In addition to activity assays, the characterization of interactions between inhibitors and the protein target has additional value. We, therefore, also include a protocol for differential scanning fluorimetry that is suitable to analyze deadenylase inhibitor interactions. This assay is particularly suitable for the characterization of binding of small, drug-like ligands to deadenylase enzymes [12]. It can be used, for example, to determine the requirement of Mg^{2+} ions for binding of inhibitors to deadenylase enzymes to support a proposed binding mode [12].

1.1 Fluorescence-Based Detection Using a Probe-Based Assay

This assay is based on a two-step reaction (Fig. 1a) [14]. In the first step, an RNA oligo containing a 5′ fluorophore is used as a substrate for the deadenylase enzyme. In the second step, the enzyme is inactivated, and a DNA oligo acts as probe. The DNA probe is complementary to the intact RNA substrate and contains a 3′ quencher. In the absence of deadenylase activity, the probe will anneal to the substrate, thereby quenching fluorescence. By contrast, in the presence of deadenylase activity, the probe will not be able to bind the converted substrate resulting in a fluorescent signal (Fig. 1a).

The short RNA oligo (15 nucleotides) that serves as the substrate contains an anchor sequence followed by nine adenosine residues. It is labeled at the 5′ end with a fluorescein (Flc, or

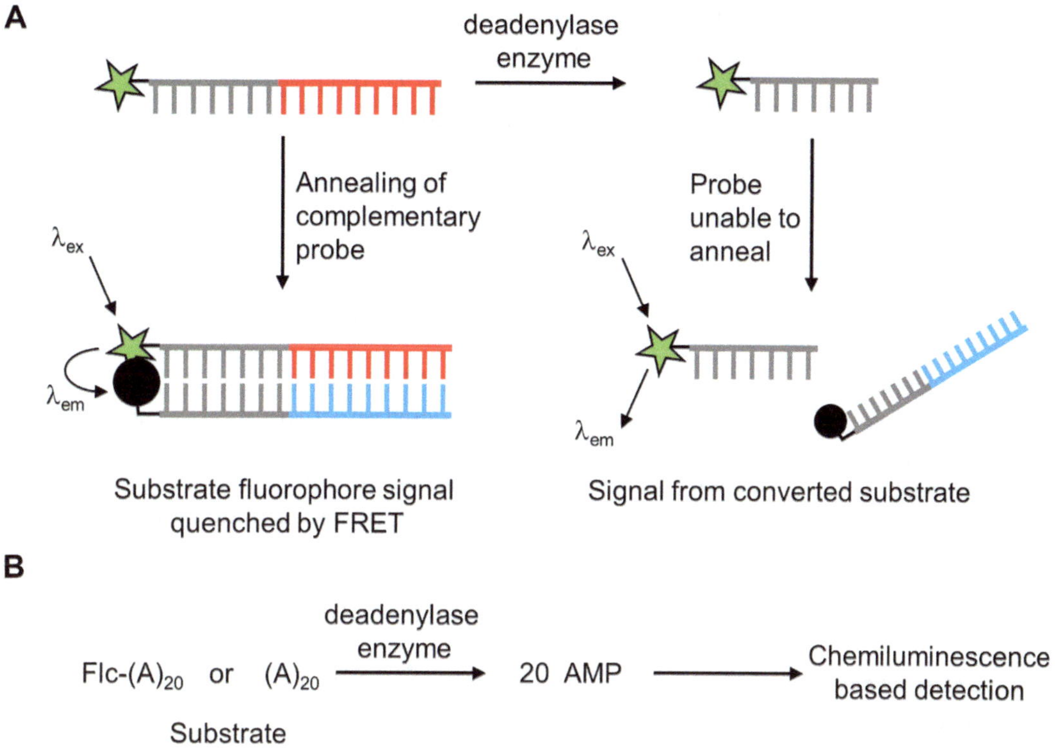

Fig. 1 Quantitative deadenylase activity assays. (**a**) Schematic of the fluorescent probe-based assay [14]. The intact oligo substrate containing a 5′ Flc (or FITC) fluorophore anneals perfectly to a probe containing a 3′ TAMRA fluorophore (or Black Hole Quencher) resulting in the absence of fluorescence due to Förster resonance energy transfer. After the turnover of the substrate and the oligo(A) tail is removed from the oligo substrate containing a 5′ Flc (or FITC) fluorophore, the converted substrate does not anneal to the probe resulting in strong Flc (FITC) fluorescence. (**b**) The AMP detection assay. Either unlabeled poly(A) or poly (A) oligonucleotides containing a 5′ Flc (or FITC) label can be used

Fluorescein isothiocyanate, FITC) fluorophore, although other fluorophores are also suitable. The reaction is terminated by adding sodium dodecyl sulfate to inactivate the deadenylase enzyme. The probe is an oligonucleotide perfectly complementary to the intact substrate and contains a 3′ tetramethyl rhodamine (TAMRA) moiety. TAMRA was selected because it effectively quenches Flc (FITC) when it is in close proximity due to Förster resonance energy transfer. Other quenchers, such as Black Hole Quencher (BHQ-1), are also suitable but do not increase signal/noise in this assay [14].

1.2 Chemiluminescence-Based Detection of AMP

A second orthogonal assay suitable for quantitative analysis of deadenylase activity is based on chemiluminescence-based detection of AMP (Fig. 1b) [12]. In this assay, which is available commercially (Promega), AMP is converted to ADP and ATP by two enzymes. The amount of ATP produced is measured by the light produced in a reaction mediated by the luciferase enzyme [17, 18].

While this reaction does not require a fluorescently labeled RNA substrate, the presence of a Flc (FITC) label does not interfere with this assay (Fig. 1c) [12]. If desired, the reaction products can be analyzed by denaturing polyacrylamide gel electrophoresis. The substrate used in the probe-based assay can also be used in the AMP detection assay. The advantage of this reaction is the assay's high sensitivity, while the reagents' cost can be a disadvantage.

When the potency of inhibitors is determined using the AMP detection assay, the resulting IC_{50} values are comparable to those obtained using the fluorescence probe-based assay [12].

1.3 Product Analysis Using Denaturing Polyacrylamide Gel Electrophoresis

While the probe-based and AMP detection assays provide direct quantitative readouts of deadenylase activity, it is occasionally desirable to carry out product analysis. The fluorescein-labeled oligonucleotide substrate used in the fluorescent probe-based assay can be directly analyzed by denaturing polyacrylamide:bisacrylamide gel electrophoresis (urea-PAGE). If a fluorescently labeled oligonucleotide substrate is used in the AMP detection assay, then the RNA reaction product can also be analyzed directly by denaturing PAGE.

1.4 Ligand Binding Analysis Using Differential Scanning Fluorimetry (Thermal Shift Assay)

A final assay discussed here is useful for the characterization of interactions between deadenylase enzymes and (small-molecule) ligands. Differential scanning fluorimetry (DSF), also known as thermal shift assay (TSA) (Fig. 2a), is based on the principle that the binding of ligands increases the conformational stability of proteins [19–21]. In this assay, the conformational stability of a protein is determined by using a fluorescent dye that has an affinity for hydrophobic regions of a protein. By measuring fluorescence during a temperature gradient (25–95 °C), the conformational stability of the protein can be represented by the melting temperature (T_m) (Fig. 2a). Ligand binding can be expressed as the difference between the melting temperature measured in the absence and presence of ligand, ΔT_m (Fig. 2b). This assay was used to support the proposed binding mode of a series of 1-hydroxy xanthine inhibitors, which require the presence of Mg^{2+} for binding Caf1/CNOT7 [12].

When a series of ligands with a similar binding mode is used, the resulting shift in melting temperature due to the presence of the ligand is proportional to the potency of the inhibitors as determined using the activity assays [12].

Together, the three assays described here are suitable for the quantitative characterization of deadenylase enzymes and are particularly suitable for the discovery and development of small drug-like inhibitors of deadenylase enzymes.

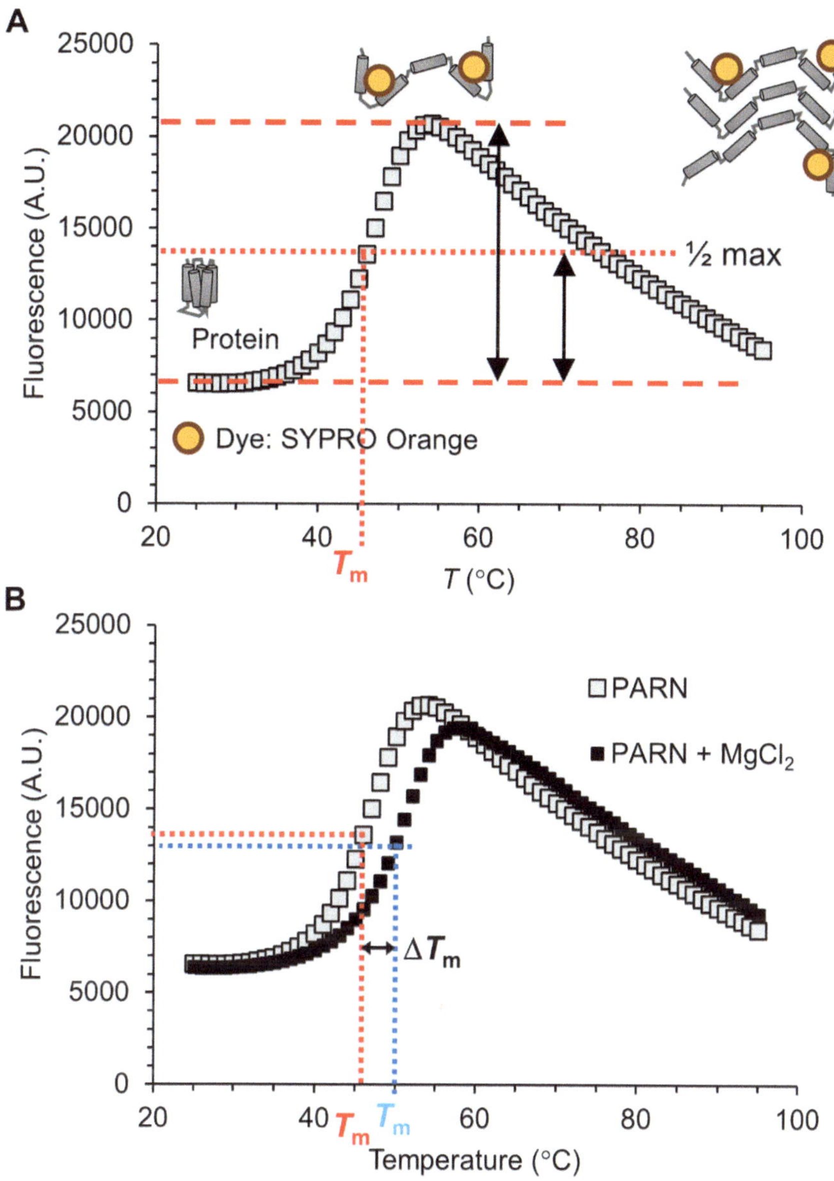

Fig. 2 Overview of differential scanning fluorimetry (DSF)/thermal shift assay (TSA). (**a**) Thermal denaturation of the deadenylase results in increased binding of the fluorescent dye SYPRO orange, which can be detected in a thermal cycler equipped with a qPCR detection system. The T_m, as a measure for the thermal stability of the protein, is determined as the temperature at the midpoint between the minimum (folded) and maximum fluorescence (unfolded). (**b**) Example of ligand binding. The addition of small molecules, such as Mg^{2+} ions or small-molecule inhibitors, increases conformational stability, which can be expressed as the difference (ΔT_m) between the melting temperatures

2 Materials

The deadenylase and thermal shift assays require highly purified deadenylase enzyme preparations. We have used the following proteins expressed in *Escherichia coli* and purified using immobilized metal-affinity chromatography: Caf1/CNOT7 [12–14], Ccr4/CNOT6L [13, 14], the BTG2-Caf1/CNOT7-Ccr4/CNOT6L nuclease complex [12, 22], the central Ccr4-Not sub--complex composed of BTG2-Caf1/CNOT7-Ccr4/CNOT6L-CNOT1(M)-CNOT9 nuclease complex [12, 23], and PARN [13, 14, 24]. Before using purified proteins for these assays, it is essential to ensure that the protein preparations do not contain contaminating nuclease activity, for example, by including nuclease-deficient variants. In addition, the specificity of the purified deadenylase enzymes and the dependence on cofactors, such as Mg^{2+} ions, can be confirmed.

2.1 Fluorescence-Based Assay

1. RNA oligonucleotide containing a 5′ fluorescein or FITC fluorophore (substrate): 5′-FITC-CCU UUC CAA AAA AAA A-3′ (HPLC grade). The oligo should be resuspended in TE [10 mM Tris–HCl, 1 mM EDTA (pH 8.0)] or ultrapure water to give a 100 μM stock solution. Store at −20 °C. Exposure to light should be minimized (*see* **Note 1**).
2. DNA oligonucleotide containing a 3′ tetramethyl rhodamine (TAMRA) fluorophore (probe): 5′-TTT TTT TTT GGA AAG G-TAMRA-3′) (HPLC grade). The oligo should be resuspended in TE [10 mM Tris–HCl, 1 mM EDTA (pH 8.0)] or ultrapure water to give a 100 μM stock solution. Store at −20 °C. Exposure to light should be minimized (*see* **Note 2**).
3. 5× Reaction buffer: 100 mM Tris–HCl, 50% (v/v) glycerol, and 5 mM β-mercaptoethanol (pH 7.9) (*see* **Note 3**).
4. 0.5 M NaCl solution.
5. 20 mM $MgCl_2$.
6. 10% (w/v) sodium dodecyl sulfate (SDS).
7. Ultrapure water (nuclease-free).
8. 96- or 384-well plate suitable for fluorescence measurements (e.g., Greiner FLUOTRAC 200 multi-well plates in 96- or 384-well format) (*see* **Note 4**).
9. Plate reader suitable for fluorescence measurements, such as BioTek Synergy HT plate reader equipped with the following filter set: 485 ± 20 nm (excitation) and 528 ± 20 nm (emission).

2.2 Chem-iluminescence-Based AMP Detection Assay

1. RNA oligonucleotide: $(A)_{20}$. The oligo should be resuspended in TE [10 mM Tris–HCl, 1 mM EDTA (pH 8.0)] or ultrapure water to give a 100 μM stock solution. Store at −20 °C (*see* **Note 5**).
2. 5× Reaction buffer: 100 mM Tris–HCl, 50% (v/v) glycerol, and 5 mM β-mercaptoethanol (pH 7.9) (*see* **Note 3**).
3. 0.5 M NaCl solution.
4. 20 mM $MgCl_2$.
5. Ultrapure water (nuclease-free).
6. AMP detection system: AMP-Glo Kit (Promega) containing the following components: AMP-Glo Reagent I, AMP-Glo Reagent II, Kinase Glo One Solution, and AMP.
7. White 96-well half-area multi-well plate (Corning Costar, Reference 3693) (*see* **Note 6**).
8. Plate reader suitable for chemiluminescence measurements such as GloMax 96 multi-well plate luminometer (Promega).

2.3 (Optional) Product Analysis Using Polyacrylamide Gel Electrophoresis

1. 20% (w/v) polyacrylamide:bisacrylamide (19:1)/50% (w/v) urea/1× TBE gel.
2. 1× TBE gel running buffer.
3. RNA gel loading buffer [95% (v/v) formamide, 0.025% (w/v) bromophenol blue, 0.025% (w/v) xylene cyanol FF, 0.025% (w/v) sodium dodecyl sulfate, and 5 mM ethylenediaminetetraacetic acid (EDTA)].
4. Thermo Fisher XCell system.
5. Fluorescence imager such as Fujifilm LAS-4000 imager equipped with an Epi-Blue illuminator (460 nm).

2.4 Differential Scanning Fluorimetry/ Thermal Shift Assay

1. SYPRO (Thermo Fisher, 5000×) (*see* **Note 7**).
2. 5× Reaction buffer: 100 mM Tris–HCl, 50% (v/v) glycerol, and 5 mM β-mercaptoethanol (pH 7.9).
3. 0.5 M NaCl solution.
4. 20 mM $MgCl_2$.
5. Ultrapure water (nuclease-free).
6. Low-profile 96-well plates with optically clear caps (*see* **Note 8**).
7. Thermal cycler with qPCR detection system (real-time PCR system) such as Agilent Mx3005P Real-Time PCR System.

3 Methods

3.1 Fluorescence-Based Assay

1. Prepare enzyme dilution. The purified proteins CNOT7, CNOT6L, and their sub-complexes, as well as PARN, can be stored in high salt conditions [20 mM Tris–HCl, 10% (v/v) glycerol, 1 mM β-mercaptoethanol, and 250 mM NaCl (pH 7.9)] to avoid aggregation at high concentrations. When setting up reactions, a 5× enzyme solution is prepared freshly by diluting enzymes in the same buffer [20 mM Tris–HCl, 10% (v/v) glycerol, 1 mM β-mercaptoethanol, 250 mM NaCl (pH 7.9)]. Typical enzyme concentrations are 0.04–0.06 μg/μL CNOT7, 0.1 μg/μL CNOT6L, 0.04–0.06 μg/μL BTG2–CNOT7–CNOT6L complex, and 5 ng/μL PARN.
2. Prepare 5× substrate dilution: 5.0 μM oligo substrate obtained by dilution of the 100 μM stock concentration with ultrapure water.
3. Optional: When determining the activity of inhibitor compounds, prepare a dilution series of the compound. A DMSO concentration of up to 10% (v/v) does not interfere with enzyme activity. Typically, compounds are dissolved in 20% (v/v) DMSO/H_2O.
4. Prepare Stop/Probe Mix: 1% (w/v) SDS, 5 μM DNA probe prepared freshly using 10% (w/v) SDS, 100 μM DNA probe solution, and ultrapure water. The stop/probe mix can be kept at room temperature in the dark until required.
5. Set up the reactions: Combine the following reagents to give a final reaction volume of 10 μL:
   - 2 μL 5× enzyme dilution.
   - 1.8 μL 5× reaction buffer.
   - 2 μL 20 mM $MgCl_2$ (or ultrapure water if Mg^{2+} dependence is tested).
   - 2.2 μL Ultrapure water.
   - 2 μL Substrate mix (*see* **Note 9**).

   Mix well, and spin briefly.
6. Incubate the reaction at 30 °C for 60 min.
7. Terminate the reaction by the addition of 10 μL stop/probe mix.
8. Incubate at 30 °C for 10 min.
9. Read fluorescence using the following settings: excitation: 485 nm; and emission: 528 nm (*see* **Notes 10** and **11**).
10. Raw data (relative fluorescence units) can be used for quantitative analysis, such as determining the K_M value or IC_{50} values. If fluorescence of a dilution series of the substrate oligonucleotide is included without the addition of the DNA probe, then data can be converted to molarity.

3.2 Chemiluminescence-Based Detection of AMP

1. Prepare enzyme dilution. The purified proteins CNOT7 and CNOT6L, their sub-complexes, and PARN can be stored in high salt conditions [20 mM Tris–HCl, 10% (v/v) glycerol, 1 mM β-mercaptoethanol, 250 mM NaCl (pH 7.9)] to avoid aggregation at high concentrations. When setting up reactions, a 5× enzyme solution is prepared freshly by diluting enzymes in the same buffer [20 mM Tris–HCl, 10% (v/v) glycerol, 1 mM β-mercaptoethanol, 250 mM NaCl (pH 7.9)]. Typically, 25 nM enzyme (CNOT7, CNOT6L, BTG2-CNOT7-CNOT6L complex, or PARN) is used.
2. Set up the reactions: Combine the following reagents to give a final reaction volume of 10 μL:
   - 2 μL 5× enzyme dilution.
   - 1.8 μL 5× reaction buffer.
   - 2 μL 20 mM $MgCl_2$ (or ultrapure water if Mg^{2+} dependence is tested).
   - 2.2 μL ultrapure water.
   - 2 μL synthetic $(A)_{20}$ oligonucleotide substrate (5 μM diluted from 100 μM stock using ultrapure water) (*see* **Notes 12** and **13**).

   Mix well, and spin briefly.
3. Incubate the reaction at 30 °C for 60 min (*see* **Note 14**).
4. Transfer 3 μL of each reaction into the white low-volume 96-well plate (*see* **Note 15**). Keep the plate in the dark (using foil lid or placing it in a dark cabinet).
5. Add 3 μL of AMP-Glo Reagent I to each reaction.
6. Incubate at room temperature for 1 h.
7. During the incubation (**step 5**), and immediately before addition, prepare a stock of detection solution using AMP-Glo Reagent II and Kinase Glo solution. AMP-Glo Reagent II needs to be diluted 1:100 (e.g., 5 μL AMP-Glo Reagent II + 495 μL of Kinase Glo Solution). Make up only what you will use (*see* **Note 16**).
8. Incubate at room temperature for 1 h.
9. Measure chemiluminescence. When using a GloMax 96 Instrument (Promega), the Kinase Glo Quick Protocol is used.
10. Raw data (relative light units) can be used for quantitative analysis such as the determination of the K_M value or IC_{50} values. Typical signals span two orders of magnitude. The inclusion of AMP solutions with known concentrations allows the conversion of relative light units to molarity.

3.3 Product Analysis Using Polyacrylamide Gel Electrophoresis

1. For gel-based detection of ribonuclease activity, deadenylase reactions were used as described in Subheading 3.1 (fluorescent probe-based assay). Reaction products can be analyzed after the measurement of fluorescence. If reactions for AMP detection use a fluorescent substrate (e.g., FITC-$(A)_{20}$), reaction products can also be analyzed by denaturing PAGE.
2. Add 12 μL RNA loading buffer to 10 μL reactions and heat for 3 min at 85 °C. Place reactions on ice until gel loading.
3. Prepare a 20% (w/v) acrylamide:bisacrylamide (19:1) gel containing 50% (w/v) urea (*see* **Note 17**).
4. Pre-run polyacrylamide gels (8 × 8 cm; Invitrogen XCell system) for 30 min at 200 V.
5. Load a small sample of the RNA mixture (3 μL) into each well (*see* **Note 18**).
6. Run gel at 200 V until loading dye has run out of the gel.
7. Visualize Flc-labeled RNA by epifluorescence, for example, using a Fujifilm LAS-4000 imager equipped with an Epi-Blue illuminator (460 nm) following the instructions of the manufacturer.

3.4 Differential Scanning Fluorimetry/Thermal Shift Assay

1. Prepare enzyme dilution. The purified proteins CNOT7, CNOT6L, and their sub-complexes, as well as PARN, can be stored in high salt conditions [20 mM Tris–HCl, 10% (v/v) glycerol, 1 mM β-mercaptoethanol, and 250 mM NaCl (pH 7.9)] to avoid aggregation at high concentrations. When setting up reactions, a 5× enzyme solution is prepared freshly by diluting enzymes in the same buffer [20 mM Tris–HCl, 10% (v/v) glycerol, 1 mM β-mercaptoethanol, and 250 mM NaCl (pH 7.9)]. Typically, 5–10 μM enzyme (CNOT7, CNOT6L, and PARN) is used (final concentration).
2. Set up the reactions: Combine the following reagents to give a final reaction volume of 20 μL:
   - 4 μL 5× enzyme dilution.
   - 3.2 μL 5× reaction buffer.
   - 2 μL 20 mM $MgCl_2$ (or ultrapure water if Mg^{2+} dependence is tested).
   - 0.1 μL SYPRO Orange (5000×).
   - 9.7 μL ultrapure water.
   - 1.0 μL ligand dissolved in 20% (v/v) DMSO.
3. Seal the plate using optically clear lids.
4. Mix thoroughly and spin briefly (30 s in a centrifuge suitable for microwell plates) to ensure all components are at the bottom of the well.

5. Transfer to thermal cycler with a qPCR detection system.
6. Run a temperature gradient from 25 to 95 °C; hold each step for 60 s.
7. Read fluorescence at each step (*see* **Note 19**).
8. To determine the T_m, data points spanning the transition from minimum to maximum fluorescence can be analyzed using nonlinear regression using scientific graphing programs such as GraphPad Prism (using the Boltzmann equation) or Systat SigmaPlot (using the five-parameter logistic curve equation) [20]. Nonlinear regression using these functions can also be carried out by Microsoft Excel using the Solver add-in package [25].

4 Notes

1. Follow instructions from the manufacturer. RNA is very stable in dried form. When dissolving RNA in a large volume, it may be preferable to aliquot batches and store the aliquots at −80 °C.
2. Follow instructions from the manufacturer. When handling large batches, it may be preferable to aliquot batches and store the aliquots at −20 °C.
3. The pH of the buffer is set at room temperature. Glycerol is not required and can be omitted.
4. The multi-well plates used for the assay should be compatible with the plate reader that is available.
5. The use of fluorescein or FITC-labeled RNA substrates, such as the oligonucleotide substrate used for assay 1, or Flc-$(A)_{20}$ does not interfere with the AMP detection assay and allows direct detection of the product analysis using denaturing PAGE if desired.
6. The multi-well plates used for the assay should be compatible with the plate reader that is available. The 96-well plate suggested here has a reduced surface area, which reduces the volume of reagents required.
7. SYPRO Orange can be obtained from various suppliers. If a reagent from another supplier is used, optimization of the amount of SYPRO Orange is required.
8. The multi-well plates used for the assay should be compatible with the qPCR detection system used.
9. The reaction is assembled in two steps when drug-like ligands are included in the deadenylase assay. First, the enzyme is incubated with the compound:

- 2 μL 5× enzyme dilution.
- 1.8 μL 5× reaction buffer.
- 2 μL 20 mM $MgCl_2$ (or ultrapure water if Mg^{2+} dependence is tested).
- 1.2 μL ultrapure water.
- 1 μL ligand (in 20% (v/v) DMSO).

Before the addition of the substrate oligonucleotide (2 μL), the enzyme/ligand mixture is pre-incubated at 4 °C for 10 min. When solutions containing DMSO are included, it is essential to mix thoroughly and spin briefly (30 s in a centrifuge suitable for microwell plates) to ensure all components are at the bottom of the well.

10. At this stage, the reaction products are stable with no signal loss for up to 7 days when the reaction products are kept at room temperature in the dark.
11. When using a BioTek Synergy HT plate reader, the following settings are used: temperature setpoint: 25; detection method: fluorescence; read type: endpoint; excitation: 485/20; emission: 528/20; optics position: top; sensitivity: 90.

    Occasionally, we found very high background values, which caused interference with fluorescence readings. We could pinpoint the source of the high background readings to some TAMRA-labeled DNA probe batches, which contained undefined contamination even when using an HPLC-purified DNA oligo. The problem was resolved by (i) changing the oligo supplier or (ii) leaving the plates at room temperature for 4–24 h before fluorescence measurement, which reduced or eliminated nonspecific background readings.
12. The reaction is assembled in two steps when drug-like ligands are included in the deadenylase assay. First, the enzyme is incubated with the compound:

- 2 μL 5× enzyme dilution.
- 1.8 μL 5× reaction buffer.
- 2 μL 20 mM $MgCl_2$ (or ultrapure water if Mg^{2+} dependence is tested).
- 1.2 μL ultrapure water.
- 1 μL ligand (in 20% (v/v) DMSO).

Before the addition of the substrate oligonucleotide $(A)_{20}$ (2 μL), the enzyme/ligand mixture is pre-incubated at 4 °C for 10 min. When solutions containing DMSO are included, it is essential to mix thoroughly and spin briefly (30 s in a centrifuge suitable for microwell plates) to ensure all components are at the bottom of the wells.

13. The presence of the fluorescein (Flc) or FITC label does not interfere with the AMP detection assay.
14. In our hands, the AMP detection assay displayed linear kinetics for over 2 h.
15. The low-volume 96-well plate can be reused to reduce costs and environmental waste. To this end, the plate should be rinsed extensively with pure water immediately after use and drained upside down overnight.
16. Kinase Glo tends to make bubbles; be careful when pipetting.
17. The gel can be prepared by transferring 5 g urea to a plastic tube. Add 1 mL 10 TBE and 5 mL 40% (w/v) acrylamide: bisacrylamide (19:1) and warm the mixture at 60 °C until all urea is dissolved (1–3 min). Adjust the volume to 10 mL and allow to cool before adding ammonium persulfate and TEMED to initiate polymerization.
18. Before loading, wells should be carefully washed to remove any precipitated urea.
19. The MX Pro software controlling the Agilent MX3005p machine has a standard SYPRO detection method option using the following filters: FAM (492 nm) and ROX (610 nm) for excitation and emission, respectively.

Acknowledgments

This work was supported by the Medical Research Council (grant G1100205), the Islamic Development Bank, the Schlumberger Foundation, and the University of Nottingham (vice-chancellor's scholarship for research excellence).

References

1. Goldstrohm AC, Wickens M (2008) Multifunctional deadenylase complexes diversify mRNA control. Nat Rev Mol Cell Biol 9: 337–344. https://doi.org/10.1038/nrm2370
2. Parker R, Song H (2004) The enzymes and control of eukaryotic mRNA turnover. Nat Struct Mol Biol 11:121–127. https://doi.org/10.1038/nsmb724
3. Collart MA, Panasenko OO (2017) The Ccr4-Not complex: architecture and structural insights. Subcell Biochem 83:349–379. https://doi.org/10.1007/978-3-319-46503-6_13
4. Wahle E, Winkler GS (2013) RNA decay machines: Deadenylation by the Ccr4-Not and Pan2-Pan3 complexes. Biochim Biophys Acta 1829:561–570. https://doi.org/10.1016/j.bbagrm.2013.01.003
5. Moon DH, Segal M, Boyraz B et al (2015) Poly (A)-specific ribonuclease (PARN) mediates 3′-end maturation of the telomerase RNA component. Nat Genet 47:1482–1488. https://doi.org/10.1038/ng.3423
6. Lardelli RM, Schaffer AE, Eggens VR et al (2017) Biallelic mutations in the 3′ exonuclease TOE1 cause pontocerebellar hypoplasia and uncover a role in snRNA processing. Nat Genet 49:457–464. https://doi.org/10.1038/ng.3762
7. Berndt H, Harnisch C, Rammelt C et al (2012) Maturation of mammalian H/ACA box snoRNAs: PAPD5-dependent adenylation and PARN-dependent trimming. RNA 18:958–

972. https://doi.org/10.1261/rna.032292.112

8. Jeske M, Temme C, Wahle E (2014) Assaying mRNA deadenylation in vitro. Methods Mol Biol 1125:297–311. https://doi.org/10.1007/978-1-62703-971-0_24
9. Cheng Y, Liu WF, Yan YB et al (2006) A nonradioactive assay for poly (a)-specific ribonuclease activity by methylene blue colorimetry. Protein Pept Lett 13:125–128. https://doi.org/10.2174/092986606775101580
10. Greiner-Stoeffele T, Grunow M, Hahn U (1996) A general ribonuclease assay using methylene blue. Anal Biochem 240:24–28. https://doi.org/10.1006/abio.1996.0326
11. He GJ, Yan YB (2012) A deadenylase assay by size-exclusion chromatography. PLoS One 7: e33700. https://doi.org/10.1371/journal.pone.0033700
12. Airhihen B, Pavanello L, Jadhav GP et al (2019) 1-Hydroxy-xanthine derivatives inhibit the human Caf1 nuclease and Caf1-containing nuclease complexes via Mg^{2+}-dependent binding. FEBS Open Biol 9:717–727. https://doi.org/10.1002/2211-5463.12605
13. Jadhav GP, Kaur I, Maryati M et al (2015) Discovery, synthesis and biochemical profiling of purine-2,6-dione derivatives as inhibitors of the human poly (A)-selective ribonuclease Caf1. Bioorg Med Chem Lett 25:4219–4224. https://doi.org/10.1016/j.bmcl.2015.07.095
14. Maryati M, Kaur I, Gopal J et al (2014) A fluorescence-based assay suitable for quantitative analysis of deadenylase enzyme activity. Nucleic Acids Res 42:e30. https://doi.org/10.1093/nar/gkt972
15. Balatsos N, Vlachakis D, Chatzigeorgiou V et al (2012) Kinetic and in silico analysis of the slow-binding inhibition of human poly (A)-specific ribonuclease (PARN) by novel nucleoside analogues. Biochimie 94:214–221. https://doi.org/10.1016/j.biochi.2011.10.011
16. Balatsos NA, Vlachakis D, Maragozidis P et al (2009) Competitive inhibition of human poly (A)-specific ribonuclease (PARN) by synthetic fluoro-pyranosyl nucleosides. Biochemistry 48: 6044–6051. https://doi.org/10.1021/bi900236k
17. Goueli SA, Hsiao K, Zegzouti H (2013) U.-S. Patent 2013/0109037 A1
18. Mondal S, Hsiao K, Goueli SA (2017) Utility of Adenosine Monophosphate detection system for monitoring the activities of diverse enzyme reactions. Assay Drug Dev Technol 15:330–341. https://doi.org/10.1089/adt.2017.815
19. Lo MC, Aulabaugh A, Jin G et al (2004) Evaluation of fluorescence-based thermal shift assays for hit identification in drug discovery. Anal Biochem 332:153–159. https://doi.org/10.1016/j.ab.2004.04.031
20. Niesen FH, Berglund H, Vedadi M (2007) The use of differential scanning fluorimetry to detect ligand interactions that promote protein stability. Nat Protoc 2:2212–2221. https://doi.org/10.1038/nprot.2007.321
21. Pantoliano MW, Petrella EC, Kwasnoski JD et al (2001) High-density miniaturized thermal shift assays as a general strategy for drug discovery. J Biomol Screen 6:429–440. https://doi.org/10.1177/108705710100600609
22. Maryati M, Airhihen B, Winkler GS (2015) The enzyme activities of Caf1 and Ccr4 are both required for deadenylation by the human Ccr4-Not nuclease module. Biochem J 469: 169–176. https://doi.org/10.1042/bj20150304
23. Pavanello L, Hall B, Airhihen B et al (2018) The central region of CNOT1 and CNOT9 stimulate deadenylation by the Ccr4-Not nuclease module. Biochem J 475:3437–3450. https://doi.org/10.1042/bcj20180456
24. Nilsson P, Virtanen A (2006) Expression and purification of recombinant poly (A)-specific ribonuclease (PARN). Int J Biol Macromol 39:95–99. https://doi.org/10.1016/j.ijbiomac.2006.02.025
25. Kemmer G, Keller S (2010) Nonlinear least-squares data fitting in Excel spreadsheets. Nat Protoc 5:267–281. https://doi.org/10.1038/nprot.2009.182

Chapter 5

A FRET-Based Assay and Computational Tools to Quantify Enzymatic Rates and Explore the Mechanisms of RNA Deadenylases in Heterogeneous Environments

Rose Irwin, Robert W. Harkness, and Julie D. Forman-Kay

Abstract

We developed a medium-throughput assay that can measure the time-dependent distribution of RNA products generated as a deadenylase degrades a polyadenosine (poly(A)) RNA tract, thereby providing insight into the mechanism of deadenylation. Importantly, this assay can be performed in both homogeneous and heterogeneous environments without relying on gel electrophoresis of RNA products or coupled enzymatic reactions that indirectly report on the RNA distribution through the detection of freed adenosine monophosphate. In parallel, we have established an open-source, Python-based command-line software package, ***deadenylationkinetics***, that can be used to numerically simulate and/or fit the datasets afforded by our assay with different deadenylation mechanisms to determine the most likely case and estimate the associated rate constants. In this chapter, we detail the implementation of our method and the quantification of poly(A) RNA binding and degradation kinetics in application to a truncated version of CNOT7 from the CCR4-NOT deadenylation complex, which serves as a model deadenylase with enhanced activity.

Key words RNA, Deadenylation, Enzyme, FRET, Enzyme kinetics

1 Introduction

Deadenylation of 3′ polyadenosine (poly(A)) mRNA tails is a major regulatory component of translation and proteostasis [1]. Previous work from our group suggests that biomolecular condensate formation can modulate mRNA deadenylase enzyme activity [2]. We aim to further understand the impact of various condensate environments on RNA deadenylation by screening deadenylase activity in conditions with different condensate-forming proteins, regulatory post-translational modifications, and disorder-associated protein variants. To quantify the impact of these numerous condensate environments on deadenylase activity, we require a high-throughput method that can be employed in both homogeneous

Eugene Valkov and Aaron C. Goldstrohm (eds.), *Deadenylation: Methods and Protocols*, Methods in Molecular Biology, vol. 2723, https://doi.org/10.1007/978-1-0716-3481-3_5,

and heterogeneous conditions, such as a phase-separated sample. Two widely used methods that are currently available for kinetic analysis of deadenylase activity are not optimal in this regard. These are (1) densitometric analysis of poly(A) RNA products resolved by gel electrophoresis and (2) colorimetric detection of liberated adenosine monophosphate (AMP) with a coupled enzyme system (e.g., AMP-Glo (Promega) or AMP Assay Kit (Abcam)).

While the gel electrophoresis method has been successfully implemented to assess deadenylase processivity [3, 4] and, in some cases, estimate steady-state enzymatic parameters [2, 5], its throughput and reliability are severely limited by the time-intensiveness of gel preparation for each desired solution condition and the difficulty in adequately resolving the bands that correspond to RNA species of different lengths for densitometric analysis. Further compounding the issue, the densitometric measurements of the band intensities rely on subjective applications of baselines and subtraction of the background, which can introduce significant error in the resulting intensities. It becomes considerably difficult to analyze heterogeneous mixtures containing multiple phases with this method, since different aliquots removed from the sample for quenching could consist of variable amounts of dilute and condensed phases, precluding accurate kinetic analyses.

In contrast to the low-throughput gel-based analysis, the colorimetric detection of AMP is well-suited to high-throughput measurements in a plate reader instrument. It has been used to qualitatively track poly(A) removal and screen for deadenylase inhibitors [6]. However, any attempts to model this type of data ultimately depend on assumptions about the kinetics of the coupled enzyme system that generates the measured signal from AMP relative to the rate of deadenylation [7–10]. As the kinetics of detection by the coupled enzymes could vary substantially across different solution conditions (e.g., salts, temperatures, and condensed and dilute phases), this can introduce substantial errors in the extracted kinetic parameters [7–10]. Finally, freed AMP is an indirect reporter of the RNA distribution, potentially limiting insight into the deadenylation mechanism that direct measurements of the amounts of each RNA species might otherwise report.

A third method, which directly reports on the RNA distribution, was initially developed by Maryati et al. [11]. In this assay, a fluorescently labeled poly(A) RNA substrate is incubated with a deadenylase, and aliquots are removed from the bulk reaction at desired time points, as is done in the PAGE method. The reaction within the extracted aliquots is quenched by the addition of SDS, which denatures the deadenylase. Added along with the SDS is a fluorescently labeled complementary DNA strand that hybridizes with the individual members of the RNA product distribution, bringing the fluorophores within proximity and leading to loss of fluorescence via quenching. As the reaction proceeds, the poly

(A) RNA distribution shifts toward smaller species with weakened affinities for the DNA probe, leading to less RNA-DNA hybridization and a concomitant increase in fluorescence as the quenching between the two fluorophores is relieved. Importantly, in addition to directly probing the time-dependent RNA product distribution, this approach is readily implemented in a plate reader format for higher throughput screening of deadenylase activity [11]. Considering the strengths of this method and the limitations of the other two approaches, we sought to adapt the Maryati assay to increase its robustness in analyses of deadenylation kinetics in heterogeneous conditions.

We introduced key modifications that allow for the use of this assay in heterogeneous *in vitro* conditions and make it easier and more robust for in-depth, quantitative analysis of deadenylase kinetics. First, we quantify changes in RNA-DNA annealing by Förster resonance energy transfer (FRET) rather than absolute fluorescence. In addition to photon emission and resonance energy transfer, there are numerous pathways by which a fluorophore can be de-excited and thereby exhibit low levels of fluorescence that do not strictly depend on the vicinity of the RNA and DNA fluorophores, such as internal conversion or intersystem crossing [12]. Therefore, we monitor and calculate FRET based on both fluorophores to ensure that signal changes depend on the proximity of the two fluorophores resulting from the specific annealing of the RNA with the DNA. Furthermore, unlike fluorescence, intrinsic FRET is not directly dependent on temperature (*see* **Note 1**), which is a highly relevant variable in both biological and condensate studies. Thus, using FRET, we have increased measurement specificity and decreased variability between conditions.

The second key modification to the Maryati method is to run identical reactions simultaneously and quench in place instead of removing and quenching aliquots from a bulk reaction. This modification can maintain consistency between quenched aliquots even with heterogeneous samples. Also, because sample spatial heterogeneity can impact fluorescence measurements, we added proteinase K, a broad-spectrum protease, to our quenching solution. This degrades proteins that contribute to the spatial heterogeneity of the sample, creating a homogenous environment for measurement. Furthermore, while not strictly necessary, the efficiency and accuracy of our method are enhanced by the use of a plate reader with a dual injection system.

Finally, we developed *deadenylationkinetics*, a flexible, Python-based package for simulating and fitting FRET-based measurements of deadenylation, enabling the extraction of RNA lengths as a function of time and quantitative insight into the mechanism of deadenylation. With this method, we were able to estimate kinetic parameters that govern the mechanism of a highly active form of human CNOT7, henceforth denoted as CNOT7X, which was designed to enhance activity by truncation of the C-terminal regulatory tail (aa 266-285) that occupies the RNA-binding site in the crystal structure [13]. These parameters enabled us to map the generated RNA species populations as a function of time and define the enzyme kinetics due to non-processive deadenylation [14].

2 Materials

2.1 Instruments

1. 96-well RNase-free polystyrene conical bottom plate(s).
2. 384-well black polystyrene plate(s) with a clear bottom.
3. Multichannel pipettor for which 20 μL is near the middle of its functional range.
4. A plate reader with fluorescence detection, temperature control, and dual-pump injection system (e.g., Synergy Neo2 Multi-Mode Assay Microplate Reader, BioTek).
5. Opaque/black 1.5 mL microcentrifuge tubes.
6. Adhesive plate seal (transparent seal, plate foil, 3 in wide packing tape, etc.).
7. Centrifuge with a rotor for plates.

2.2 Reagents

1. Reaction buffer: 25 mM sodium phosphate, 50 mM sodium chloride (NaCl), 2 mM magnesium chloride ($MgCl_2$), 2 mM fresh dithiothreitol (DTT), and 5% (v/v) glycerol (pH 7.4) (*see* **Note 2**).
2. Quenching buffer: 25 mM sodium phosphate, 50 mM NaCl, 2 mM fresh DTT, 5% (v/v) glycerol, 30 mM ethylenediaminetetraacetic acid (EDTA), 0.6 mg/mL Proteinase K, and 3% (w/v) sodium dodecyl sulfate (SDS) (pH 7.4) (*see* **Note 3**).
3. Purified deadenylase in reaction buffer.
4. Single-strand RNA substrate oligonucleotide with a 5′--conjugated donor fluorophore solubilized in nuclease-free water (e.g., 5′ [6FAM]-CCUUUCC$(A)_{18}$; *see* **Note 4**).
5. Single-strand DNA oligonucleotide that is complementary to the RNA substrate with a 3′-conjugated acceptor fluorophore solubilized in nuclease-free water (*see* **Note 5**) (e.g., 5′ $(T)_{38}$GGAAAGG-[TAMRA]; *see* **Note 6**).

3 Methods

One species will be titrated, while the others will be held at constant concentration (Fig. 1a).

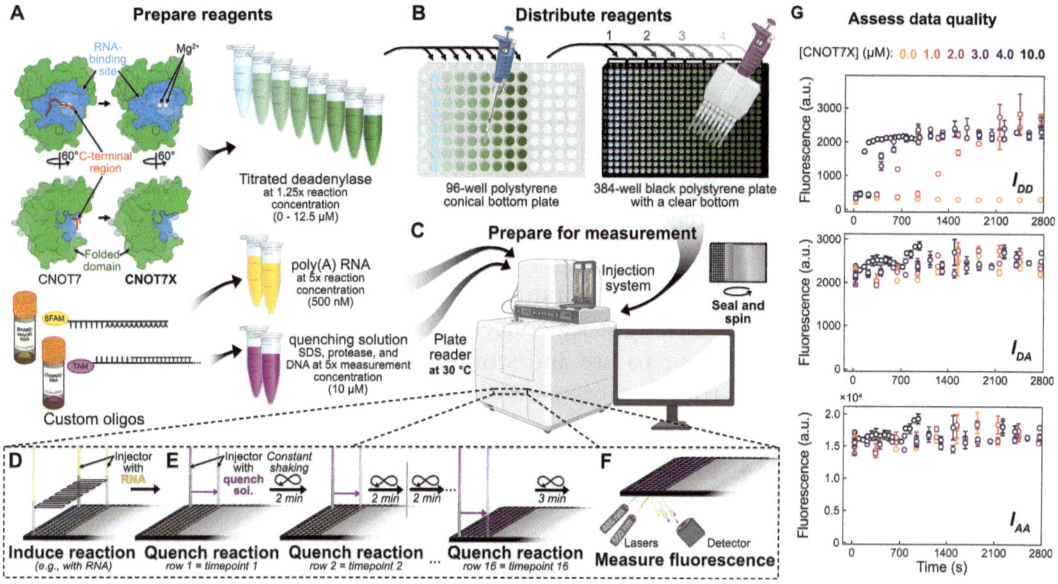

Fig. 1 Step-by-step outline of deadenylation assay setup and data acquisition by FRET. (**a**) Prepare reagents: Major reagents used in this method include a purified deadenylase, poly(A) RNA substrate, and complementary DNA. We used, respectively, a C-terminal truncated form of CNOT7 that leaves the RNA-binding site perpetually open for RNA binding, a custom, fluorescently labeled RNA substrate (5′ [6FAM]-CCUUUCC $(A)_{18}$), and a custom, fluorescently labeled DNA (5′ $(T)_{38}$GGAAAGG-[TAMRA]). For a deadenylase titration, prepare sufficient volume of deadenylase at 1.25× the reaction concentration. Prepare enough RNA in the reaction buffer and DNA in the quenching buffer with SDS and protease for use at each enzyme concentration and in priming of the injection system. (**b**) To efficiently distribute the titrated deadenylase into the 384-well plate, first pipet equal volumes into eight adjacent wells in a 96-well plate with a single channel pipettor. Then, use an 8-channel pipettor to distribute each condition into three columns of 16 wells (for triplicates of 16 timepoints). We typically run the assay several times, and only two conditions are tested at once. (**c**) Prepare for measurement: Prime the injection system in the plate reader with the appropriate buffer and then with the oligo solutions. In addition, seal the 384-well plate and briefly spin to bring all liquid to the base of the well and remove bubbles that could interfere with fluorescence measurements. Heat the plate reader to the measurement temperature (e.g., 30 °C). Finally, generate a method with the plate reader software to perform steps (**d**–**f**). (**d**) Induce reaction: Inject the 5× RNA solution into each well such that replicate points are induced sequentially. (**e**) Immediately after RNA is dispensed, quench the first row with quenching solution. Then, wait the appropriate amount of time while the plate is continuously shaking. For a ~30 min run, the shaking period should be 2 min. Repeat this process with each row until all rows are quenched. After the final quenching, wait an additional 3 min to allow for complete equilibration of injected DNA. (**f**) Measure fluorescence for I_{DD} (donor emission after donor excitation), I_{DA} (acceptor emission after donor excitation), and I_{AA} (acceptor emission after acceptor excitation). (**g**) Assess data quality: To make sure no experimental errors occurred, such as reagent omission and fluorophore quenching, plot the I_{DD}, I_{DA}, and I_{AA} data. Error bars are standard deviation of triplicates

3.1 Preparation of Reaction Materials and Plate Setup

3.1.1 Prepare DNA, RNA, and Deadenylase for the Reaction

1. For the titrated species (to be used as a pre-aliquoted reagent), prepare (*reaction volume* – *injection volume*) × # *time points* × # *technical replicates* × 1.2 (to account for dead volume) µL at 1.25× the reaction concentration of each titration point in reaction buffer (to account for dilution to the working concentration; *see* **Note 7**).

2. For the constant concentration species, which will be added to the reactions via the plate reader injector, prepare *injection volume* × # *time points* × # *technical replicates* × 1.2 (to account for dead volume) μL for each condition in the titration (*see* **Note 7**). Extra induction and quenching solution, approximately 1.5–2× the nominal volumes, should be prepared to prime the injection system if using a plate reader equipped with this functionality.
3. Prepare the induction solution by diluting the RNA to the working stock concentration with reaction buffer immediately prior to use and store on ice in the dark. Opaque microcentrifuge tubes and/or covered ice buckets work well for RNA storage until use.
4. Prepare the quenching solution by diluting the complementary DNA to working stock concentration with quenching buffer immediately prior to use and store in the dark. Do not store on ice as the SDS in the quenching buffer will begin to precipitate. If this occurs due to cooler ambient temperatures, submerge the buffer container in warm water and swirl periodically to homogenize the solution.

3.1.2 Prepare the Reaction Plate

1. Aliquot the titrated species into a sterile 96-well polystyrene plate such that each condition is divided evenly down a column (Fig. 1b). Once aliquoted, seal the plate with an adhesive film and store at 4 °C or on ice. If liquid condenses onto the adhesive film, spin the plate briefly before removing the seal.
2. Using a multichannel pipettor, pipet 20 μL from the 96-well plate into the 384-well black polystyrene plate (*see* **Note 8**). Once aliquoted, seal the plate with an adhesive film.
3. Prior to measurement, spin the 384-well plate at ~1500 × *g* for 2–5 min at reaction temperature to remove bubbles and bring the plate to the reaction temperature (Fig. 1c).

3.2 Assay Induction, Quenching, and Fluorescence Measurements

3.2.1 Set Up the Plate Reader

1. Calibrate the instrument (i.e., adjust the detector gain) before running the assay such that low concentrations of RNA/DNA can be detected but high concentrations do not saturate the detector. To do this, set up some wells with the highest and lowest concentrations of RNA only, DNA only, and RNA with DNA. Include all reagents except for the deadenylase. Using these samples, measure "dummy" FRET values to find the gain that provides optimal signal-to-noise for all cases.
2. Consider the following points when configuring the assay to your specific setup. It is possible to perform this assay manually (i.e., without an injection system) although this will impinge on method efficiency. However, the following description for plate reader use includes salient points to consider in a manual setup as well:

- Incubate the plate at the reaction temperature (e.g., 30 °C).
- Inject the induction (RNA) solution into each well (Fig. 1d).
  - Because the well volume of the 384-well plate is ~40 μL, 5 μL of induction solution was injected into each well, which already contained 20 μL of aliquoted solution. This left space for 5 μL of the quenching solution to be added later, for a final volume of 30 μL.
  - Induction solution should be injected such that replicate wells are induced sequentially. For example, the Synergy Neo2 plate reader injects row-by-row, so the plate is set up such that replicates are in the same row and the time points are arrayed down the columns.
- Inject the quenching solution at regular time intervals (Fig. 1e). For example, to monitor deadenylation for 30 min, we quenched a well every ~120 s (*see* **Note 9**).
  - Quenching should occur systematically such that each replicate of one condition is quenched consecutively. Because of the row-by-row injection method inherent to the Synergy Neo2 plate reader, quenching is, in turn, also carried out in this way, with minimal deadtime between quenching of each replicate reaction. Thus, each of the replicates within a given row can be treated as one time point.
  - The first row should be quenched immediately after the induction solution is added to all wells in order to measure a point as close as possible to $t = 0$. It may be necessary to manually determine the time between induction and quenching for the first point if the plate reader does not report this value.
- Constantly shake the plate at a low speed and large amplitude (e.g., 180 cycles per min, 6 mm).
- Allow the plate to shake for at least 2 min after the final time point is quenched. This ensures complete sample homogenization and consistent fluorescence measurements.
- Measure the fluorescence of all quenched time points after the reaction is complete (Fig. 1f):
  - Donor fluorescence after donor excitation, I_{DD} ($\lambda_{6FAM,ex} = 485$ nm, $\lambda_{6FAM,em} = 528$ nm).
  - Acceptor fluorescence after donor excitation, I_{DA} ($\lambda_{6FAM,ex} = 485$ nm, $\lambda_{TAM,em} = 590$ nm).
  - Acceptor fluorescence after acceptor excitation, I_{AA} ($\lambda_{TAM,ex} = 550$ nm, $\lambda_{TAM,em} = 590$ nm).

3.2.2 Review Data Quality

After data collection, review the data to ensure no issues arose during the experiment (Fig. 1g). Variation between technical replicates should be minimal and the change in fluorescence in I_{DD} should be consistent across all conditions, indicating that the RNA injection volume was uniform.

The I_{AA} data indicate the stability of the acceptor fluorophore over the course of the assay, while the no-enzyme condition will indicate the stability of both acceptor and donor. Any changes unattributed to the acceptor fluorophore may result from issues with the donor fluorophore.

We observed a slight, linear increase in acceptor fluorescence during longer assay runs; we attribute this to the loss of reducing agent, which leads to increased relaxation via the triplet state in early timepoints and an apparent increase in I_{AA} over time. In this case, I_{AA} can be used to normalize I_{DA}. However, unexpected, large changes are an indication of an issue with the reagents.

3.3 Organize Data for Analysis

1. Calculate relative FRET efficiency according to

$$E_{\text{rel.}} = \frac{I_{DA}}{I_{DD} + I_{DA}} \tag{1}$$

where I_{DD} and I_{DA} denote 6FAM fluorescence, that is, the donor fluorophore, after 6FAM excitation (485–528 nm), and TAMRA fluorescence, that is, the acceptor fluorophore, after 6FAM excitation (485–590 nm), respectively (*see* **Note 10**).

2. Create a tabular CSV file with all the FRET data modeled after the example in the *deadenylationkinetics* package "data" folder). This will be the data input file. The following columns should be included, with entries for each data point:
   - `Time` (s)
   - `FRET` (arbitrary units, a.u.)
   - `Error` (error for fitting averaged FRET values; if not averaged, set error to 0)
   - `Rna` (RNA concentration in molar)
   - `Enzyme` (enzyme concentration in molar)
   - `Replicate` (the data replicate number, should be 1 for all data points corresponding to the first replicate, 2 for the second, etc.)
   - `IDD`
   - `IDA`
   - `IAA`

3.4 Fit Assay Data with the deadenylationkinetics Package

The *deadenylationkinetics* package enables the user to simulate and fit the FRET data afforded by our assay in a straightforward manner. In what follows, we describe the usage of the version of deadenylationkinetics as of publication. Future versions of *deadenylationkinetics* will have additional features and content than are defined below; please check the GitHub repository for the most current version and updated instructions for implementation.

We have assumed that the modeling of the experimental FRET data can be performed in three main steps. First, the reaction kinetics are simulated according to a system of rate equations that govern the binding of the deadenylase to the RNA and the subsequent conversion of the initial substrate into AMP and RNA species with shorter poly(A) tails (*see* **Note 11**). In the second part of the modeling procedure, the total amount of each RNA species at each time point (*see* **Note 12**) is annealed to the DNA probe strand according to a set of two-state equilibria, resulting in the formation of a series of hybrid RNA-DNA duplexes (*see* **Note 13**). Finally, FRET values are calculated using the time-dependent fraction of RNA annealed to the DNA strand and are compared with the experimental data. The model parameters are subsequently incremented until optimal kinetic values are found (*see* **Note 14**).

Importantly, this three-step procedure can quantitatively account for the experimental data. As the reaction proceeds and the product distribution tends toward shorter RNAs with weaker affinities for the DNA strand, the fraction of annealed RNA decays, leading to a decrease in the simulated FRET values.

3.4.1 Workspace Setup

1. To begin, download and install the latest version of Python 3. Python 3 versions and installation instructions can be found at https://www.python.org/downloads/.
2. Open a terminal (MacOS or Linux) or command prompt (Windows) and create and activate a virtual environment for installing *deadenylationkinetics* with the command,

```
python -m venv /path_to_new_virtual_environment  (MacOS/Linux)
```

or

```
c:\>python -m venv c:\path\to\myenv (Windows)
```

This will create an isolated Python 3 installation for analyzing deadenylation data that will not interfere with the system Python 3. More information about virtual environments can be found at https://docs.python.org/3/library/venv.html.

3. Download the code repository for *deadenylationkinetics* from https://github.com/robharkness/deadenylationkinetics, in the form of a ZIP file, and unzip within a directory of choice (*see* **Note 15**). There should now be a directory named *deadenylationkinetics* that contains a set of subdirectories: "data," "scripts," and "analysis."
   - The "data" directory contains example files like a CSV data file for CNOT7X and a YAML configuration file that is read by *deadenylationkinetics* to initiate fits and simulations.
   - The "scripts" directory contains utility scripts that are used by *deadenylationkinetics* and can be used to process the experimental data into FRET data.
   - The "analysis" directory contains the program files for executing *deadenylationkinetics* to fit and simulate the data for the assay using different deadenylation mechanisms.
4. To install, change the current directory to *deadenylationkinetics* and type the command

```
pip install -e .
```

into the terminal or command prompt. More information on the package can be found at https://github.com/robharkness/deadenylationkinetics.

3.4.2 Configure Fitting Parameters

Copy and update the YAML configuration file for fitting deadenylation data generated by the user. This configuration file is the main route for interacting with the *deadenylationkinetics* package. It contains sections for file names (input data and output plots), experimental parameters (e.g., enzyme, RNA, and DNA concentrations), modeling parameters (i.e., fitting and simulation, which depend on the deadenylation model of choice, and parameter error analysis), and data plotting. All inputs are specified as `key:value` pairs that are read by *deadenylationkinetics.*

The YAML file can be opened and edited in a text editor to update its parameters. For example, to enter the sample name for a given dataset, type the desired text into the value field after the sample name key (e.g., `Sample name: CNOT7X`). Provide values for *all* of the following keys:

- `Sample name:` unique identifier.
- `Data file to fit:` name of a CSV file with data to be fit.
- `Output plot file:` name of a PDF file for plot output.
- `Optimal fit parameter file:` name of a CSV file for parameter output.
- `Experimental parameters:`

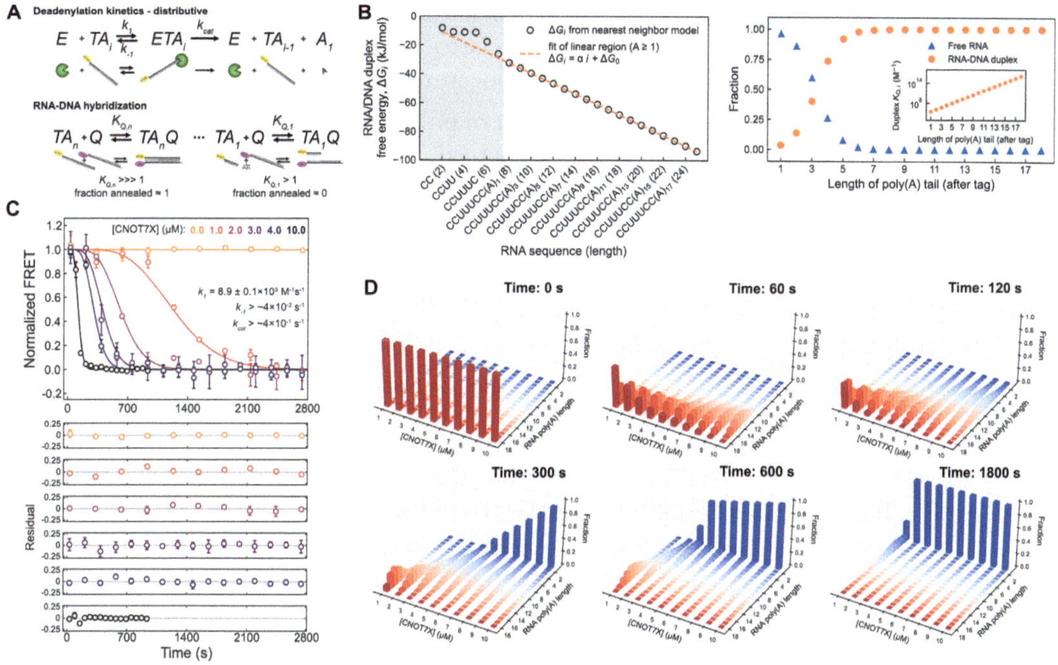

Fig. 2 Modeling mRNA deadenylation with the *deadenylationkinetics* package. (**a**) Schematics depicting the distributive model used to describe deadenylation kinetics for CNOT7X and the RNA-DNA hybridization thermodynamics. *Upper:* Terms shown are *E* for enzyme, TA_i, TA_{i-1}, or TA_n for RNA substrate with a tag *T* and poly(A) tail *A* of length *i*, *i*–1, or *n*, ETA_i for the enzyme-RNA complex where the RNA has poly(A) tail length of *i*, A_1 for AMP removed by the enzyme, k_1 and k_{-1} for the rate constants of enzyme-RNA association and dissociation, and k_{cat} for the rate constant of removing one adenosine base per catalytic step. *Lower:* Terms shown are *Q* for the complementary DNA oligo, TA_1Q for the RNA-DNA hybrid where the RNA has 1 A, $K_{Q,1}$ for the corresponding RNA-DNA binding affinity, TA_nQ for the RNA-DNA complex where the RNA has poly (A) tail length of *n*, and $K_{Q,n}$ for the associated RNA-DNA binding affinity. When the poly(A) tail is long (i.e., contains many As), the RNA-DNA binding affinity is very tight, and the fraction of the RNA annealed is close to 1. If the poly(A) tail is short (i.e., contains few bases), then the RNA-DNA binding affinity is weak, and the fraction of RNA annealed to DNA is close to 0. (**b**) *Left:* Free energy of RNA-DNA hybridization for RNA of length *i*, ΔG_i, in kJ/mol (black circles). The linear region of the plot, that is, where $A \geq 1$, is fit as a function of total RNA length (orange dashed line) where $\alpha = -3.6$ kJ/mol and $\Delta G_0 = -3.7$ kJ/mol. Points in the gray box lie outside the linear region and were not included in fitting. *Right:* The fraction of RNA bound to DNA (orange circles) and the fraction of free RNA (blue triangles) modeled based on the association constant $K_{Q,i}$ between the DNA oligo and RNA with the tag sequence 5′ CCUUUCC and poly(A) tail of length *i* ($i \in (1, 2, 3, \ldots, n)$). *Inset:* The values for $K_{Q,i}$ calculated based on ΔG_i values generated from the linear fit of nearest neighbor free energy values for RNA-DNA hybridization [12] for the RNA sequence 5′ $CCUUUCC(A)_{1-18}$. (**c**) *Upper* Averaged, normalized FRET for RNA degraded in the presence of 0–10 μM CNOT7X with the kinetic parameters extracted from fits of the data, where k_{-1} and k_{cat} are estimates of the minimum values. *Lower* Residuals of the fits for each concentration of CNOT7X. Error bars are standard deviation of triplicates. (**d**) Populations of RNA species with poly(A) length of 1–18 for CNOT7X concentrations 1–10 μM based on the starting material of 100 nM RNA with poly(A) length of 18 As at the time points 0, 60, 120, 300, 600, and 1800 s

- `Enzyme:/Value:` list of enzyme concentration(s) (molar). Use square brackets to define the start and end of the list, with individual concentrations separated by commas.
- `RNA:/Value:` list of poly(A) RNA concentration(s) (molar), defined identically to the enzyme concentrations.
- `QT:/Value:` DNA probe strand concentration (molar).
- `n:/Value:` length of the poly(A) tail in the substrate RNA (an integer).
- `Temperature:/Value:` reaction temperature (Kelvin).

• `Modeling parameters:`

- `Fit:` `True` to fit the data, or `False` to simulate FRET data with the input parameters.
- `Use individual replicates:` `True` to fit individual FRET replicates sequentially, or `False` to average the FRET replicates and fit the average obtained at each enzyme and RNA concentration as a function of time.
- `Minimizer:` for specifying which optimization algorithm that is used by *deadenylationkinetics* to fit the data; *deadenylationkinetics* implements the highly flexible Python package *lmfit* to perform the minimization, which has a number of minimization algorithms to choose from (see https://lmfit.github.io/lmfit-py/).

  `Initial:` for specifying the initial optimization algorithm to be used when fitting multiple FRET replicates sequentially, as defined by the "`Use individual replicates`" option above. We recommend the Nelder–Mead simplex algorithm (`nelder`) as a robust method for finding optimal parameters in the first fit. Other options can be found at https://lmfit.github.io/lmfit-py/fitting.html.

  `Subsequent:` for specifying the optimization algorithm to be used in the fits of the additional FRET replicates once optimal parameters are found from the fit of the first replicate. We recommend the least-squares approach (`leastsq`) if the data are highly reproducible and thus are expected to have similar optimal parameter sets, as it proceeds to the minimum more rapidly than the Nelder–Mead approach.

- `Kinetic model`: for specifying the mechanism by which the deadenylase degrades the poly(A) RNA (Fig. 2a). Additional mechanisms will be added and outlined in further revisions of ***deadenylationkinetics***, please check the GitHub repository for updates and associated instructions.

  `Processive`: The deadenylase binds and completely degrades the full-length RNA without dissociating from any intermediate species, leading to the absence of free intermediate RNA lengths. This is summarized in Eq. 2 as follows:

$$\begin{aligned} E + TA_n \underset{k_{-1}}{\overset{k_1}{\rightleftharpoons}} ETA_n &\overset{k_{cat}}{\rightarrow} ETA_{n-1} + A_1 \\ ETA_{n-1} &\overset{k_{cat}}{\rightarrow} ETA_{n-2} + A_1 \\ &\vdots \\ ETA_2 \overset{k_{cat}}{\rightarrow} ETA_1 + A_1 &\underset{k_{-1}}{\overset{k_1}{\rightleftharpoons}} E + TA_1 \end{aligned} \quad (2)$$

  where E, TA_n, ETA_i $(i \in (1, 2, 3, \ldots, n))$, and A_1, are the deadenylase enzyme, the full-length RNA, the enzyme complexes with RNA strands of length i that contain the tag sequence T (e.g., 5′-CCUUUCC…) that promotes specific annealing with the DNA oligo in the detection portion of the assay, and freed AMP, respectively, and k_1, k_{-1}, and k_{cat} are the RNA binding, dissociation, and catalytic rate constants, respectively.

  `Distributive`: The deadenylase dissociates from each RNA product after the catalytic step, leading to free intermediate-length RNAs that are then re-bound and cleaved by the enzyme according to Eq. 3:

$$\begin{aligned} E + TA_n \underset{k_{-1}}{\overset{k_1}{\rightleftharpoons}} ETA_n &\overset{k_{cat}}{\rightarrow} E + TA_{n-1} + A_1 \\ E + TA_{n-1} \underset{k_{-1}}{\overset{k_1}{\rightleftharpoons}} ETA_{n-1} &\overset{k_{cat}}{\rightarrow} E + TA_{n-2} + A_1 \\ &\vdots \\ E + TA_2 \underset{k_{-1}}{\overset{k_1}{\rightleftharpoons}} ETA_2 &\overset{k_{cat}}{\rightarrow} E + TA_1 + A_1 \end{aligned} \quad (3)$$

- `Fit parameters`: for defining parameters that are used in modeling the RNA deadenylation kinetics and the annealing between the RNA product distribution and the DNA oligo (Fig. 2a). For each parameter in the fitted kinetic model (`k1`, `km1`, `kcat`, `dGo`, and `alpha`), set values for the following:

`Value`: the starting fit value.

`Units`: the parameter units, which are used to generate the final optimal parameter values of the CSV file.

`Vary`: `True` if the parameter should be varied in the fit, or `False` to hold the parameter constant (*see* **Note 16**).

`Minimum`: A minimum value to constrain the fit; for example, rate constants cannot assume negative values and so the minimum for these should be set to zero.

Instructions for determining the hybridization parameters `dGo` and `alpha` can be found in Subheading 3.4.3. Generally, these values will not be varied for fitting.

- `Error estimation`: for estimating errors in fitted kinetics parameters and to assess correlations between them.
  - `Monte Carlo`: performs fits of synthetic data to obtain the standard deviation of the fitted parameters as their errors.

    `Run`: set to `True` or `False` to turn on or off.

    `Iterations`: an integer, must be at least ~200–300 for accurate assessment of fitted parameter uncertainties.
  - `Error surfaces`: generates pairwise parameter correlation surfaces to assess the degree of correlation between each of the fitted parameters, that is, how robustly can each parameter be defined, or whether only limits can be set on parameter values:

    `Run`: set as defined in the "`Monte Carlo`" option.

    `Parameter range factor`: a multiple of the optimal parameter value that sets the range to be explored in the correlation analysis; we recommend 50–100 to explore several orders of magnitude around the optimal parameter value.

    `Points`: an integer that sets the number of parameter values to be assessed within the range defined by the parameter range factor.
- `Plot parameters`: for generating several types of plots as the output of simulations or fits by *deadenylationkinetics*.
  - `Plot best fit`: `True` or `False` to turn on or off plotting the best fit of the model of choice to the assay FRET data (Fig. 2c).
  - `Plot residuals`: as above, except for plotting fit residuals calculated as $\mathrm{FRET}_{\mathrm{calc},x,y,z} - \mathrm{FRET}_{\mathrm{experiment},x,y,z}$ where x, y, and z index the FRET value at a given time point, enzyme concentration, and RNA substrate concentration, respectively (Fig. 2c).

- `Plot 2D population bars:` `True` yields 2D bar plots of enzyme and RNA species fractions as a function of time at each total enzyme and RNA concentration.
- `Plot 3D population bars:` set as above, if `True` generates 3D bar plots as a function of reaction time where the x, y, and z axes are total enzyme concentration, poly(A) tail length, and species fraction (Fig. 2d).

3.4.3 Calculate Hybridization Parameters

Hybridization parameters, denoted as `dGo` and `alpha` in the YAML file, are necessary to describe the binding affinity between the RNA species derived from the enzymatic activity of the deadenylase and the DNA oligo, as the change in the fraction of RNA annealed to the DNA is ultimately what underlies the shape of the measured FRET data (*see* **Note 13**).

1. Determine the free energy of binding to the DNA probe strand for each RNA species as a function of the total RNA strand length (including the tag sequence T).
   - The constants ΔG_0 and α can be calculated and fixed using known values from the literature for RNA-DNA hybridization (*see* Banerjee, et al. [15]), or can be determined from titration or thermal melting experiments performed with the DNA probe and a set of RNA strands of different lengths.
   - It is possible to allow these parameters to be optimized in the fit, though we do not recommend this as they can be highly correlated with the kinetic parameters, leading to large uncertainties.
2. To estimate the hybridization constants ΔG_0 and α, plot the free energies of DNA binding as a function of total RNA length (Fig. 2b). This should yield a plot with a linear region (i.e., where $A \geq 1$) that can be fit by the linear equation:

$$\Delta G_i = \alpha \cdot i + \Delta G_0 \tag{4}$$

   where ΔG_i is the free energy of binding of an RNA strand with a poly(A) tail of total length i to the DNA probe, ΔG_0 (`dGo`) approximates the free energy of binding between the RNA tag sequence that remains after the poly(A) tail has been removed, which we denote as T, and the DNA oligo, and α (`alpha`) is the free energy gain per adenosine A added to the poly(A) tail sequence.
   - Within *deadenylationkinetics*, ΔG_0 and α are used to calculate the set of ΔG_i, which are converted into RNA-DNA duplex affinities, $K_{Q,\,i}$, according to

$$K_{Q,i} = e^{\frac{-\Delta G_i}{RT_{\text{exp.}}}} = \frac{[TA_iQ]}{[TA_i][Q]} \tag{5}$$

where R is the ideal gas constant, $T_{\text{exp.}}$ is the experimental temperature, and TA_i, TA_iQ ($i \in (1, 2, \ldots, n)$), and Q are the free RNA species, the DNA-bound RNA species, and the free DNA probe strand, respectively (Fig. 2b). Thus, the variation in $K_{Q,\,i}$ with RNA length, as defined by Eqs. 4 and 5, controls the fraction of each RNA species bound to the DNA probe and thus the proximity of the fluorophores, leading to decreases in FRET as the RNA distribution tends toward smaller species.

- It is not necessary to supply estimates of the FRET values for converting the RNA-DNA duplex populations into the measured FRET signal, as this is performed internally by *deadenylationkinetics* (*see* **Note 14**).

3.4.4 Run a Simulation or Fit with deadenylationkinetics

1. In a terminal or command prompt, change the current directory to

```
path_to_folder/deadenylationkinetics/data/yourdatafolder/
```

2. From here, issue the command

```
../../analysis/main.py configuration_file.yaml
```

and *deadenylationkinetics* will run and perform the tasks as outlined in the configuration file. Later versions of *deadenylationkinetics* will be updated with simpler execution; please check the GitHub repository for the most recent version.

The outputs of *deadenylationkinetics* include (assuming all plotting and error analysis options are set to `True`) the following:

- A PDF file with the following plots:
  - FRET data with fits, raw and normalized (Fig. 2c).
  - Fit residuals for each FRET curve collected at a given enzyme and RNA concentration (Fig. 2c).
  - 2D RNA population plots showing the RNA distributions produced over time for a given condition.
  - 3D population plots of the RNA distributions as a function of time, as defined above (Fig. 2d).
- A CSV file with the fitted parameter values and their associated errors as estimated by the *lmfit* package (*see* **Note 17**).

- A CSV file of the Monte Carlo error analysis results that contains the average of the resultant parameter values and their associated standard deviations.
- A PDF file with plots of the Monte Carlo parameter distributions.
- A CSV file of the parameter correlation results containing each parameter pair, their values, and the associated fit quality.
- A PDF file containing contour plots of pairwise parameter correlation surfaces for assessing parameter robustness.

4 Notes

1. FRET is dependent on the Förster radius R_0 of two fluorophores and the distance between them in solution, r:

$$E = \frac{1}{1 + \left(\frac{r}{R_0}\right)^6} \quad (6)$$

with

$$R_0^6 = \frac{9000 \ln(10) \kappa^2 Q_D J}{128 \pi^5 n^4 N_A} \quad (7)$$

where κ depends on the relative orientation between the two dyes, Q_D is the quantum efficiency of the donor fluorophore, J is the overlap integral between donor emission and acceptor absorption spectra, n is the refractive index of the medium, and N_A is Avogadro's number [12].

2. While not strictly necessary, use of a reducing agent (e.g., DTT or β-mercaptoethanol) is highly recommended to enhance photostability of the fluorophores [16]. If using DTT (1–3 mM) or β-mercaptoethanol (BME, ~10 mM), fresh reagent is necessitated at the time of assay setup.
3. After injection, the working concentration of the quenching buffer components will be 25 mM sodium phosphate pH 7.4, 50 mM NaCl, 2 mM DTT, 5 mM EDTA, 0.1 mg/mL Proteinase K, and 0.5% SDS.
4. Our RNA substrate was purchased from Bio Basic Inc. (Canada) post-HPLC purification, delivered in lyophilized aliquots, and stored at −20 °C until use.
5. Our DNA oligo was obtained from Sigma (USA) post-desalting. We then further purified in-house by loading on a DNAPac PA-100 BioLC 9 × 50 mm guard and semi-prep column (Thermo Scientific) and eluted with a sodium chloride gradient. DNA was then precipitated by adding 1/10 volume

of 3 M sodium acetate pH 5.2 and two volumes of ice-cold 100% ethanol to the DNA sample. The sample was mixed and stored at −20 °C for 1 h. After centrifugation at 13,000 × *g* for 15 min, the pellet was gently washed twice with 70% ethanol, then left in the dark to dry, and finally stored at −20 °C.

6. A DNA oligo longer than the full-length RNA, that is, containing more T bases than there are complementary As, is not necessary since the binding affinity, to first-order approximation, depends on the length of the RNA poly(A) tail. It is most important to start with oligo lengths and concentrations that, when degraded, shift from highly associated with the DNA to highly unassociated, in order to observe a substantial change in FRET.

7. For an 8-point CNOT7X titration run in triplicate with 16 time points and individual reaction volumes of 25 μL, we prepared 20 μL × 16 timepoints × 3 replicates × 1.2 = ~1200 μL of each CNOT7X concentration. Because the pre-aliquot volume was 20 μL, and we titrated from 1 to 10 μM, the working stock concentrations of CNOT7X were 1.25–12.5 μM (1.25×). Note, a no-enzyme condition (i.e., 0 μM) is necessary for quality control. For the DNA/RNA, we prepared 5 μL × 16 timepoints × 3 replicates × 8 titration points × 1.2 = 2300 μL of working stock at 5× reaction concentration. The DNA working stock was 10 μM because our measurement concentration was 2 μM and the injection volume was 5 μL. The reaction concentration of RNA was 100 nM for our experiments, so the working RNA stock concentration was 500 nM.

8. To balance efficiency and precision/accuracy, we aliquot one condition into three adjacent 16-well columns (Fig. 1b) and run six columns at a time (two conditions in triplicate or three conditions in duplicate). One can run as many or as few conditions as desired but should consider the following:
   - How long will the reaction run?
   - How much time can be saved by running multiple conditions simultaneously?
   - How long will it take to induce each reaction?
   - How much time will pass before the first time point can be quenched?
   - If the reaction is not run at room temperature, can the plate be kept at the proper temperature during induction/quenching, or does it need to be removed?
   - Is the time required to quench the reaction long enough such that the reactions might change temperature during quenching?

9. Since different deadenylases have distinct deadenylation kinetics, the precise time sampling interval will need to be established in each case with a pilot experiment. In addition, if a very long RNA oligo is implemented in the assay, it will be necessary to run the reaction for a much longer period of time in order to allow for the deadenylase to remove enough of the poly(A) tail such that the binding affinity for the DNA probe decreases to the point at which the FRET values begin to transition from the high to the low FRET baseline.

10. The relative FRET efficiency is sufficient in this situation because the assay is used to measure deadenylation rates as opposed to determining quantitative distances between fluorophores. However, it is recommended that one uses a FRET index calculation that includes both donor and acceptor fluorescence (as opposed to ratios of donor fluorescence in the absence and presence of the acceptor, or acceptor photobleaching) because it provides a more reproducible and reliable measurement [17]. If one would like to determine the absolute FRET efficiency, one must take channel crosstalk into consideration when choosing fluorophores and substrate/quencher concentrations. One will also need to make additional measurements outside of assay runs, for example, donor absorbance at the acceptor excitation wavelength, etc.

11. *Poly(A) RNA deadenylation kinetics:* Here, we consider the distributive model of deadenylation as an example, since previous work demonstrated that the *Schizosaccharomyces pombe* CAF1, which is homologous to CNOT7 [13], operates by this mechanism, and we also found this to be the case for CNOT7X. In distributive deadenylation, the enzyme E (CNOT7X in this instance) initially binds to the full-length RNA substrate, denoted TA_n (containing the tag sequence T for annealing to the DNA, and the poly(A) tail of length n, as outlined in the protocol above):

$$E + TA_n \underset{k_{-1}}{\overset{k_1}{\rightleftharpoons}} ETA_n \tag{8}$$

where the forward and reverse rate constants k_1 and k_{-1} dictate the association of E and TA_n and the dissociation of ETA_n, respectively. Following the formation of ETA_n, an RNA product of length $n - 1$ and one molecule of AMP, A_1, is generated. We assume that this occurs with no reverse reaction; that is, the cleaved RNA product and AMP cannot be spontaneously re-ligated by E,

$$ETA_n \xrightarrow{k_{cat}} E + TA_{n-1} + A_1 \tag{9}$$

where k_{cat} is the rate constant for catalysis of TA_n, leading to $TA_{n-1} + A_1$ and product release. Intermediate RNA lengths begin to accumulate upon cleavage of the first TA_n strand, thus creating a pool of different-length substrates for CNOT7X to interact with and subsequently cleave into even shorter strands. In the absence of other information, we assume that CNOT7X's respective association, dissociation, and catalysis energetics for these intermediate length RNA species are identical to those of the full-length RNA. Thus, for the binding equilibria,

$$\begin{aligned} E + TA_{n-1} &\underset{k_{-1}}{\overset{k_1}{\rightleftharpoons}} ETA_{n-1}, \\ E + TA_{n-2} &\underset{k_{-1}}{\overset{k_1}{\rightleftharpoons}} ETA_{n-2}, \\ &\vdots \\ E + TA_1 &\underset{k_{-1}}{\overset{k_1}{\rightleftharpoons}} ETA_1, \end{aligned} \tag{10}$$

and for the catalytic steps,

$$\begin{aligned} ETA_{n-1} &\overset{k_{cat}}{\rightarrow} E + TA_{n-2} + A_1, \\ &\vdots \\ ETA_2 &\overset{k_{cat}}{\rightarrow} E + TA_1 + A_1. \end{aligned} \tag{11}$$

We can recast these kinetic schemes in the form of rate equations:

$$\frac{d[E]}{dt} = -k_1[E]\sum_{i=1}^{n}[TA_i] + k_{-1}\sum_{i=1}^{n}[ETA_i] + k_{cat}\sum_{i=2}^{n}[ETA_i] \tag{12}$$

$$\frac{d[ETA_1]}{dt} = k_1[E][TA_1] - k_{-1}[ETA_1]$$

$$\frac{d[ETA_i]}{dt} = k_1[E][TA_i] - k_{-1}[ETA_i] - k_{cat}[ETA_i], \quad i > 1$$

$$\frac{d[TA_i]}{dt} = -k_1[E][TA_i] + k_{-1}[ETA_i] + k_{cat}[ETA_{i+1}], \quad 1 \le i < n$$

$$\frac{d[TA_n]}{dt} = -k_1[E][TA_n] + k_{-1}[ETA_n]$$

$$\frac{d[A_1]}{dt} = k_{cat}\sum_{i=2}^{n}[ETA_i]$$

where it is assumed that the smallest possible species are A_1 and TA_1; once all other lengths have been converted entirely to TA_1 and A_1, the reaction is terminated; that is, the deadenylase requires at least two adenosine moieties to be present in the tail in order to remove an A_1. We further assume that A_1 cannot bind to CNOT7X, though the species TA_1 retains the same binding capacity as all other RNA lengths. Note that A_1 is a special case since one A_1 molecule is generated by cleavage of all species with a poly(A) tail, that is, $i > 1$. Thus, in the limit that all initial TA_n is converted to TA_1 and A_1, and in the absence of binding of E to A_1, $[A_1] = (n - 1)[TA_n]$, where the factor of $n - 1$ takes into account that a single A remains within the TA_1 strand. The time-dependent concentrations are obtained by numerically integrating this system of equations over the experimental timespan starting from an initial condition that matches the beginning of an experiment, namely that the RNA is entirely full length (the concentrations of all other, shorter RNA species are zero) and that the total amount of enzyme and RNA added to the system is unbound at $t = 0$.

12. The total amount of each RNA species corresponds to the sum of the concentrations of the free and any enzyme-bound amounts. The summation approximates the experimental quenching of the reaction where the deadenylase is denatured and degraded, thus releasing any bound RNA.

13. *RNA-DNA hybridization:* Following simulation of the time evolution of the RNA distribution, the formation of the RNA-DNA hybrid duplexes, giving rise to the variation in FRET values observed experimentally, is modeled according to a series of two-state equilibria:

$$\begin{array}{c} TA_n + Q \overset{K_{Q,n}}{\rightleftharpoons} TA_nQ, \quad K_{Q,n} = \frac{[TA_nQ]}{[TA_n][Q]} \\ TA_{n-1} + Q \overset{K_{Q,n-1}}{\rightleftharpoons} TA_{n-1}Q, \quad K_{Q,n-1} = \frac{[TA_{n-1}Q]}{[TA_{n-1}][Q]} \\ \vdots \\ TA_1 + Q \overset{K_{Q,1}}{\rightleftharpoons} TA_1Q, \quad K_{Q,1} = \frac{[TA_1Q]}{[TA_1][Q]} \end{array} \qquad (13)$$

where Q is the DNA probe strand defined as in the protocol, and $K_{Q,\ i}$ ($i \in (1,2,\ldots,n)$) is the set of RNA-DNA affinity constants for the product distribution. The total concentrations of $TA_{i,\ \text{tot.}}$ and $Q_{\text{tot.}}$ of the RNA and DNA, respectively, in terms of free and hybridized forms, are given by

$$\begin{array}{c} TA_{n,\text{tot.}} = [TA_n] + [TA_nQ] \\ TA_{n-1,\text{tot.}} = [TA_{n-1}] + [TA_{n-1}Q] \\ \vdots \\ TA_{1,\text{tot.}} = [TA_1] + [TA_1Q] \\ Q_{\text{tot.}} = [Q] + \sum_{i=1}^{n} [TA_iQ]. \end{array} \qquad (14)$$

Note that each member of the set of $TA_{i,\ \mathrm{tot.}}$ is equal to $[TA_i]$ + $[ETA_i]$ from the kinetic simulations at each time point, as the enzyme is destroyed upon quenching of the reaction, and thus, the enzyme-bound RNA becomes free and can participate in duplex formation within this part of the model. This system of equations is solved for the unknown $[TA_i]$, $[Q]$, and $[TA_iQ]$ by passing the RNA product distribution from the kinetic simulations as the set of $TA_{i,\ \mathrm{tot.}}$ at each desired time point, in addition to the total amount of capture strand $Q_{\mathrm{tot.}}$ and estimates of $K_{Q,\ i}$ into a numerical system of equations solver in Python. To compare the simulations with experiment, the amounts of each hybrid duplex that contribute to the FRET signal are calculated as fractions of the total initial RNA according to

$$P_i = \frac{[TA_iQ]}{TA_{n,\mathrm{tot.}}} \tag{15}$$

14. *Calculation of FRET values:* The FRET signal can be approximated by assuming that the FRET intensity is given by a population-weighted average of the maximum change in FRET expected for the two fluorophores and an offset corresponding to the minimum value observed when the reaction is complete

$$F_{\mathrm{obs.}} = F_0 + \Delta F \sum_{i=1}^{n} P_i \tag{16}$$

where $F_{\mathrm{obs.}}$ is the observed FRET intensity at each time point, F_0 is the FRET offset at $t = \infty$, ΔF is the total change in FRET, and $\sum_{i=1}^{n} P_i$ is the sum of the fraction of each RNA species annealed to the DNA. In practice, values for F_0 and ΔF at each experimental condition (e.g., enzyme and total initial RNA concentration) are determined internally by *deadenylationkinetics* through an algorithm that treats these as "baseline" parameters that map the annealed RNA populations onto the experimental FRET values.

15. The installation process will be updated so that the user can install *deadenylationkinetics* via the standard route by issuing the command `pip install deadenylationkinetics`, which will install the package from the Python PiPy package repository directly, as opposed to using the downloaded GitHub repository. Check the GitHub page for further developments.

16. Allowing certain parameters to vary while others remain constant can be used to assess parameter correlations, test the influence of a given parameter on the simulated or fitted data, or constrain the fit if the minimization becomes "lost" with all parameters varying at once.

17. In this case, the RNA association rate constant k_1 is well-defined, while only boundaries can be set for k_{-1} and k_{cat}. Robust values for these rate constants can be obtained from experiments performed with [RNA] > [deadenylase], as in standard Michaelis–Menten enzyme experiments. We set [CNOT7X] > [RNA] for illustrative purposes, to ensure rapid consumption of the RNA substrate and to demonstrate the types of curve shapes that can be expected from the assay.

References

1. Wiederhold K, Passmore LA (2010) Cytoplasmic deadenylation: regulation of mRNA fate. Biochem Soc T 38:1531–1536
2. Kim TH, Tsang B, Vernon RM et al (2019) Phospho-dependent phase separation of FMRP and CAPRIN1 recapitulates regulation of translation and deadenylation. Science 365: 825–829
3. Viswanathan P, Ohn T, Chiang Y-C et al (2004) Mouse CAF1 can function as a Processive Deadenylase/3′–5′-Exonuclease in vitro but in yeast the Deadenylase function of CAF1 is not required for mRNA Poly (A) Removal*. J Biol Chem 279:23988–23995
4. Martı^nez J, Ren Y-G, Nilsson P et al (2001) The mRNA cap structure stimulates rate of Poly(A) Removal and amplifies processivity of degradation*. J Biol Chem 276:27923–27929
5. Webster MW, Stowell JAW, Tang TTL et al (2017) Analysis of mRNA deadenylation by multi-protein complexes. Methods 126:95–104
6. Airhihen B, Pavanello L, Jadhav GP et al (2019) 1-Hydroxy-xanthine derivatives inhibit the human Caf1 nuclease and Caf1-containing nuclease complexes via Mg2+-dependent binding. Febs Open Bio 9:717–727
7. Kruger NJ (1995) Errors and artifacts in coupled spectrophotometric assays of enzyme activity. Phytochemistry 38:1065–1071
8. Acker MG, Auld DS (2014) Considerations for the design and reporting of enzyme assays in high-throughput screening applications. Perspect Sci 1:56–73
9. Wang Y, Wang G, Moitessier N et al (2020) Enzyme kinetics by isothermal titration calorimetry: allostery, inhibition, and dynamics. Front Mol Biosci 7:583826
10. McClure WR (1969) Kinetic analysis of coupled enzyme assays. Biochemistry 8:2782–2786
11. Maryati M, Kaur I, Jadhav GP et al (2014) A fluorescence-based assay suitable for quantitative analysis of deadenylase enzyme activity. Nucleic Acids Res 42:e30–e30
12. Valeur B, Berberan-Santos MN (2012) Molecular fluorescence: principles and applications, 2nd edn. Wiley-VCH Verlag GmbH & Co. KGaA, Weinheim
13. Petit A-P, Wohlbold L, Bawankar P et al (2012) The structural basis for the interaction between the CAF1 nuclease and the NOT1 scaffold of the human CCR4–NOT deadenylase complex. Nucleic Acids Res 40:11058–11072
14. Webster MW, Chen Y-H, Stowell JAW et al (2018) mRNA Deadenylation is coupled to translation rates by the differential activities of Ccr4-Not nucleases. Mol Cell 70:1089–1100.e8
15. Banerjee D, Tateishi-Karimata H, Ohyama T et al (2020) Improved nearest-neighbor parameters for the stability of RNA/DNA hybrids under a physiological condition. Nucleic Acids Res 48:12042–12054
16. Glembockyte V, Cosa G (2017) Redox-based Photostabilizing agents in fluorescence imaging: the hidden role of intersystem crossing in geminate radical ion pairs. J Am Chem Soc 139:13227–13233
17. Berney C, Danuser G (2003) FRET or no FRET: a quantitative comparison. Biophys J 84:3992–4010

Chapter 6

Measuring Poly-Adenosine Tail Length of RNAs by High-Resolution Northern Blotting Coupled with RNase H Cleavage

René M. Arvola and Aaron C. Goldstrohm

Abstract

The poly-adenosine, or poly(A) tail, plays key roles in controlling the stability and translation of messenger RNAs in all eukaryotes, and, as such, facile assays that can measure poly(A) length are needed. This chapter describes an approach that couples RNase H-mediated cleavage of an RNA of interest with high-resolution denaturing gel electrophoresis and northern blot-based detection. Major advantages of this method include the ability to directly measure the abundance of any RNA and the length of its poly(A) tail without amplification steps. The assay provides high specificity, sensitivity, and reproducibility for accurate quantitation using standard molecular biology equipment and reagents. Overall, the high-resolution northern blotting approach offers a cost-effective means of poly(A) RNA analysis that is especially useful for small numbers of transcripts and comparisons between experimental conditions or time points.

Key words Northern blot, RNase H, Poly-adenosine tail, Poly(A)

1 Introduction

Nearly all eukaryotic messenger RNAs, and many noncoding RNAs, are modified by the addition of a non-templated 3′ poly-adenosine tail [1]. The poly(A) tail is initially established with an average length of 250 nucleotides in vertebrates. Thereafter, poly (A) tail length is dynamically modulated by enzymatic removal of adenosine nucleotides from the 3′ end via deadenylase enzymes (i.e., deadenylation) or, in some cases, extension via specialized polymerases.

Regulation of poly(A) tail length has emerged as a critical control point for translation and mRNA decay [1]. Long poly (A) tails generally promote translation and stability. Shortening of the poly(A) tail by deadenylases initiates mRNA degradation pathways. These processes act on all poly-adenylated mRNAs and can be differentially regulated by cis-acting RNA sequences and their

Eugene Valkov and Aaron C. Goldstrohm (eds.), *Deadenylation: Methods and Protocols*, Methods in Molecular Biology, vol. 2723,
https://doi.org/10.1007/978-1-0716-3481-3_6,

corresponding RNA-binding factors [2–4]. For example, members of the Pumilio family of sequence-specific RNA-binding proteins bind to RNAs that contain one or more Pumilio response elements [4–7]. Pumilio recruits specific deadenylase enzymes to the PRE-containing mRNA to promote the shortening of its poly (A) tail, reducing translation, and accelerating mRNA decay.

Given the important role of poly(A) length in gene regulation, effective approaches to measure tail length of specific RNAs are essential. Ideally, such an approach should be quantitative with a wide dynamic range, high sensitivity, and reproducibility. The technique should be able to monitor poly(A) dynamics under various experimental conditions. For example, it could measure poly (A) length changes in response to perturbations in cis-acting RNA elements, trans-acting factors, and components of RNA decay pathways. Moreover, the kinetics of poly(A) length changes could be monitored over time in response to stimuli and stress, during the cell cycle, and throughout development and differentiation.

Multiple approaches have been developed to analyze poly (A) tails of individual mRNAs [8–11] and globally [8, 12–18], each with their inherent strengths and limitations. This chapter focuses on the use of northern blot-based detection of poly (A) tail length [4, 5, 19]. Northern blotting has long been used to detect, visualize, and measure the levels of a specific transcript (Fig. 1) [20]. As an overview, purified RNA—from any source—is first separated by gel electrophoresis. Typically, denaturing formaldehyde–agarose gel electrophoresis is used for size separation of RNAs up to ~10,000 nucleotides [20, 21], followed by transfer of the RNA from gel to a membrane (e.g., nitrocellulose or nylon) by blotting. Alternatively, denaturing urea–polyacrylamide gel electrophoresis can achieve higher resolution size separation of RNA molecules with lengths less than ~1000 nucleotides [22, 23]. To detect the specific RNA, a radiolabeled antisense nucleic acid probe, which is the reverse complement of the target RNA, is incubated with the membrane to enable specific hybridization to the transcript of interest. The unbound probe is subsequently removed by washing the membrane several times. Finally, the RNA is detected using autoradiography. Traditionally, X-ray film was used to visualize the signal on the northern blot. Though the film can be quantitated, the linear range is limited. Better still, phosphorimaging can detect and quantify the radioactive signal. The migration of the transcript of interest can be compared to RNA size standards included in the analysis. The transcript's poly(A) tail length distribution can thus be determined (Fig. 1). The poly (A) tails of transcripts can be removed by treating the RNA sample with RNase H and a thymidine DNA oligonucleotide before electrophoresis to create the so-called pA_0 control sample (Fig. 1). RNase H acts exclusively to degrade RNA within an RNA:DNA

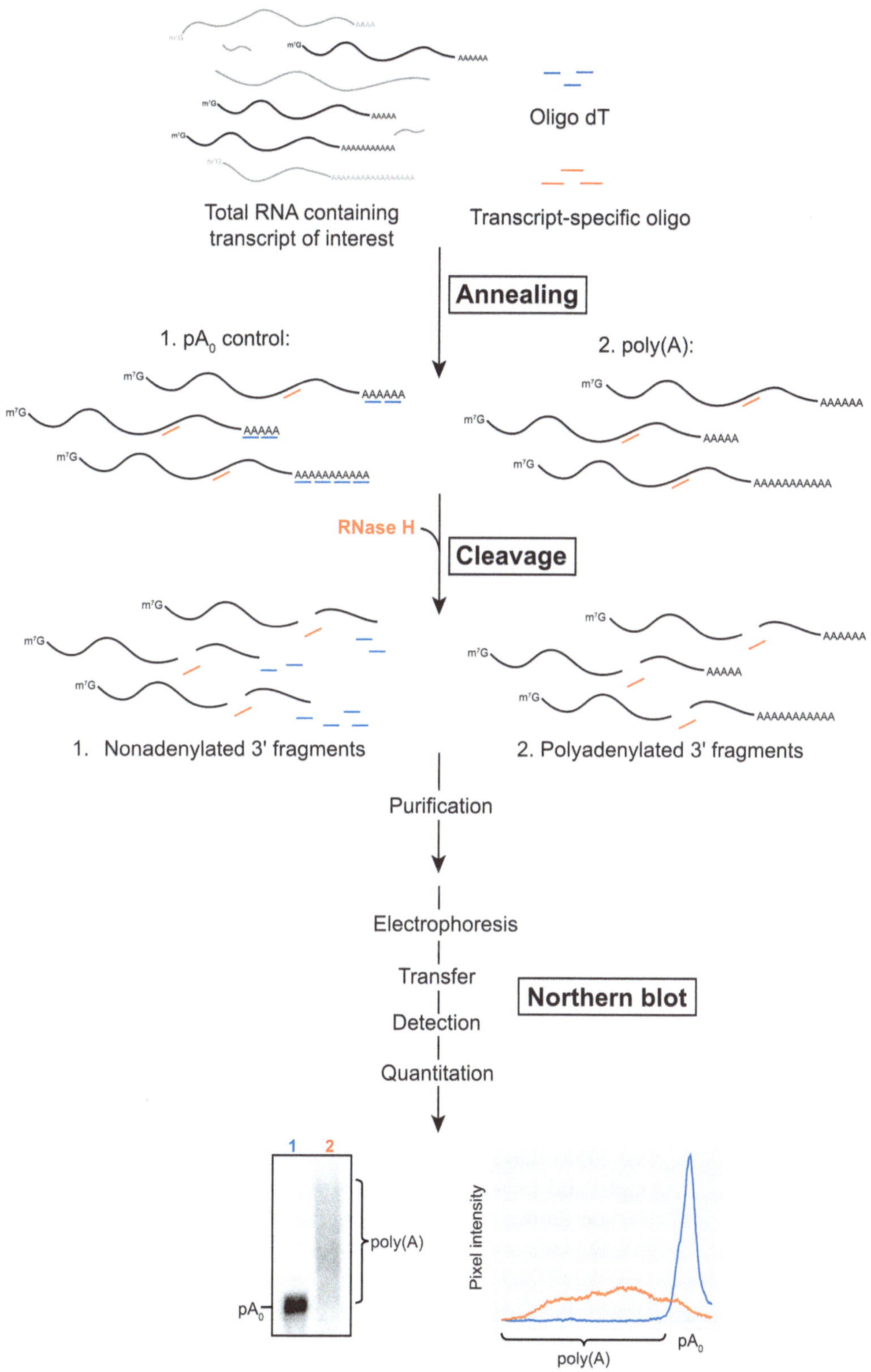

Fig. 1 Schematic describing the workflow of RNase H cleavage followed by high-resolution northern blotting. Briefly, a transcript-specific oligo is annealed to a transcript of interest within total RNA. The DNA:RNA hybrid is then cleaved by RNase H to generate shorter 3′ fragments suitable for high-resolution northern blotting. For

hybrid. In this manner, the size of an RNA that lacks its poly(A) tail can be compared to the native RNA.

Northern blotting offers several advantages. The first advantage is its simplicity, making it accessible to most molecular biologists. Second is its directness, in that the hybridization of the probe directly to the transcript of interest enables detection and quantitation without an intermediary amplification step. Third, northern blotting resolves RNA intermediates with different sizes, providing important insight into RNA metabolism and regulation. Fourth is its sensitivity and broad quantitation range when using a phosphorimager. Fifth is the inherent versatility of northern blotting; nearly any transcript can be analyzed, including transcripts from endogenous genes or exogenous reporter genes. The nature of the probe is also versatile. The method described here details the use of antisense RNA probes generated by in vitro transcription, which enables high specific activity labeling using radioactive nucleotides. Alternative types of probes can be used, including synthetic DNA oligonucleotides. Different detection methods can also be used for northern blotting. The method described here uses radioactivity-based detection; alternatively, colorimetric, chemiluminescence, or fluorescence detection methods are commercially available.

Northern blotting also has inherent limitations that need consideration. Probing is typically limited to one target RNA at a time. The format of the gel electrophoresis also limits the number of experimental conditions tested. The necessary reagents and most equipment are available commercially and are common in molecular biology labs. However, some more specialized equipment is less common (e.g., hybridization oven, UV crosslinker) and can be quite expensive (e.g., phosphorimager). The use of radioactivity necessitates proper licensing, training, safety precautions, equipment, and facilities. Moreover, the resolution of northern blotting is limited by the capabilities of gel electrophoresis and blotting, and single-nucleotide resolution of tail length is not feasible. Poly (A) tail lengths will be more readily resolved for shorter transcripts. To overcome this limitation, longer transcripts can be cleaved into smaller fragments using RNase H guided by an antisense DNA oligonucleotide (Fig. 1). Northern blot analysis of the resulting 3′ cleavage product of the transcript can better resolve the poly(A) tail length distribution (Fig. 2). In this chapter, we provide a protocol for DNA oligonucleotide-directed RNase H cleavage and both high- and low-resolution northern blotting. We offer suggestions for designing and optimizing this method for measuring poly (A) tails of specific transcripts.

Fig. 1 (continued) measurement of poly(A) tails, two conditions are used: cleavage with (1) oligo dT in addition to transcript-specific oligo to generate a non-adenylated (pA_0) control 3′ fragment and (2) only transcript-specific oligo to generate polyadenylated (poly(A)) fragments

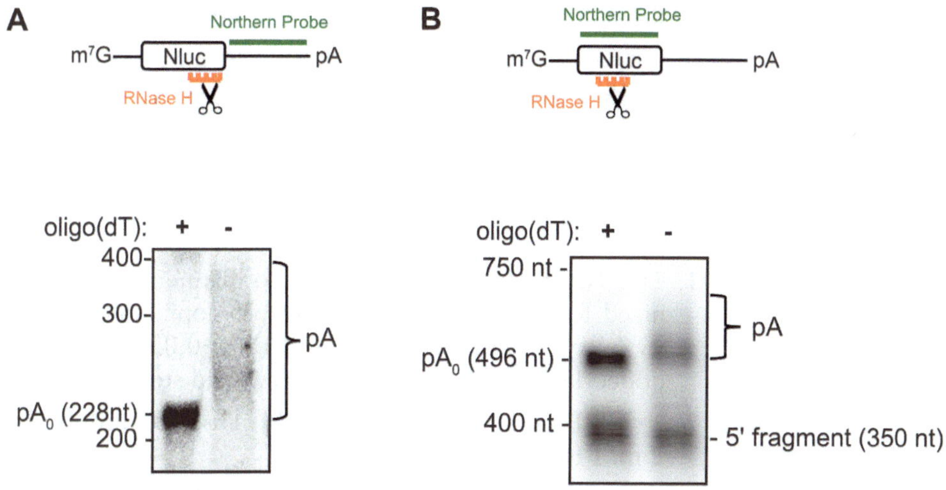

Fig. 2 (**a**) RNase cleavage strategy to generate a small 3′ fragment for optimal resolution of poly(A) RNA. In this approach, the cleavage oligo is positioned at the end of the open reading frame and the northern blot probe detects this 3′ fragment. (**b**) RNase cleavage strategy to generate both 5′ and 3′ fragments of resolvable lengths. In this approach, the cleavage oligo is positioned in the middle of the ORF and the northern probe spans both fragments. The 5′ fragment can be used as an internal control

2 Materials

Prepare all reagents using ultrapure, nuclease-free distilled, deionized water (ddH_2O). Use ribonuclease (RNase) free reagents and plasticware. Maintain sterile conditions and wear disposable gloves to help avoid RNase contamination. Avoid heating RNA samples in the presence of divalent metal ions, which catalyze the hydrolysis of RNA.

2.1 RNase H Cleavage

1. RNA sample: 10 μg (*see* **Note 1**).
2. Gene-specific antisense cleavage oligodeoxynucleotide at 20 pmol/μL (see below for design considerations) (*see* **Notes 2–4**).
3. Thymidine oligodeoxynucleotide (dT_{16}): 500 ng/μL (*see* **Note 5**).
4. 3× annealing buffer: 600 mM KCl and 3 mM ethylenediaminetetraacetic acid (EDTA).
5. Cleavage buffer: 120 mM Tris–HCl (pH 8.0) and 168 mM $MgCl_2$.
6. RNase H enzyme: 5000 units/mL.
7. RNase-free ultrapure ddH_2O.

8. Recombinant RNase Inhibitor (RNaseIn): 40 units/μL.
9. EDTA-EGTA mixture (pH 8.0): 120 mM EDTA, 120 mM ethylene glycol-bis(β-aminoethyl ether)-N,N,N′,N′-tetraacetic acid (EGTA).
10. Incubator for microfuge tubes.
11. RNA purification spin columns.

2.2 Electrophoresis and Northern Blotting

2.2.1 High-Resolution Urea–Polyacrylamide Gel

1. 2× RNA gel loading buffer with formamide: 88% (v/v) deionized formamide, 0.025% (w/v) bromophenol blue, 0.025% (w/v) xylene cyanol FF, 10 mM EDTA, 0.025% (w/v) sodium dodecyl sulfate (SDS). Filter-sterilize and store at −20 °C.
2. 1× TBE running buffer: 89 mM Tris, 89 mM boric acid, 2 mM EDTA.
3. 0.5× TBE transfer buffer (diluted from 1×).
4. 0.25% (w/v) methylene blue dissolved in ddH_2O.
5. RNA molecular weight markers (*see* **Note 6**).
6. Urea–polyacrylamide gel: 8 M urea, 1× TBE, 5% (w/v) polyacrylamide. We use commercial precast gels (14 cm wide × 9 cm tall × 1 mm thick) (*see* **Notes 7** and **8**).
7. Vertical gel electrophoresis apparatus.
8. Wet electroblotting apparatus.
9. Electrophoresis power supply.

2.2.2 Lower-Resolution Formaldehyde–Agarose Gel

1. 10× MOPS buffer (pH 7.0): 0.2 M 3-(N-Morpholino)propanesulfonic acid hemisodium salt (MOPS), 0.05 M sodium acetate, 10 mM EDTA. Filter-sterilize, and do not autoclave.
2. 3× northern RNA sample buffer: 9% formaldehyde, 69% formamide, 13.8 mM MOPS (pH 7), 3.3 mM sodium acetate, and 0.6 mM EDTA.
3. 10× northern RNA loading dye: 50% (v/v) glycerol, 100 mM EDTA (pH 8.0), 0.25% (w/v) bromophenol blue, and 0.25% (w/v) xylene cyanol FF.
4. Ethidium bromide solution: 1 μg/μL in ddH_2O (*see* **Note 9**).
5. RNA molecular weight markers. RNA markers are available from commercial suppliers.
6. Formaldehyde-agarose gel: 1% (w/v) agarose in 1× MOPS (20 mM MOPS (pH 7.0), 5 mM sodium acetate, and 1 mM EDTA) with 1.48% formaldehyde. Formaldehyde vapors are toxic, and all steps involving formaldehyde should be performed in a fume hood. Carefully melt the agarose-dissolved ddH_2O in a microwave oven and then cool to 70 °C before adding formaldehyde and the MOP buffer to a final

concentration of 1×. Mix the solution and cast in the electrophoresis gel tray (*see* **Note 7** for gel percentage considerations).

7. Cellulose filter paper.
8. Horizontal gel electrophoresis apparatus.
9. Electrophoresis power supply.
10. Transfer reservoir for downward capillary transfer. We use the Whatman TurboBlotter apparatus (*see* **Note 10**).
11. UV crosslinker with emission wavelength of 254 nm.

2.3 In Vitro Transcription of Radioactively Labeled Probes

See **Notes 11–13** for considerations and alternatives pertaining to the design and detection format of the probe.

1. Ribonucleotide triphosphates: 10 mM each of ATP, CTP, GTP, and CTP.
2. T7 RNA polymerase enzyme: 15 U/μL from a commercial source.
3. 10× T7 transcription buffer, supplied with T7 RNA polymerase.
4. DNase 1: 2 U/μL, RNase-free.
5. Radioactively labeled α-^{32}P UTP: 800 Ci/mmol, 10 mCi/mL, and 250 μCi.
6. G25 Sephadex gel filtration medium (superfine): 50% slurry in ddH_2O.
7. Microspin columns in microfuge tubes.
8. Microcentrifuge with rotor and lid.
9. Calibrated Geiger–Müller counter.
10. EDTA-EGTA mixture (pH 8.0): 0.25 mM EDTA and 0.25 mM EGTA.

2.4 Northern Blotting

1. 20× SSC buffer: 3 M NaCl, 300 mM sodium citrate–HCl (pH 7.0).
2. Positively charged nylon transfer membrane with 0.45 μm diameter pore size.
3. Hybridization buffer. We use ULTRAhyb from ThermoFisher Scientific with proprietary composition. Alternatively, a hybridization buffer can be made from individual components as previously described [24].
4. 10% (w/v) SDS in ddH_2O.
5. Northern wash buffers:

   Wash Buffer #1: 2× SSC, 0.1% (w/v) SDS.

   Wash Buffer #2: 0.1× SSC, 0.1% (w/v) SDS.

6. Hybridization oven with integral rotisserie and appropriately sized glass hybridization bottles with leak-proof lids.

2.5 Autoradiography

1. Phosphorimager.
2. Phosphorimager screen with light-proof cassette.
3. Data analysis software. We use ImageQuant TL from Cytiva Life Sciences.

3 Methods

3.1 RNase H Cleavage

RNA samples should be thawed and kept on ice while assembling the reaction. *See* **Note 14** for reaction optimization considerations.

1. Assemble RNA-DNA oligo annealing reactions:

   x μL RNA sample (10 μg).

   10 μL 3× annealing buffer.

   1 μL of 20 pmol/μL cleavage oligo (20 pmol).

   19-*x* μL ddH_2O.

   Total: 30 μL.

   For pA_0 control, assemble another reaction that includes 1500 ng of oligo dT_{16}, adjusting final reaction volume as necessary.
2. Denature the RNA by heating the annealing reaction to 70 °C for 3 min. Then, cool samples on ice for at least 10 min to allow annealing.
3. Prepare cleavage reactions:

   10 μL of cleavage buffer.

   1.2 μL of 40 U/μL RNasin.

   17.8 μL of ddH_2O.

   1 μL of 5 U/μL RNase H enzyme.

   Total: 30 μL.
4. For each 30 μL annealing reaction, add 30 μL of the cleavage reaction mixture.
5. Incubate at 37 °C for 1 h (*see* **Note 14**).
6. Chelate divalent metal ions by adding 60 μL of EDTA-EGTA mixture.
7. Purify RNA using spin columns following the manufacturer's instructions and eluting in 30 μL ultrapure RNase-free ddH_2O (*see* **Note 15** for alternative purification methods).

3.2 Gel Electrophoresis and Blotting

3.2.1 High-Resolution Urea–Polyacrylamide Gel Electrophoresis and Electroblotting

1. Assemble the urea–polyacrylamide gel in the electrophoresis apparatus with 1× TBE buffer and pre-run at a constant 200 V until the current reaches 24–25 mA.
2. Prepare samples by combining 15 μL total RNA (5 μg) with 15 μL of 2× RNA gel loading buffer with formamide.
3. Prepare the RNA molecular weight marker by mixing it with 15 μL of 2× RNA gel loading buffer with formamide.
4. Heat denature the samples and marker at 75 °C for 10 min.
5. Turn off the electrophoresis power supply.
6. Before loading the samples into the gel, flush the gel wells thoroughly with the running buffer using a syringe or pipette.
7. Carefully load each RNA sample into the gel wells using a pipette with a fine-gauge pipette tip.
8. Resume electrophoresis at 200 V. Stop the electrophoresis once the bromophenol blue dye in the samples has migrated ~ 2⁄3 the length of the gel or until the desired resolution is reached (*see* **Note 7**). On a 5% polyacrylamide gel, bromophenol blue and xylene cyanol migrate at the same rates as 35 and 140 nt RNA molecules, respectively.
9. Soak the positively charged nylon membrane in 0.5× TBE for 5 min. Handle the membrane with forceps and avoid touching it with gloved hands.
10. Briefly wet two sheets of cellulose filter paper in 0.5× TBE.
11. Set up wet electroblotting apparatus. On the side of the electroblotting cassette facing the positive electrode, first layer the mesh support and then cellulose filter paper, followed by the positively charged nylon membrane. Remove air bubbles between the membrane and cellulose filter paper by rolling a serological pipet over the membrane.
12. Carefully remove the gel from between glass (or plastic) plate casing. Layer the gel onto the positively charged nylon membrane. Next, layer a piece of cellulose filter paper on top of the gel, followed by a final piece of mesh support. This direction is important so that the negatively charged nucleic acid is transferred toward the positive electrode and onto the nylon membrane. Assemble the cassette. The final order of the assembled stack in the cassette should be as follows:
    - Positive electrode side of the cassette.
    - Support mesh.
    - Cellulose filter paper.
    - Positively charged nylon membrane.
    - Urea–polyacrylamide gel.
    - Cellulose filter paper.

- Mesh support.
- Negative electrode side of the cassette.

13. Insert the cassette into the tank of the electroblotting apparatus.
14. Fill the tank with 0.5× TBE, prechilled at 4 °C.
15. Connect the apparatus to the power supply and apply 30 V for 45 min. Perform the transfer at 4 °C in a cold room. If possible, mix the running buffer during electroblotting to aid in cooling. Moreover, some electroblotting apparatuses have an ice pack that can be added to the tank to aid cooling.
16. Stop the power supply and remove the cassette.
17. Disassemble the cassette and remove the nylon membrane using forceps. Wrap the membrane in clear plastic wrap.
18. Crosslink the RNA to the membrane in a UV crosslinker at 120 J/cm^2 UV ($\lambda = 254$). The membrane is now ready for probing or can be stored at 4 °C.
19. To visualize the RNA molecular weight markers on the blot, stain them with methylene blue before proceeding to the northern blot probing steps. Cut the strip of the blot containing the RNA marker, taking care to mark the orientation of the ladder relative to the remainder of the blot. Stain the RNA marker by soaking the strip in 0.25% (w/v) methylene blue dissolved in ddH$_2$O for 15 min. Wash the ladder portion of the blot several times with ddH$_2$O until the RNA marker bands are seen, and the background signal on the strip is only faintly blue.

3.2.2 Formaldehyde–Agarose Gel Electrophoresis and Capillary Transfer Blotting

1. Prepare samples:

   2.4 μL 10× northern RNA loading dye.

   8 μL 3× northern RNA sample buffer.

   x μL RNA sample (5 μg).

   Optional: 1 μL of 1 μg/μL ethidium bromide (0.04 μg/μL final).

   11.6-*x* μL ddH$_2$O.

   Total: 24 μL.
2. Prepare RNA molecular weight marker in 1× northern RNA loading dye with 1× northern sample buffer.
3. Heat the samples and RNA marker at 70 °C for 10 min.
4. Assemble the formaldehyde–agarose gel in the horizontal electrophoresis apparatus and fill the tank with 1× MOPS running buffer to a level where the buffer covers the surface of the gel. All steps involving formaldehyde should be conducted in a chemical fume hood.

5. Before loading the gel, flush the wells thoroughly with 1× MOPS running buffer using a pipette or syringe.
6. Carefully load the samples into the wells of the gel.
7. Electrophorese at 5 V/cm of gel width. Stop electrophoresis once the bromophenol blue dye has reached ~ ⅔ the length of the gel or until the desired size resolution is achieved.
8. Soak the positively charged nylon membrane in ddH_2O for 5 min and then in 20× SSC for 5 min.
9. Prepare the transfer reservoir by placing a stack of cellulose paper sheets at a height equivalent to the height of the reservoir tank.
10. Wet two sheets of cellulose filter paper with 20× SSC and add on top of the stack.
11. Add the positively charged nylon membrane to the top of the stack. Use a serological pipet or roller to remove bubbles.
12. Carefully place the formaldehyde–agarose gel to the top of the stack. Hold the fragile gel carefully in gloved hands, allowing the middle of the gel to touch the membrane. Slowly move hands apart to lower each gel side onto the nylon membrane. If bubbles form, carefully lift that part of the gel and lower it again. Adding a few milliliters of 20× SSC buffer to the top of the nylon membrane can help eliminate air bubbles.
13. Wet and add two sheets of cellulose filter paper and place them on top of the stack, rolling out any bubbles.
14. Add a dry wick of cellulose filter paper to the top of the stack so that the middle contacts the transfer stack and the edges are submerged in the 20× SSC buffer in the transfer reservoir. It may help to roll the top of the wick to contact the stack evenly. If the capillary transfer is working, you should see the dry regions of the wick start to absorb liquid slowly. The order of the stack from the bottom of the stack to the top should be as follows:
    - Stack of dry cellulose filter paper within the reservoir (bottom of the stack).
    - Two wetted sheets of cellulose filter paper.
    - Nylon membrane.
    - Agarose gel.
    - Two wetted sheets of cellulose filter paper.
    - Wick (top of the stack).
15. Allow blotting transfer to proceed for 16–24 h.
16. Remove the blot from the transfer apparatus and wrap it in clear plastic wrap.

17. Crosslink the RNA to the membrane in a UV crosslinker at 120 J/cm^2 UV ($\lambda = 254$ nm). The membrane is now ready for probing or can be stored at 4 °C.
18. If ethidium bromide was included in the samples in **step 1**, then the RNA molecular weight markers can be visualized on the blot with a handheld UV light and subsequently marked with a pencil. Alternatively, the RNA molecular weight markers on the blot can be visualized by staining with methylene blue, as described in Subheading 3.2.1, **step 16**.

3.3 Antisense Radiolabeled RNA Probe Generation

1. Set up in vitro transcription reaction as follows:

   x μL of linear DNA template (1 μg) (*see* **Note 13**).

   1 μL of 10 mM ATP.

   1 μL of 10 mM CTP.

   1 μL of 10 mM GTP.

   2.5 μL 10× T7 transcription buffer.

   15.3-*x* μL ultrapure ddH_2O.

   2 μL 12.5 μM α^{32}P UTP (800 Ci/mmol, 10 mCi/mL) (1 μM, 10–20 μCi in final reaction).

   0.2 μL of 1 mM unlabeled UTP.

   2 μL T7 RNA polymerase enzyme (15 U/μL).

   25 μL total.
2. Incubate transcription reaction at 37 °C for 10 min.
3. Add 1 μL (2 U) of DNase 1 and incubate at 37 °C for 10 min.
4. To chelate divalent metal ions, add 1 μL 0.25 M EDTA–0.25 M EGTA mix and incubate at 37 °C for 10 min.
5. Pipette 1 mL of G25 Sephadex 50% slurry to a microspin column in a 1.7 mL collection tube.
6. Centrifuge the microspin column for 1 min at 1000 g and empty the eluted liquid from the collection tube. Repeat this step twice.
7. Carefully add the RNA probe from **step 4** to the top of the G25 Sephadex microspin column.
8. Cap the microspin column, place it in the centrifuge rotor, and add the rotor cap. Centrifuge at 1000 g for 1 min.
9. Collect the eluate containing the radiolabeled RNA probe. Discard the microspin column containing the unincorporated radioactive nucleotide.

3.4 Probing and Imaging of the Northern Blot

1. Prehybridize the northern blot in a hybridization bottle by incubating with rotation for at least 30 min at 68 °C in the hybridization oven with enough hybridization buffer to cover the blot. We use 6–8 mL for a 12 × 8 cm blot (*see* **Note 16**).
2. Measure the radioactivity of the RNA probe by counting 1 μL in a microfuge tube with a Geiger–Müller counter set at 100× sensitivity range.
3. Add 2.5–7.5 million counts per minute of the probe to the hybridization bottle containing the northern blot and hybridization buffer (*see* **Note 17**).
4. Hybridize the RNA probe to the northern blot by incubating with rotation in the hybridization oven at 68 °C for 16–24 h.
5. Wash the blot twice with Wash buffer #1 (2× SSC, 0.1% SDS) in the hybridization oven with rotation at 68 °C for 15 min/wash.
6. Wash the blot twice with Wash buffer #2 (0.1× SSC, 0.1% SDS) in the hybridization oven with rotation at 68 °C for 15 min/wash.
7. Wrap the blot in plastic and expose it to a phosphor screen in its light-proof exposure cassette. Exposure time will vary from several hours to several days and may need to be optimized for specific RNAs.
8. Scan the phosphor screen using a phosphorimager.
9. Visualize and quantitate the northern blot image using phosphorimager analysis software.

4 Notes

1. Because this method is based on directly detecting a target RNA, it requires more total RNA than amplification-based approaches. While we routinely use 10 μg of total RNA, we have successfully used as little as 1 μg of total RNA to detect an exogenously expressed reporter mRNA. The amount of RNA loaded on the gel will depend on the amount recovered after the RNase H cleavage assay and spin column purification. The anticipated copy number of a transcript of interest is one significant factor to consider, as the abundances of different RNAs can vary; transcript number per cell between genes can vary as much as ~1000-fold [25]. Moreover, the abundance of an RNA can vary widely depending on cellular context and conditions. Thus, the amount of total RNA used for this approach may need to be determined experimentally for the specific transcript being detected. This amount will depend on factors such as the abundance of that RNA and the probe's sensitivity. Endogenous RNAs of lower abundance may require more total

RNA input for detection, while overexpressed reporters generally require less material.

2. RNase H cleaves the phosphodiester backbone of RNA within an RNA:DNA hybrid. The position of the gene-specific reverse complement DNA oligo used in the RNase H cleavage reaction will determine the length of the cleavage products and will depend on the desired resolution of the assay. For example, it is often best for the 3′ cleavage product to be as short as possible to allow for higher resolution while maintaining hybridization to the antisense probe (as in Fig. 2a). Alternatively, the experiment may benefit from measuring the 5′ and 3′ cleavage products, as in Fig. 2b (the former serving as an internal control). In this case, the oligo can be positioned to generate cleavage products that are far enough apart to be resolved but close enough to be detected on the same gel.
3. Design parameters of the cleavage oligo follow general primer design guidelines. As general guidelines, we aim for a DNA oligo that is 18–24 base pairs long with a melting temperature of 58–60 °C and as close to 50% GC content as possible. Oligos should be evaluated for propensity to form competing secondary structures or homodimers using one of many available online tools. We use the OligoAnalyzer from IDT (https://www.idtdna.com/pages/tools/oligoanalyzer) for oligos designed in SnapGene or Benchling, but other tools such as Primer3 (https://primer3.ut.ee/) are also commonly used. It may also be beneficial to search the oligo sequence on Primer-BLAST (https://www.ncbi.nlm.nih.gov/tools/primer-blast/) for potential off-target binding sites on other transcripts, which may be undesirable. Longer oligos will confer more specificity but may also be more prone to be highly structured. Moreover, since RNA within the hybrid is degraded, this length will be subtracted from the final cleavage product(s).
4. While cleavage oligo design for established reporter RNAs is straightforward, additional considerations exist for detecting endogenous RNAs in this assay. This assay may also reveal the presence of alternatively processed RNA isoforms, such as those produced by alternative splicing and/or alternative cleavage and polyadenylation site utilization. These products would be evident by multiple bands in the non-adenylated pA_0 control (*see* **Note 5**). If multiple isoforms are detected with lengths that cannot be sufficiently resolved, then isoform-specific cleavage oligos and antisense probes would be required.
5. As a crucial control, a pA_0 control reaction is set up using oligo dT in addition to the transcript-specific oligo to generate a non-adenylated 3′ fragment of the RNA of interest. Combined with RNA markers, this control enables the measurement of

poly(A) tail length for the RNA of interest. While this method can be used to validate the polyadenylated status of an RNA (or poly(A) tail-lacking control), it can also be used as a reference point when quantitating the poly(A) tail lengths for a given RNA. To do this, the blot image can be analyzed using image analysis software. We use ImageQuant TL software from Cytiva Biosciences. To quantitate poly(A) signal at different lengths, a histogram can be generated by integrating the radioactive signal across the lane of each sample, thereby plotting the distribution of polyadenylated RNA (as demonstrated in Fig. 1) [4, 19].

6. To measure poly(A) tail length, RNA molecular weight markers with known lengths are used. RNA molecular weight markers are available from multiple commercial sources and should be selected based on the desired size range necessary for the resolution and detection of the RNA of interest (Fig. 3). These RNA markers can be detected via staining with methylene blue followed by imaging and comparison alongside the

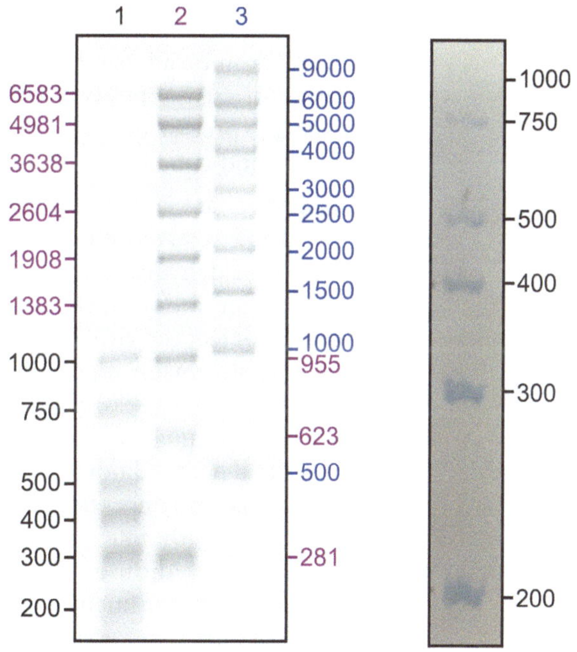

Fig. 3 *Left*: Resolution of three sets of RNA molecular weight markers on a denaturing formaldehyde 1.25% agarose gel. RNA is visualized via the inclusion of ethidium bromide in the sample. Lane 1: RNA Century-Plus Markers (Invitrogen), Lane 2: RNA Markers (Promega), and Lane 3: Millennium RNA Markers (Ambion). *Right*: Resolution of RNA Century markers on a methylene blue-stained nylon membrane transferred from an 8 M urea and 5% polyacrylamide gel. Sizes (length in nucleotides) of the RNA markers are indicated on the sides

blot. Alternatively, suppose ethidium bromide is used in the sample. In that case, the ladder can be detected on the transferred blot via UV transillumination and then later compared alongside the autoradiography image of the probed blot.

7. Gel composition and percentage are important considerations for the optimal resolution of the RNA of interest. We routinely use 8 M urea, 5% polyacrylamide gels to resolve fragments of ~200–400 nucleotides in length with poly(A) tails ranging up to 250 nucleotides. We use formaldehyde–agarose gels ranging from 0.85% to 2% agarose to resolve RNA fragments >400 nucleotides with tails of up to 250 nucleotides. Low melting temperature agarose can produce gels with a higher percentage of agarose, enabling the resolution of small polyadenylated RNAs. Still, our experience is that polyacrylamide is more effective in this case. A helpful guide for choosing gel percentage can be found here: https://www.idtdna.com/pages/education/decoded/article/running-agarose-and-polyacrylamide-gels.
8. Established protocols exist for pouring urea–polyacrylamide gels [22, 26, 27]. Pouring your own gels affords flexibility in the composition and resolution of RNAs. The percentage of the gel can be adjusted as necessary to resolve the desired size range of transcripts best. Alternatively, we use commercially available precast urea–polyacrylamide gels, which offer convenience and consistency.
9. The abundant large and small ribosomal RNAs (rRNA) can be visualized as a readout of RNA integrity and transfer efficiency for formaldehyde–agarose gels and blots. We routinely add a low concentration of ethidium bromide in the sample preparation to detect the rRNA in the gel before and after the transfer and on the nylon membrane after blotting. This provides a facile way to assess RNA integrity and transfer efficiency. While ethidium bromide in the sample at the low concentration described here did not affect migration, it is not recommended to use higher concentrations, as this could impede migration. Do not add ethidium bromide to the gel mix when casting, as this creates a diffuse fluorescent signal throughout the gel. Alternatively, the membrane can be stained with methylene blue (or a similar dye) after transfer but before incubation in a hybridization buffer to visualize rRNA bands. In our experience, the dye is subsequently removed upon incubation with a hybridization buffer at high temperatures, though we have not evaluated whether this affects sensitivity.
10. Using a commercially available transfer reservoir, we describe a downward capillary transfer (absorbent stack on the bottom, wick on top). However, upward capillary transfer (wick on

bottom, absorbent stack on top) can be done using a rectangular glass pan as the reservoir, with the transfer stack assembled on a glass plate that partially covers the pan reservoir (so that it is elevated from the reservoir). Paper towels can be used as a substitute for cellulose filter paper, though they may not be as absorbent.

11. We use antisense RNA probes synthesized by in vitro transcription with a radiolabeled nucleotide. This enables the generation of probes with high specific activity (i.e., radioactivity per unit mass) and sequence specificity. As an alternative for detecting very abundant transcripts, antisense synthetic DNA oligo probes can be generated by 5′ end labeling with γ-^{32}P ATP using polynucleotide kinase (PNK), as we have done to detect 18S rRNA [5] and 7SL noncoding RNA [19].
12. We use radiation-based detection due to the high sensitivity. In the case of safety concerns regarding the use of radiation, other non-radioactive detection methods exist that others have demonstrated to be sensitive and effective, such as an IR-dye-based approach [28]. These methods also circumvent the issue of short half-lives of signal when using radioactive ^{32}P isotopes.
13. For our antisense RNA probe, we generate linear in vitro transcription templates using PCR with a T7 promoter appended to the reverse primer (so that the transcribed probe will be the reverse complement of the target RNA). While we generally aim for northern probes to be ~500 base pairs, this approach may necessitate shorter probes depending on cleavage product (s) length. For example, we have successfully used probes complementary to shorter 3′UTRs (down to 187 bp). Longer probes enable increased sensitivity due to the incorporation of more radioactive nucleotides. If multiple cleavage products are to be detected, the probe will need to span both 5′ and 3′ products (as in Fig. 2b), or multiple probes will be necessitated.
14. Parameters of the RNase H cleavage protocol may be modified to optimize conditions for difficult templates. The denaturation time and temperature may need to be optimized for highly structured RNAs. The reaction time may need to be extended if incomplete cleavage is observed. The optimal incubation time should ensure complete cleavage while minimizing potential nonspecific degradation of the RNA. RNase H enzyme units could be increased to enhance cleavage efficiency with short reaction times. Potassium in the annealing buffer aids in the hybridization of the DNA oligo to the target RNA. Magnesium is necessarily excluded from the denaturation/hybridization step because divalent metals catalyze RNA hydrolysis at elevated temperatures. Still, magnesium must be subsequently added to the RNase H reaction because it is an

essential cofactor for RNase H. Chelation by EDTA and EGTA and RNA purification are included to remove RNase H and eliminate divalent metal ions to prevent degradation of RNA during subsequent steps.

15. Alternatively, phenol–chloroform, followed by precipitation with isopropanol/ethanol, or magnetic bead-based purification, can clean up RNA after RNase H cleavage.
16. The volume of the hybridization buffer can be adjusted depending on the size of blot so long as the blot is fully covered when rotating; however, increasing the hybridization volume will dilute the radiolabeled probe.
17. The counts per minute estimate the activity of the radiolabeled probe and can be adjusted depending on the abundance of the target, required detection sensitivity, and intended exposure time. Alternatively, the counts per minute can be measured using a scintillation counter.

Acknowledgments

This work was supported by the National Institute of General Medical Sciences, National Institutes of Health grant R01GM105707 to A.C.G. R.M.A. was supported by fellowship DGE 1256260 from the National Science Foundation. We thank Dr. Katherine McKenney and Robert Connacher for their comments and suggestions. We thank Dr. Daniel Seay and Dr. Nathan Blewett for the technical information.

References

1. Passmore LA, Coller J (2022) Roles of mRNA poly(A) tails in regulation of eukaryotic gene expression. Nat Rev Mol Cell Biol 23:93–106
2. Raisch T, Valkov E (2022) Regulation of the multisubunit CCR4-NOT deadenylase in the initiation of mRNA degradation. Curr Opin Struct Biol 77:102460
3. Goldstrohm AC, Wickens M (2008) Multifunctional deadenylase complexes diversify mRNA control. Nat Rev Mol Cell Biol 9: 337–344
4. Goldstrohm AC, Hook BA, Seay DJ, Wickens M (2006) PUF proteins bind Pop2p to regulate messenger RNAs. Nat Struct Mol Biol 13: 533–539
5. Arvola RM, Chang C-T, Buytendorp JP et al (2020) Unique repression domains of Pumilio utilize deadenylation and decapping factors to accelerate destruction of target mRNAs. Nucleic Acids Res 48:1843–1871
6. Van Etten J, Schagat TL, Hrit J et al (2012) Human Pumilio proteins recruit multiple deadenylases to efficiently repress messenger RNAs. J Biol Chem 287:36370–36383
7. Enwerem III, Elrod ND, Chang C-T et al (2021) Human Pumilio proteins directly bind the CCR4-NOT deadenylase complex to regulate the transcriptome. RNA 27:445–464
8. Brouze A, Krawczyk PS, Dziembowski A, Mroczek S (2022) Measuring the tail: methods for poly(A) tail profiling. Wiley Interdiscip Rev RNA 14:e1737
9. Sallés FJ, Strickland S (1995) Rapid and sensitive analysis of mRNA polyadenylation states by PCR. PCR Methods Appl 4:317–321
10. Sallés FJ, Strickland S (1999) Analysis of poly (A) tail lengths by PCR: the PAT assay. Methods Mol Biol 118:441–448
11. Jänicke A, Vancuylenberg J, Boag PR et al (2012) ePAT: a simple method to tag

adenylated RNA to measure poly(A)-tail length and other 3′ RACE applications. RNA 18: 1289–1295

12. Chang H, Lim J, Ha M, Narry Kim V (2014) TAIL-seq: genome-wide determination of poly (A) tail length and 3′ end modifications. Mol Cell 53:1044–1052
13. Subtelny AO, Eichhorn SW, Chen GR et al (2014) Poly(A)-tail profiling reveals an embryonic switch in translational control. Nature 508:66–71
14. Lim J, Lee M, Son A et al (2016) mTAIL-seq reveals dynamic poly(A) tail regulation in oocyte-to-embryo development. Genes Dev 30:1671–1682
15. Liu Y, Zhang Y, Wang J, Lu F (2022) Transcriptome-wide measurement of poly (A) tail length and composition at subnanogram total RNA sensitivity by PAIso-seq. Nat Protoc 17:1980–2007
16. Legnini I, Alles J, Karaiskos N et al (2019) FLAM-seq: full-length mRNA sequencing reveals principles of poly(A) tail length control. Nat Methods 16:879–886
17. Workman RE, Tang AD, Tang PS et al (2019) Nanopore native RNA sequencing of a human poly(A) transcriptome. Nat Methods 16:1297–1305
18. Coon SL, Li T, Iben JR et al (2021) Single-molecule polyadenylated tail sequencing (SM-PAT-Seq) to measure polyA tail lengths transcriptome-wide. Methods Enzymol 655: 119–137
19. Weidmann CA, Raynard NA, Blewett NH et al (2014) The RNA binding domain of Pumilio antagonizes poly-adenosine binding protein and accelerates deadenylation. RNA 20:1298–1319
20. Alwine JC, Kemp DJ, Stark GR (1977) Method for detection of specific RNAs in agarose gels by transfer to diazobenzyloxymethyl-paper and hybridization with DNA probes. Proc Natl Acad Sci U S A 74:5350–5354
21. Rio DC (2015) Denaturation and electrophoresis of RNA with formaldehyde. Cold Spring Harb Protoc 2015:219–222
22. Green MR, Sambrook J (2021) Separation of RNA according to size: electrophoresis of RNA through denaturing urea polyacrylamide gels. Cold Spring Harb Protoc 2021:prot101766. https://doi.org/10.1101/pdb.prot101766
23. Rio DC, Ares M Jr, Hannon GJ, Nilsen TW (2010) Polyacrylamide gel electrophoresis of RNA. Cold Spring Harb Protoc 2010:db.prot5444
24. He SL, Green R (2013) Northern blotting. Methods Enzymol 530:75–87
25. Schwanhäusser B, Busse D, Li N et al (2011) Global quantification of mammalian gene expression control. Nature 473:337–342
26. Albright LM, Slatko BE (2001) Denaturing polyacrylamide gel electrophoresis. Curr Protoc Nucl Acid Chem Appendix 3:Appendix 3B
27. Summer H, Grämer R, Dröge P (2009) Denaturing urea polyacrylamide gel electrophoresis (urea PAGE). J Vis Exp 29(32):1485. https://doi.org/10.3791/1485
28. Miller BR, Wei T, Fields CJ et al (2018) Near-infrared fluorescent northern blot. RNA 24: 1871–1877

Chapter 7

An RNA-Ligation-Based RACE-PAT Assay to Monitor Poly(A) Tail Length of mRNAs of Interest

Fabienne Mauxion and Bertrand Séraphin

Abstract

In eukaryotes, a non-templated poly-adenosine (poly(A)) tail is added co-transcriptionally to almost every messenger RNA (mRNA). The length of this poly(A) tail changes during the lifetime of mRNAs and has been shown in many circumstances to be an important factor controlling transcript fates. Yet, the measure of the length of this homogenous nucleotide sequence is technically challenging, making it difficult to assess its dynamic variation. In this chapter, we describe an RNA-ligation-based RACE-PAT (*R*apid *A*mplification of *c*DNA *E*nd—*P*oly(*A*) *T*ail) assay to monitor the poly(A) tail length of mRNAs. In the first step, an RNA oligonucleotide is ligated to mRNA 3′ ends providing an anchoring site to prime cDNA synthesis, avoiding the bias introduced by oligo(dT)-derived primers. Afterward, reverse transcription is performed with an anchor primer with a unique 5′ extension. The choice of the oligonucleotide 3′ end at this step allows further flexibility to amplify modified tails, for example, by uridylation. Next, short DNA fragments encompassing the poly(A) tails are amplified by Polymerase Chain Reaction (PCR) using as forward primer, a transcript-specific primer hybridizing close to the transcript polyadenylation signal, and as reverse primer, an oligonucleotide corresponding to the 5′ extension of the primer used for cDNA synthesis, ensuring that only cDNAs are amplified. The resulting DNA fragments are then visualized after size fractionation by electrophoresis. This method does not provide exact nucleotide count and composition but has the advantage of allowing the processing of many samples in parallel at a low cost.

Key words Poly(A) tail, Polyadenylation, Deadenylation, RT-PCR, RNA ligation, Uridylation

1 Introduction

A stretch of adenosines is added co-transcriptionally to the 3′ end of most mRNAs in eukaryotes except for the transcripts encoding replication-dependent histones in mammals. The initial size of this poly(A) tail varies between species (e.g., about 200 adenosines in mammals vs. 70 in yeast) but is universally bound by poly(A)-binding proteins. Poly(A) tails bound by poly(A)-binding proteins dynamically control many aspects of the life of mRNAs from nuclear export, translation efficiency, storage, and decay [1, 2]. The length of the poly(A) tail usually shortens in the

Eugene Valkov and Aaron C. Goldstrohm (eds.), *Deadenylation: Methods and Protocols*, Methods in Molecular Biology, vol. 2723, https://doi.org/10.1007/978-1-0716-3481-3_7,

cytoplasm during the lifetime of mRNAs. However, cytoplasmic re-adenylation may occur in particular contexts, such as during early development, inflammation, or synaptic plasticity [3]. Thus, poly(A) tail length is a dynamic landmark of mRNA 3′ ends.

However, the evaluation of poly(A) tail length is challenging. Techniques such as northern blot are often not adapted as the size difference between polyadenylated and deadenylated transcripts is generally too limited to generate a visible shift in migration during electrophoresis, which is further exacerbated when multiple alternatively spliced mRNA isoforms coexist. For simple mRNA populations, this problem has been bypassed in the past by treating RNA samples with RNase H and antisense oligonucleotides targeting the mRNA 3′ UTR to reduce the size of the transcript under study [4]. Nevertheless, northern blotting remained tedious and required a large amount of starting RNA material.

With the advent of high-throughput sequencing, methods to determine poly(A) tail dynamic at the transcriptome level were developed. Strategies were implemented to adapt to this end the initial short-read sequencing technologies, such as the one developed by Illumina, that are not suited for sequencing long homogenous nucleotide stretches. Results of the first analyses of poly(A) tail sequencing using these specifically developed strategies, such as PAL-seq [5] or TAIL-seq [6], challenged previous models correlating poly(A) tail length to mRNA translatability. They revealed furthermore frequent modifications of the poly(A) tail by uridylation [6] or less frequently by guanylation and cytidilation [7]. These early results raised interest in sequencing transcriptome-wide mRNA poly(A) tails despite some limitations, especially with longer poly(A) tails. Thereby, new methods to sequence them by third-generation sequencing techniques not using amplification by polymerases are currently in development [8]. However, these assays are costly and may not be suitable when many conditions have to be tested. Further, some methods still involve oligo(dT) selection and/or oligo(dT) priming and thus may distort the representation of the poly(A) tail population.

To visualize poly(A) tail length variations for mRNAs of interest, Reverse Transcription—Polymerase Chain Reaction (RT-PCR)-based methods that have been developed to amplify mRNA 3′ extremity, including the poly(A) tail remain an alternative of choice. These strategies circumvent requirements for large quantities of biological material required for northern analyses, are independent of transcript length or the presence of alternative splicing, and allow, for a target mRNA, the rapid processing of many samples in parallel at limited cost. The distinct methods differ mainly in the strategy used to prime cDNA synthesis for reverse transcription. In LM-PAT (Ligation-Mediated-PAT) [9] and ePAT (extension PAT) [10], reverse transcription is primed by oligo(dT)-derived anchors. These assays are likely to miss very short poly

(A) tails and result thus in a distorted presentation of poly(A) tail distribution. Moreover, when primers ending with oligo(dT) are used for amplification, those may hybridize to internal parts of the poly(A) sequence at each cycle, artifactually generating ever shorter products that are preferentially amplified as the PCR reaction proceeds. To avoid these possible biases, we used ligation of an RNA adapter to the 3′ end of mRNAs to prime reverse transcriptase polymerization. This robust method offers also the possibility to be adapted for transcripts carrying posttranscriptionally added non-A residues at the transcript 3′ end. When coupled to transcriptional arrest using actinomycin D treatment in the case of endogenous transcripts [11] or tetracycline-regulated promoters to drive the expression of reporter transcripts [12], this RNA-ligation-based RACE-PAT assay allowed visualizing the kinetics of mRNA poly(A) tail shortening for transcripts of interest. In our hands, it was successfully used to decipher the function of the BTG/Tob factors [11, 13, 14] or specific subunits of the CCR4–NOT complex [12] in regulating mRNA deadenylation.

The RACE-PAT assay is presented in Fig. 1. The first step involves ligating an RNA oligonucleotide (Fig. 1a) to the 3′ end of each molecule in a total RNA sample (Fig. 1b). To avoid multiple ligations to the same molecule, a dideoxycytidine blocks the RNA oligonucleotide at its 3′ extremity. Following ligation, cDNA synthesis is primed using a reverse transcriptase RT primer with a 3′ sequence complementary to the ligated 3′ RNA oligonucleotide containing an additional T at the 3′ end (Fig. 1a; note that the RT primer sequence is shown 3′ to 5′ to indicate complementarity with the other primers) to favor cDNA synthesis from RNA ending with adenosine. (This additional T is optional and may be substituted by another nucleotide, for example, an A if one aims to analyze uridylated transcripts, or omitted if one wants to detect the tails of the whole population of the mRNA of interest.) In the final step, PCR amplification is performed with a primer pair consisting of a PCR primer specific for the transcript of interest and the PCR-all primer corresponding to the 5′ extremity of the RT primer (Fig. 1a, b). All enzymatic reactions can be performed in a thermocycler. The results are then visualized after size fractionation and detection of the products. A deadenylated transcript control can be generated by digesting total RNA with RNase H and oligo $(dT)_{18}$.

A representative example of results generated with the RACE-PAT technique is shown Fig. 1c. In this experiment, human HEK293 cells expressing constitutively the Tet-Off regulator [15] were treated with siRNAs to deplete different subunits of the main deadenylase CCR4–NOT complex. They were then transfected with a plasmid reporter encoding a ß-globin gene under the control of the tetracycline-regulated promoter [15]. A transcriptional pulse-chase was performed [16]. The poly(A) tail length of the β-globin mRNA was assayed using the RACE-PAT assay and

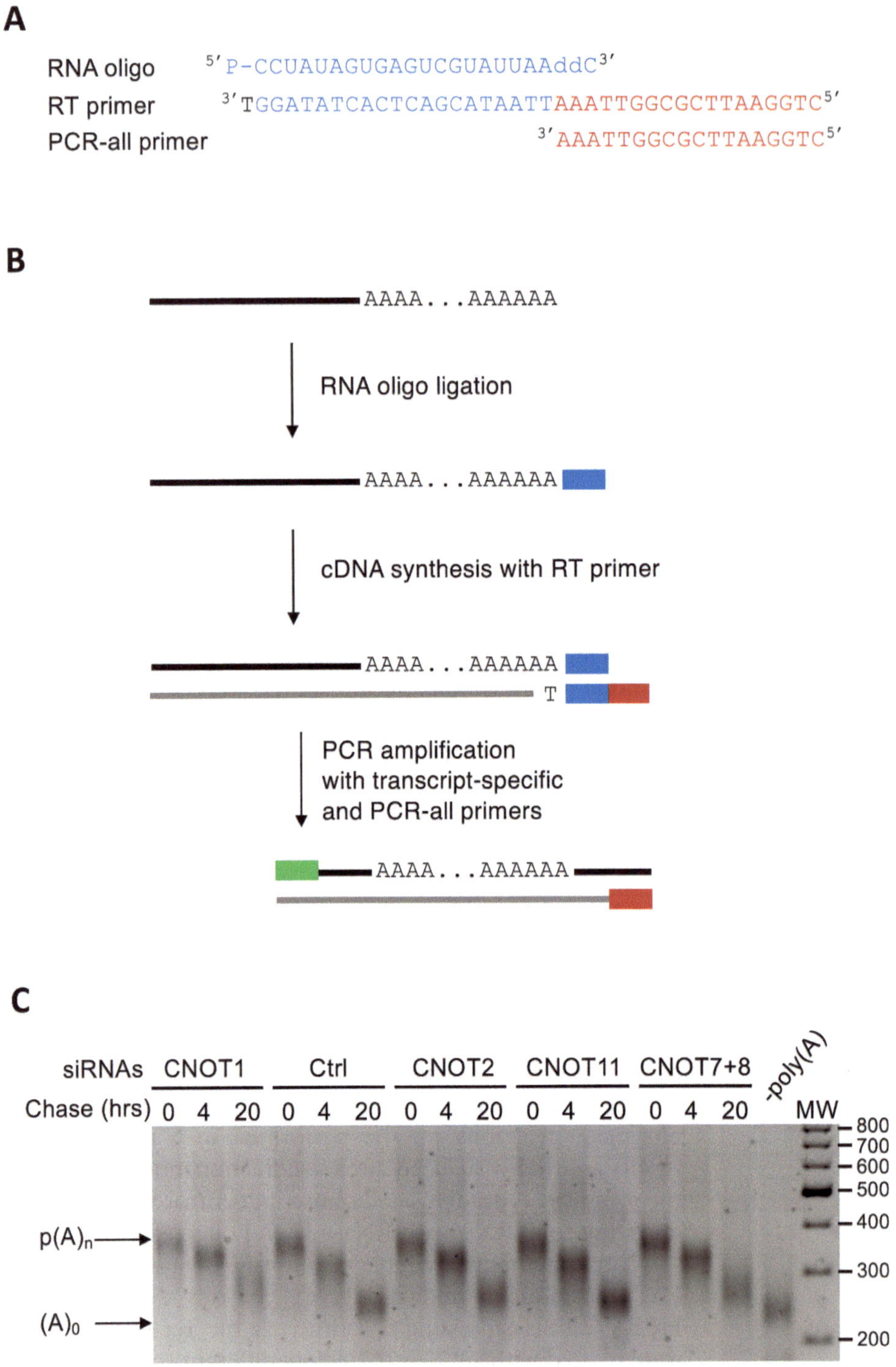

Fig. 1 (**a**) Sequences of oligonucleotides used in the technique. The oligonucleotides are colored and aligned to visualize their complementarity. Note that the RT and PCR-all primer sequences are shown 3′ to 5′ to indicate complementarity with the other primers. (**b**) Schematic representation of the RACE-PAT assay. The different steps are depicted with oligonucleotides colored as in A. The transcript-specific primer is shown in green. (**c**) Poly(A) tail shortening of a β-globin mRNA reporter. HEK293 cells expressing the Tet-Off regulator stably were first transfected with the indicated siRNAs and re-transfected 24 h later with the siRNAs and a

detected after electrophoresis on a 3% agarose gel stained with Ethidium Bromide. This experiment demonstrated impaired deadenylation upon depletion of the CNOT1 scaffold subunit of the CCR4–NOT complex, upon simultaneous depletion of the deadenylases CNOT7 and CNOT8, and to a lesser extent upon depletion of CNOT2, but not upon depletion of the CNOT11 subunit.

2 Materials

2.1 RNA Extraction

1. The protocol was established with 1 μg total RNA extracted from human cells in culture using the NucleoSpin RNA kit (*see* **Note 1**).
2. Ultraviolet (UV) spectrophotometer (e.g., NanoDrop ND1000).
3. (*optional*) 3 M sodium acetate (pH 5.2).
4. (*optional*) Cold absolute ethanol kept at −20 °C.
5. (*optional*) Cold 80% ethanol kept at −20 °C.

2.2 (Optional) RNase H: Oligo $(dT)_{18}$ Treatment

1. RNase H buffer 4×: 80 mM Tris–HCl (pH 7.5), 40 mM $MgCl_2$, 2 mM EDTA, 200 mM NaCl, 4 mM DTT, 120 μg/ml BSA.
2. oligo$(dT)_{18}$: 1 μg/μl.
3. RNase H: 2 U/μl.
4. Stop Buffer: 20 mM EDTA, 0.4 mM LiCl.
5. Heating block for 65 °C and 30 °C incubation.
6. PCI: Phenol:Chloroform:Isoamyl alcohol solution 25:24:1 by volume (caution, this solution is toxic, perform extraction under a fume hood and discard wastes according to local regulations).
7. 3 M sodium acetate (pH 5.2).
8. Cold absolute ethanol kept at −20 °C.
9. Cold 80% ethanol kept at −20 °C.

Fig. 1 (continued) plasmid expressing a β-globin mRNA reporter expressed under a tetracycline-regulated promoter. Immediately after transfection, doxycycline (1 ng/ml) was added to the medium to block transcription of the reporter. Two days later, cells were washed, and a 4-h-transcriptional-pulse was performed before the re-addition of doxycycline (2 μg/ml). Chase times indicate hours (h) after the second doxycycline addition. The poly(A) tail length of the reporter was analyzed by RACE-PAT. An RNA sample treated with oligo$(dT)_{18}$ and RNase H was used as a marker for the migration of the deadenylated (-poly(A)) β-globin mRNA

2.3 3′ RNA Oligonucleotide Ligation

1. RNA oligonucleotide: 50 mM with sequence shown in Fig. 1a. The RNA oligonucleotide is 5′ phosphorylated (*see* **Note 2**) and blocked at the 3′ end by a dideoxycytidine.
2. T4 RNA Ligase 1 (ssRNA ligase): 10,000 units/ml (New England BioLabs, *see* **Note 2**).
3. T4 RNA Ligase Reaction Buffer 10× (supplied by the manufacturer).
4. Adenosine-5′-Triphosphate (ATP): 10 mM (supplied by the manufacturer).
5. PEG 8000 1× (supplied by the manufacturer).
6. Heating block for 37 °C and 70 °C incubation.

2.4 cDNA Synthesis by Reverse Transcription

1. Reverse Transcriptase RT primer: 50 μM with sequence shown in Fig. 1a (note that the RT primer sequence is shown 3′ to 5′).
2. RevertAid H Minus Reverse Transcriptase: 200 U/μl (*see* **Note 3**).
3. RT Reaction Buffer (5×) supplied by manufacturer.
4. dNTP Mix: dATP, dGTP, Dctp, and dTTP; each at 10 mM.
5. RNasin Ribonuclease Inhibitor: 40 units/μl.
6. Heating block for 42 °C and 70 °C incubation.

2.5 PCR Amplification

1. Transcript-specific primer: 100 μM. This primer should hybridize between 200 and 300 base pairs upstream of the polyadenylation site of the transcript of interest (*see* **Note 4**).
2. PCR-all primer: 100 μM with sequence shown in Fig. 1a (note that the sequence is shown 3′ to 5′).
3. *Taq* DNA Polymerase: 5000 units/ml (New England BioLabs, *see* **Note 5**).
4. ThermoPol Buffer 10× (New England BioLabs, *see* **Note 5**).
5. dNTP Mix: dATP, dGTP, Dctp, and dTTP; each at 10 mM.
6. 6× DNA Gel Loading Dye (preferably without xylene cyanol FF): 20 mM Tris–HCl (pH 8.0), 0.25% bromophenol blue, 30% glycerol.
7. Thermocycler instrument.

2.6 Product Fractionation, Detection, and Quantification

We routinely used agarose gel electrophoresis and Ethidium Bromide staining (*see* **Note 6**).

1. Agarose.
2. 10× Tris-Borate-EDTA running buffer (TBE): 0.89 M Tris base, 0.89 M boric acid, 0.02 M EDTA.

3. Ethidium Bromide (EtBr): 10 mg/ml (caution, EtBr is classified as a potent mutagen; precautions should be taken for handling and waste discarded according to local regulations).
4. DNA molecular weight marker.
5. System for preparing horizontal agarose gel.
6. An electrophoresis power supply.
7. A UV transilluminator and camera allowing imaging with numeric digitalization.

3 Methods

3.1 RNA Extraction

1. Extract total RNA using the NucleoSpin RNA kit following recommendations from the manufacturer (*see* **Note 1**).
2. Measure sample concentration with a UV spectrophotometer. If the sample concentration is below 250 ng/μl, concentrate the sample by ethanol precipitation as follows.
3. Add 0.1 volume of 3 M sodium acetate (pH 5.2).
4. Add 2.5 volumes of cold absolute ethanol, mix by inverting the tube, and incubate for 1 h at −80 °C or overnight at −20 °C.
5. Centrifuge 20 min at 16,000 g at 4 °C; carefully remove the supernatant.
6. Wash the pellet to lower salt concentration by adding 500 μl cold 80% ethanol, centrifuge for 10 min at 16,000 g at 4 °C, and carefully remove the supernatant.
7. Air dry the pellet at room temperature and redissolve it in RNase-free distilled, deionized water to obtain a concentration of 250–1000 ng/μl.

3.2 (Optional) RNase H: Oligo $(dT)_{18}$ Treatment

RNase H digestion in the presence of oligo $(dT)_{18}$ removes mRNA poly (A) tails. A fraction of the RNA sample is treated and handled in parallel with undigested RNA samples. The digested sample thus provides a marker for a deadenylated transcript (e.g., Fig. 1c). This requires sufficient quantities of RNA.

1. To an appropriate volume of RNase-free water to complete the final reaction volume to 20 μl, add 5 μg total RNA.
2. Add 5 μl RNase H buffer 4×.
3. Add 1 μg oligo $(dT)_{18}$ (1 μg/μl).
4. Incubate 15 min at 65 °C to denature RNAs, spin briefly and leave 5 min at room temperature to allow hybridization of oligo $(dT)_{18}$ to mRNAs.
5. Add 1 μL RNase H (2 U/μl).
6. Incubate 1 h at 30 °C.

7. Add 200 μl Stop Buffer.
8. To remove the RNase H enzyme, extract by adding 200 μl PCI (perform extraction under a fume hood), mix to form the emulsion, centrifuge 4000 g for 1 min, recover the aqueous supernatant avoiding pipetting the interphase that contains proteins.
9. To remove oligo $(dT)_{18}$ and concentrate the sample, precipitate with ethanol as described in Subheading 3.1.

3.3 3′ RNA Ligation

Set up a 10 μl reaction as follows:

1. Prepare a Mix for the appropriate number of reactions (*see* **Note 7**) containing per reaction:

   1 μl T4 RNA Ligase Reaction Buffer 10×.

   1 μl 10 mM ATP.

   1 μl RNA ligation oligonucleotide (50 μM).

   2 μl PEG 8000 1×.

   1 μl T4 RNA Ligase 1 (10,000 units/ml).
2. Distribute 6 μl of Mix in an adapted reaction tube (i.e., adapted to the thermocycler if the reaction is performed in this device).
3. Add 1 μg total RNA.
4. Add an appropriate volume of RNase-free water to complete the final volume to 10 μl.
5. Incubate 2 h at 25 °C.
6. Stop the reaction by incubating for 10 min at 70 °C.

3.4 cDNA Synthesis by Reverse Transcription (RT)

1. Pipet 5 μl ligation reaction in an adapted reaction tube (keep remaining ligation reaction at −20 °C for additional experiments if needed).
2. Add 5 μl RNase-free water.
3. Add 2 μl RT primer (50 μM).
4. Incubate for 5 min at 65 °C, centrifuge briefly, and put on ice.
5. Prepare a Mix for the appropriate number of reactions (*see* **Note 7**), containing per reaction:

   4 μl RT Buffer 5×.

   1 μl RNasin (40 U/μl).

   2 μl dNTP Mix (10 mM each).

   1 μl RevertAid H Minus Reverse Transcriptase (200 U/μl).
6. Distribute 8 μl of Mix to each reaction tube.
7. Incubate 1 h at 42 °C.
8. Stop reaction by incubating 10 min at 70 °C.

3.5 PCR Amplification

Set up on ice a 50 μl PCR reaction as follows:

1. Prepare a Mix for the appropriate number of reactions (*see* **Note 7**), containing per reaction:

   5 μl ThermoPol Buffer 10×.

   1 μl dNTP Mix (10 mM each).

   0.25 μl Transcript-specific PCR primer (100 μM; 0.5 μM final concentration).

   0.25 μl PCR-all primer (100 μM; 0.5 μM final concentration).

   0.25 μl *Taq* DNA Polymerase (5000 units/ml).

   42.25 μl nuclease-free water.

2. Distribute 49 μl of the Mix in an adapted reaction tube.
3. Add 1 μl of cDNA from the RT reaction (keep the remaining RT reaction at −20 °C).
4. Spin briefly and transfer to a thermocycler block preheated to 95 °C.
5. Thermocycling conditions are initial denaturation of 30 s at 95 °C; 30 to 40 cycles as follows (*see* **Note 8**): denaturation of 15 s at 95 °C, annealing of 15 s at 60 °C, extension of 15 s at 68 °C; final extension for 5 min at 68 °C and hold at 12 °C.
6. For electrophoresis on agarose gel (*see* **Note 6**), add 10 μl 6× DNA Gel Loading Dye.

3.6 Electrophoresis

If necessary, this step can be adapted to more sensitive detection methods (*see* **Note 6**).

1. Pour a 3% agarose gel containing ethidium bromide at 0.5 μg/ml final concentration. This agarose concentration is suitable to fractionate PCR amplicons from 200 to 500 base pairs, corresponding reciprocally to deadenylated and polyadenylated transcripts.
2. Load 15 μl of the PCR reactions.
3. Run in 1× TBE migration buffer containing ethidium bromide (0.5 μg/ml final concentration).
4. Visualize the gel using a UV transilluminator and generate a numeric file (*see* **Note 9**).

4 Notes

1. The protocol described here uses 1 μg total RNA and is suitable for medium to high-abundance transcripts; for transcripts of low abundance, more sensitive detection methods might be considered, as discussed in **Note 6**.

2. A pre-adenylated RNA oligonucleotide can be used. Instead of T4 RNA ligase 1, use T4 RNA ligase 2, truncated K227Q, and truncated KQ (New England BioLabs) following the manufacturer's recommendation.
3. Other Reverse Transcriptase enzymes will also work. In such case, follow the manufacturer's recommendation to perform the cDNA synthesis step.
4. 3′ untranslated regions are often A-U rich and might be more susceptible to polymorphism than the coding sequence. These features must be considered when choosing the specific primer sequence. It is advisable, thus, to verify the primer by performing validation experiments in conditions in which sufficient amounts of mRNA are available. Those assays may include control reactions in which the ligase and/or the reverse transcriptase are omitted to ensure that products do not derive from nonspecific amplifications. Treatment with RNase H and oligo $(dT)_{18}$ (*see* Subheading 3.2) is also useful to confirm amplicon size for the deadenylated transcript.
5. These DNA Polymerase and Reaction Buffer are used because they yield high product levels at limited cost.
6. Alternative size fractionation systems (acrylamide gel electrophoresis, BioAnalyzer (Agilent)), as well as alternative staining reagents (e.g., SYBR Green, a fluorescently labeled or radioactive PCR-all primer) can be considered to increase sensitivity in case of medium-to-low abundant transcripts.
7. The method is suitable for the analysis of many samples at a same time. To reduce pipetting steps and accompanying errors, it is advisable to prepare a Mix. To avoid problems of missing Mix for the final sample due to pipetting errors, we recommend planning an additional ½ reaction when preparing the Mix.
8. To minimize bias due to less efficient polymerization of longer poly (A) sequences at each PCR cycle, perform the minimum number of PCR cycles that are needed to visualize the amplicons.
9. The digitalized image can be analyzed with software like ImageJ.

Acknowledgments

We acknowledge our team members for support. The work in the Séraphin lab was supported by funding from the Ligue Contre le Cancer (Equipe Labellisée 2020) [to B.S.], the Agence Nationale pour la Recherche grant ANR-21-CE12-0003 [to F.M.]. This work of the Interdisciplinary Thematic Institute IMCBio+, as part of the ITI 2021-2028 program of the University of Strasbourg,

CNRS and Inserm, was supported by IdEx Unistra (ANR-10-IDEX-0002), by SFRI-STRAT'US (ANR 20-SFRI-0012), and EUR IMCBio (ANR-17-EURE-0023) under the framework of the French Investments for the Future Program [to B.S.].

References

1. Tudek A, Schmid M, Jensen TH (2019) Escaping nuclear decay: the significance of mRNA export for gene expression. Curr Genet 65: 473–476. https://doi.org/10.1007/s00294-018-0913-x
2. Passmore LA, Coller J (2022) Roles of mRNA poly (A) tails in regulation of eukaryotic gene expression. Nat Rev Mol Cell Biol 23:93–106. https://doi.org/10.1038/s41580-021-00417-y
3. Weill L, Belloc E, Bava FA, Mendez R (2012) Translational control by changes in poly (A) tail length: recycling mRNAs. Nat Struct Mol Biol 19:577–585. https://doi.org/10.1038/nsmb.2311
4. Muhlrad D, Decker CJ, Parker R (1995) Turnover mechanisms of the stable yeast PGK1 mRNA. Mol Cell Biol 15:2145–2156. https://doi.org/10.1128/MCB.15.4.2145
5. Subtelny AO, Eichhorn SW, Chen GR, Sive H, Bartel DP (2014) Poly (A)-tail profiling reveals an embryonic switch in translational control. Nature 508:66–71. https://doi.org/10.1038/nature13007
6. Chang H, Lim J, Ha M, Kim VN (2014) TAIL-seq: genome-wide determination of poly (A) tail length and 3′ end modifications. Mol Cell 53:1044–1052. https://doi.org/10.1016/j.molcel.2014.02.007
7. Lim J, Kim D, Lee YS, Ha M, Lee M, Yeo J, Chang H, Song J, Ahn K, Kim VN (2018) Mixed tailing by TENT4A and TENT4B shields mRNA from rapid deadenylation. Science 361:701–704. https://doi.org/10.1126/science.aam5794
8. Brouze A, Szczepan Krawczyk P, Dziembowski A, Mroczek S (2022) Measuring the tail: methods for poly (A) tail profiling. Wiley Interdiscip Rev RNA 26:e1737. https://doi.org/10.1002/wrna.1737
9. Sallés FJ, Strickland S (1999) Analysis of poly (A) tail lengths by PCR: the PAT assay. Methods Mol Biol 118:441–448. https://doi.org/10.1385/1-59259-676-2:441
10. Jänicke A, Vancuylenberg J, Boag PR, Traven A, Beilharz TH (2012) ePAT: a simple method to tag adenylated RNA to measure poly (A)-tail length and other 3′ RACE applications. RNA 18:1289–1295. https://doi.org/10.1261/rna.031898.111
11. Mauxion F, Faux C, Séraphin B (2008) The BTG2 protein is a general activator of mRNA deadenylation. EMBO J 27:1039–1048. https://doi.org/10.1038/emboj.2008.43
12. Mauxion F, Prève B, Séraphin B (2013) C2ORF29/CNOT11 and CNOT10 form a new module of the CCR4-NOT complex. RNA Biol 10:267–276. https://doi.org/10.4161/rna.23065
13. Stupfler B, Birck C, Séraphin B, Mauxion F (2016) BTG2 bridges PABPC1 RNA-binding domains and CAF1 deadenylase to control cell proliferation. Nat Commun 7:10811. https://doi.org/10.1038/ncomms10811
14. Amine H, Ripin N, Sharma S, Stoecklin G, Allain FH, Séraphin B, Mauxion F (2021) A conserved motif in human BTG1 and BTG2 proteins mediates interaction with the poly (A) binding protein PABPC1 to stimulate mRNA deadenylation. RNA Biol 18:2450–2465. https://doi.org/10.1080/15476286.2021.1925476
15. Gossen M, Bujard H (1992) Tight control of gene expression in mammalian cells by tetracycline-responsive promoters. Proc Natl Acad Sci U S A 89:5547–5551. https://doi.org/10.1073/pnas.89.12.5547
16. Xu N, Loflin P, Chen CY, Shyu AB (1998) A broader role for AU-rich element-mediated mRNA turnover revealed by a new transcriptional pulse strategy. Nucleic Acids Res 26: 558–565. https://doi.org/10.1093/nar/26.2.558

Chapter 8

Dissecting the Role of the Ccr4–Not Deadenylase Complex in Pluripotency and Differentiation

Xiukun Wang, Qing Chen, Brad Lackford, and Guang Hu

Abstract

In this chapter, we describe methods and assays to examine the role of the Ccr4–Not mRNA deadenylase complex in regulating mouse embryonic stem cell (ESC) pluripotency and differentiation. We present the following procedures: sgRNA design and cloning, the culture of mouse ESCs, transfection and colony picking, genotyping, generation of *Cnot3* conditional deletion ESCs, assessment of *Cnot3* deletion efficiency, and examination of the impact on ESC maintenance and differentiation. The above approach and protocols can also be readily applied to study other Ccr4–Not subunits in ESC fate regulation, thereby facilitating the systematic dissection of Ccr4–Not function in stem cells and development.

Key words Embryonic stem cells, Pluripotency, Differentiation, Ccr4–Not, Poly(A)-tail, Deadenylation

1 Introduction

The poly(A)-tail influences mRNA maturation, export, stability, and translation and therefore plays many important roles in the life of an mRNA (reviewed in [1]). The shortening or removal of mRNA poly(A)-tails is catalyzed by dedicated exonucleases called deadenylases [2]. Among all the deadenylases in eukaryotes, the Ccr4–Not complex plays a dominant role in poly(A)-length determination and gene expression [3–5]. Ccr4–Not was originally identified in yeast and was later found well-conserved in all eukaryotes. In recent studies, Ccr4–Not has gradually emerged as an essential player in gene regulation and has been implicated in various physiological and developmental processes, such as heart development, germ cell development, immune cell differentiation, energy metabolism, osteoporosis, and cancer (reviewed in [6, 7]). Importantly, Ccr4–Not has also been shown to regulate stem cell self-renewal and differentiation by modulating mRNA poly(A)-tail length and stability in pluripotent stem cells [8, 9], germline stem

Eugene Valkov and Aaron C. Goldstrohm (eds.), *Deadenylation: Methods and Protocols*, Methods in Molecular Biology, vol. 2723, https://doi.org/10.1007/978-1-0716-3481-3_8,

cells [10], cardiac progenitors [11], and somatic cell reprogramming [12]. These findings strongly suggest that Ccr4–Not-mediated poly(A)-tail regulation is likely an essential regulatory step in controlling cell fate maintenance and transitions.

At the molecular level, the core Ccr4–Not complex comprises nine to eleven subunits in different organisms [13], which form distinct functional modules. Ccr4–Not can regulate gene expression in almost every step, from transcription initiation, elongation, mRNA export, and degradation to mRNA translation [1, 6]. However, it is not fully understood how it controls specific physiological or cellular phenotypes and how each subunit contributes to the complex function.

In this chapter, we use the CNOT3 subunit as an example and present a genetic approach to examine and characterize its function in mouse embryonic stem cell self-renewal and differentiation. We first describe the procedures for mouse ESC culture. We then explain the experimental steps to generate a Cnot3 conditional deletion mouse ESC line using CRISPR-mediated genome-editing tools. Finally, we demonstrate assays to examine the impact of Cnot3 deletion on ESC maintenance and differentiation using the above cell line. The approach and procedures illustrated in this chapter can be readily applied to study all Ccr4–Not subunits in regulating ESC pluripotency and differentiation. Thus, they will greatly facilitate the systematic dissection of Ccr4–Not function in pluripotent stem cell fate determination and help to uncover Ccr4–Not-mediated posttranscriptional regulation in stem cells and development.

2 Materials

2.1 sgRNA Oligos

1. Cnot3 sgRNA-d1F: /5phos/cacc GGTTGATAGAGGAAGT GGGCC.
2. Cnot3 sgRNA-d1R: /5phos/aaac GGCCCACTTCCTCTAT CAACC.
3. Cnot3 sgRNA-d2F: /5phos/cacc GTGGCTCACGGTGAGT TGGGA.
4. Cnot3 sgRNA-d2R: /5phos/aaac TCCCAACTCACCGTGA GCCAC.
5. Cnot3 sgRNA-LoxP1F: /5phos/caccGGAGAGGAGGTCGG TTGGCA.
6. Cnot3 sgRNA-LoxP1R: /5phos/aaacTGCCAACCGACCTC CTCTCC.
7. Cnot3 sgRNA-LoxP2F: /5phos/caccGCCTAAATGTTTAAG CGTAA.

8. Cnot3 sgRNA-LoxP2R: /5phos/aaacTTACGCTTAAACATTTAGGC.
9. pX330 sequencing: AGGGATGGTTGGTTGGTGGG.

2.2 Single-Stranded Oligodeoxynucleotides (ssODNs)

1. Cnot3-EcoRILoxP1-ssoligo: attcagaagggcctgaggacaaaggactgaggagaggaggtcggttggcaGAATTCataacttcgtataatgtatgctatacgaagttatgggaaatccacactacaaaagttgaccccgaaacagggcatcctggagag.
2. Cnot3-BamHILoxP2-ssoligo: cccaacctcaatttgcaattagttggatcagcctaaatgtttaagcgtaaGGATCCataacttcgtataatgtatgctatacgaagttatgggaactaggacaatcaattccttccgaatgtcagctaatcatgaaactg.

2.3 Genotyping Primers

1. Cnot3 sgRNA-d1-geno-F: GGGCCTTTCCTTGATCCCTG.
2. Cnot3 sgRNA-d2-geno-F: GAAGGGCAGAAATGAGGGCT.
3. Cnot3 sgRNA-d2-geno-R: CACTCACATCCTTGTCGCCT.
4. Cnot3 LoxP1 F-1: ACCCATGGGGGAGAAGGTTA.
5. Cnot3 LoxP1 R-1: CAGGAAGCAGCACTCAAGGA.
6. Cnot3 LoxP2 F-1: TAGGAGGCAGTTTGGCATGG.
7. Cnot3 LoxP2 R-1: GGCTACACAGCGAGATGCT.

2.4 RT-qPCR Primers

1. Cnot3-F: AGAGGCCGATCTACAGATAGTGA.
2. Cnot3-R: GACAGGCTTGGAGCCATTT.
3. Pou5f1-F: TCTTTCCACCAGGCCCCCGGCTC.
4. Pou5f1-R: TGCGGGCGGACATGGGGAGATCC.
5. Sox2-F: TTCCGCAGCTGTCGTTTCGCTG.
6. Sox2-R: TTCAAAGAGATACAAGGGAATTGGGA.
7. Fgf5-F: AAGTAGCGCGACGTTTTCTTC.
8. Fgf5-R: CTGGAAACTGCTATGTTCCGAG.
9. Pax6-F: TAGCCCAGTATAAACGGGAGTG.
10. Pax6-R: CCAGGTTGCGAAGAACTCTG.
11. Sox17-F: GATGCGGGATACGCCAGTG.
12. Sox17-R: CCACCTCGCCTTTCACCTTTA.
13. Foxa2-F: CGAGCTAAAGGGAGCACCT.
14. Foxa2-R: TAATGGTGCTCGGGCTTC.
15. Emoes-F: CGGCAAAGCGGACAATAACA.
16. Emoes-R: GGAGCCAGTGTTAGGAGATTC.
17. Cdx2-F: GCGGCTGGAGCTGGAGAAGGAGTT.
18. Cdx2-R: CGGCGGCTGTGGAGGCTGTTGT.
19. T-F: AAGAACGGCAGGAGGATGT.
20. T-R: TCACGAAGTCCAGCAAGAAA.
21. Gapdh-F: GCCAGCCTCGTCCCGTAGACA.

22. Gapdh-R: CAACAATCTCCACTTTGCCACTGC.
23. Blimp1-F: AGCATGACCTGACATTGACACC.
24. Blimp1-R: CTCAACACTCTCATGTAAGAGGC.
25. Dppa3-F: AGGCTCGAAGGAAATGAGTTTG.
26. Dppa3-R: TCCTAATTCTTCCCGATTTTCG.
27. Prdm14-F: ACAGCCAAGCAATTTGCACTA.
28. Prdm14-R: TTACCTGGCATTTTCATTGCTC.
29. Nanos3-F: CACTACGGCCTAGGAGCTTGG.
30. Nanos3-R: TGATCGCTGACAAGACTGTGG.
31. mGata3-F: TTATCAAGCCCAAGCGAAG.
32. mGata3-R: TGGTGGTGGTCTGACAGTTC.
33. Krt8-F: AGTTCGCCTCCTTCATTGAC.
34. Krt8-R: GCTGCAACAGGCTCCACT.
35. Esrrb-F: CCTCCTCGAACTCGGTCA.
36. Esrrb-R: CGATTCATGAAATGCCTCAA.

2.5 Antibodies

1. CNOT1: ProteinTech 14276-1-AP.
2. CNOT2: ProteinTech 10313-1-AP.
3. CNOT3: Abnova H00004849-M01.
4. CNOT7: Cell Signaling 86665.
5. CNOT8: Thermo PA5-13451.
6. RAN: BD Biosciences 610340.

2.6 Reagents and Equipment

1. 0.1% Gelatin (w/v): Add gelatin (from porcine skin) in water. Dissolve the powder and sterilize the solution by autoclaving.
2. Dulbecco's phosphate-buffered saline (DPBS): 200 mg/L KCl, 200 mg/L KH_2PO_4, 8000 mg/L NaCl, 2160 mg/L Na_2HPO_4-$7H_2O$.
3. Opti-ME I Reduced-Serum Medium (Thermo).
4. 0.05% Trypsin-EDTA (w/v).
5. 4-Hydroxy-Tamoxifen (4-OHT): make 1 mM 4-OHT solution in DMSO. Aliquot and store at −80 °C until use.
6. TE buffer: 10 mM Tris-HCl, pH 8, 1 mM EDTA.
7. DMSO (cell culture grade).
8. Direct PCR lysis buffer: 50 mM KCl, 10 mM Tris–HCl, pH 8, 2.5 mM $MgCl_2$, 0.45% NP-40, 0.45% Tween 20. Add Proteinase K to 100 μg/ml before use.
9. Human plasma fibronectin.
10. 2-Mercaptoethanol.

11. ESGRO Complete PLUS Clonal Grade Medium: Millipore.
12. M10 medium: Dulbecco's Modified Eagle Medium (DMEM), supplemented with 10% ES cell-qualified Fetal Bovine Serum (FBS, v/v), 1 mM Sodium Pyruvate, 0.1 mM MEM nonessential amino acids, 0.1 mM 2-Mercaptoethanol, and 1000 U/ml ESGRO LIF (Millipore).
13. EpiLC medium: 1% (v/v) N2-2 supplement (Thermo), 2% (v/v) B-27 supplement (Thermo), 48% (v/v) NeuroBasal medium (Thermo), 1% (v/v) Knockout Serum Replacement (Thermo), 48% DMEM/F12 basal medium, activin A (20 ng/ml), and bFGF (12 ng/ml).
14. PGC-LC medium: GMEM (Thermo) with 15% Knockout Serum Replacement (Thermo), 0.1 mM MEM nonessential amino acids, 1 mM sodium pyruvate, 0.1 mM 2-mercaptoethanol, 2 mM L-glutamine, 500 ng/ml BMP4, 1000 U/ml ESGRO LIF, 100 ng/ml SCF, 500 ng/ml BMP8b, and 50 ng/ml EGF.
15. Ampicillin (100 μg/ml) Luria-Bertani (LB) Broth Agar plates.
16. Lipofectamine 2000 (Thermo).
17. NuPAGE LDS Sample Buffer (or equivalent).
18. BbsI restriction enzyme (includes enzyme and digestion buffer).
19. Calf Intestinal alkaline phosphatase (CIP).
20. Taq DNA polymerase.
21. 96-well plate aluminum seal.
22. Quick DNA ligation kit (includes ligase and quick ligation buffer).
23. Chemically *E. coli* Competent Cells.
24. RNA miniprep kit.
25. Reverse transcription kit (includes enzyme, primers, and reaction buffer).
26. SYBR Green Real-Time PCR Master Mix.
27. DNA gel extraction kit.
28. Tissue culture plates: Tissue-culture treated 6-well, 12-well, 96-well flat-bottom plates, round-bottom 96-well plates, low-cell-binding U-bottom 96-well plate.

    AggreWell-400 plate: Stemcell Technologies.
29. 12-channel multichannel pipette.
30. Cryo-preservation vials.
31. Cell freezing containers.
32. Dry Block Incubators.
33. DNA agarose gel electrophoresis system.

34. Protein gel electrophoresis system.
35. Western blot protein transfer system.
36. Orbital shaker incubator (for bacteria culture).
37. Table-top micro-centrifuge and swinging-bucket centrifuge.
38. PCR and real-time PCR thermal cyclers.
39. Inverted phase contrast microscope for cell culture.
40. Tissue culture biosafety cabinet.
41. Tissue culture CO_2 incubator.
42. E14-CreERT2 cells: kindly provided by Dr. Shaun Cowley [14].
43. pX330-U6-Chimeric_BB-CBh-hSpCas9 (hereinafter referred to as pX330, [15]).

3 Methods

3.1 sgRNA Design and Cloning

1. Design two sets of sgRNA oligos using the webtool https://portals.broadinstitute.org/gppx/crispick/public: Cnot3 sgRNA-d1F, Cnot3 sgRNA-d1R and Cnot3 sgRNA-d2F, Cnot3 sgRNA-d2R. These sgRNAs will be used for the first round of targeting to introduce the deletion of exon-1 to exon-6 in one allele of *Cnot3* (*see* **Note 1**).
2. Design two additional sets of sgRNA oligos using the same webtool: Cnot3 sgRNA-LoxP1F, Cnot3 sgRNA-LoxP1R and Cnot3 sgRNA-LoxP2F, Cnot3 sgRNA-LoxP2R. These sgRNAs will introduce the two LoxP sites and create the second conditional deletion allele.
3. Synthesize the sgRNAs commercially. Note that the sgRNA oligos need to be 5′-phosphorylated.
4. Resuspend synthesized oligos to 100 μM in TE buffer.
5. Anneal each pair of sgRNAs on a thermal cycler: Mix the two oligos (2 μl each) and 2 μl 10× BbsI digestion buffer in a 20 μl reaction. Incubate in a thermal cycler at 95 °C for 5 min and then ramp down to 25 °C at 5 °C/min. For ligation (see below), take 1 μl of annealed oligos and dilute 1:200 in TE.
6. Digest the pX330 plasmid with BbsI: Mix 1 μg of pX330, 2 μl 10× Buffer, 10 U BbsI, 5 U CIP in a 20 μl reaction, and incubate at 37 °C for 1 h.
7. Separate digested pX330 by Agarose gel electrophoresis using 1% agarose gel.
8. Gel-purify the pX330 vector using a DNA Gel Extraction kit.
9. Ligate the annealed sgRNA oligos with the digested pX330: Mix 50 ng of gel-purified digested pX330, 1 μl annealed oligo

(1:200 dilution), 5 μl 2× Quick ligation buffer, 1 μl quick ligase in a 10 μl reaction. Incubate at room temperature for 10 min.

10. Use 5 μl of the ligation reaction to transform chemically competent *E. coli* cells, and plate transformed cells on ampicillin-LB Agar plates.
11. Pick 3–6 colonies and grow in ampicillin (100 μg/ml) LB media in 37 °C bacteria shaker incubator overnight.
12. Extract plasmids from the culture by DNA Miniprep kit.
13. Sequence the plasmids by Sanger sequencing using the pX330 sequencing primer to identify clone(s) with the correct sgRNA sequence. Store the correct pX330-sgRNA plasmid at −20 °C.

3.2 Mouse ESC Culture

1. Add 0.1% gelatin solution (enough to thoroughly cover the tissue culture plates and incubate at room temperature for ≥30 min. Gelatin-coated plates can be stored with the gelatin solution in the tissue culture CO_2 incubator for later use.
2. Plate mouse ESCs (such as the E14-creERT2 cells) at 2–3 × 10^5 cells/cm^2 in gelatin-coated plates in the ESGRO medium and culture the cells at 37 °C with 5% CO_2 in a tissue culture incubator (*see* **Note 2**).
3. Refresh the medium every day.
4. Passage mouse ESCs when they reach ~80% confluence (*see* **Note 3**).
5. Rinse cells with DPBS.
6. Dissociate cells by adding 0.05% Trypsin-EDTA (e.g., 0.25 ml Trypsin/well in a 6 well-plate). Ensure that the trypsin solution covers the entire surface of the culture vessel.
7. Incubate at room temperature for ~5 min or until the cells detach from the culture plate.
8. Add M10 medium and dissociate cells into single-cell suspension by pipetting up and down for 5–10 times. Use ≥3 times the volume of Trypsin-EDTA to ensure that trypsin is fully inactivated by the FBS in M10.
9. Transfer the cell suspension into a microfuge tube or 15 ml conical tube. Centrifuge at 300 g for 5 min to collect cells.
10. Resuspend cells as a single-cell suspension in a fresh ESGRO medium.
11. Count the cells and plate at 2–3 × 10^5 cells/cm^2 in new gelatin-coated plates.
12. To freeze ESCs as stock, trypsinize and dissociate the cells as described above in **steps 5–8**.
13. Centrifuge to collect the cells at 300 g for 5 min, and resuspend cells in M10 to ~1 × 10^6/ml.

14. Prepare 2× freezing medium: 20% DMSO, 20% ES-FBS, 60% M10.
15. Add an equal volume of the 2× freezing medium to the cell suspension and mix well.
16. Aliquot 0.5–1 ml mixed cell suspension in cryo-preservation vials and freeze the cells in cell freezing containers in a −80 °C freezer overnight.
17. Transfer the frozen cell stocks to liquid nitrogen freezers for long-term storage.

3.3 Mouse ESC Transfection and Colony Picking

1. For transfection in a 12-well plate, assemble the transfection mixture of 1 μg DNA and 3 μl lipofectamine 2000 in 100 μl Optim-MEM according to the manufacturer's instructions.
2. Plate 2 × 10^5 ESCs (e.g., E14-creERT2) in one well of a gelatin-coated 12-well plate in ESGRO and add the transfection mixture. Culture the cells overnight.
3. Passage the transfected cells the next day. Plate 4000–6000 cells in one 6 cm gelatin-coated plate in ESGRO (*see* **Note 4**) and continue to culture the cells for 7–10 days with daily medium refreshment. The colonies will be ready to be picked when visible to the naked eye.
4. To pick colonies, aliquot 25 μl 0.05% Tripsin-EDTA/well in a 96-well round-bottom plate using a multichannel pipette.
5. Remove the medium from the 6 cm plate and add 5 ml DPBS.
6. Under an inverted phase contrast microscope, gently scrape and detach an individual colony with the tip of a 10 μl pipet tip. Aspirate the detached colony into the pipet tip in 5 μl volume, and dispense it into one well of the 96-well round-bottom plate from **step 4**.
7. After picking enough clones (*see* **Note 5**), keep the 96-well round-bottom plate at room temperature for an additional 5 min to allow cell dissociation by trypsin.
8. Neutralize trypsin by adding 150 μl M10 medium to each well using a multichannel pipette (*see* **Note 6**). Dissociate the colony into single-cell suspension by pipetting up and down for 5–10 times.
9. Transfer the entire cell suspension to a new gelatin-coated flat-bottom 96-well plate. Culture the cells for 3–4 days with daily medium change in ESGRO, until most of the wells become confluent.
10. Rinse cells in the 96-well plate with DPBS.
11. Add 25 μl 0.05% Trypsin-EDTA in each well and incubate at room temperature for 5 min.

12. While waiting, aliquot 100 μl/well fresh ESGRO medium in two gelatin-coated flat-bottom 96-well plates.
13. To the trypsin-digested cells from **step 11**, add 100 μl/well M10 and pipette up and down 5–10 times to dissociate cells into a single-cell suspension.
14. Aliquot 50 μl of cell suspension/well to each of the two gelatin-coated plates with fresh ESGRO medium from **step 12**.
15. Culture the cells for 2 days with the daily medium change in ESGRO. Now there are two duplicated plates: one for genotyping and one for stock.

3.4 Mouse ESC Genotyping in 96-Well Plates

1. For PCR genotyping, remove the medium from one 96-well plate from Subheading 3.2, **step 15**, and add 150 μl/well direct PCR lysis buffer.
2. Seal the plate with an aluminum seal and incubate at 55 °C on a dry block incubator for 30 min.
3. Inactivate Proteinase K in the lysis buffer by incubating the plate at 95 °C for 5 min.
4. Centrifuge the plate at 1000 × g for 5 min to collect content to the bottom.
5. Set up 20 μl PCR reactions in a 96-well PCR plate: in each well, add 2 μl cell lysate from **step 6**, PCR primers (0.5 μM each), 0.4 mM dNTP, 1× PCR buffer, 0.5 U Taq polymerase, and H_2O as needed.
6. Run the reactions on a thermal cycler using the following program: 94 °C 4 min, (94 °C 20 s, 55 °C 20 s, 72 °C 30 s) × 35 cycles, 72 °C 5 min.
7. Load the PCR reactions on 1% agarose gel and examine the PCR products by electrophoresis. Identify clones that have both the wild-type and deletion allele.
8. Expand the candidate clones from **step 9** from the stock 96-well plate.
9. Rinse the wells with DPBS, add 25 μl/well 0.05% Trypsin-EDTA, and incubate at room temperature for 5 min.
10. While waiting, aliquot 1 ml/well fresh ESGRO medium in a gelatin-coated 12-well plate.
11. Add 100 μl of M10 to the trypsin-digested cells from **step 11**. Pipette up and down 5–10 times to dissociate cells into a single-cell suspension.
12. Transfer the entire cell suspension to the 12-well plate from **step 12**, and culture the cells for 2 days or until they become ~80% confluent.
13. Expand the cells further or freeze the cells down (*see* Subheading 3.1) as needed.

3.5 Generation of *Cnot3* Conditional Deletion ESCs (See Note 7)

Please see Fig. 1a for the *Cnot3* conditional knockout (Cnot3-cKO) strategy (*see* **Note 8**).

1. Design and synthesize single-stranded oligodeoxynucleotides (ssODNs) using the web tool https://www.idtdna.com/pages/tools/alt-r-crispr-hdr-design-tool: Cnot3-EcoRI-LoxP1-ssoligo and Cnot3-BamHILoxP2-ssoligo. They will be used as homologous recombination donors to introduce the two LoxP sites (*see* **Note 9**).
2. Design and synthesize PCR genotyping primers and RT-qPCR primers.
3. Target the first allele by transfecting E14-CreERT2 ESCs with pX330-sgRNAs-d1 and pX330-sgRNA-d2 (0.5 μg of each plasmid) using lipofectamine 2000 as described in Subheading 3.3.
4. Pick 24–48 colonies as described in Subheading 3.3.
5. Genotype the clones as described in Subheading 3.4. Set up two PCR reactions for each clone: In one reaction, use the Cnot3 sgRNA-d1-geno-F and Cnot3 sgRNA-d2-geno-R primers, which generate a 427 bp PCR product for the wild-type Cnot3 allele. In the second reaction, use the Cnot3 sgRNA-d2-geno-F and Cnot3 sgRNA-d2-geno-R primers, which generate a 259 bp PCR product for the Cnot3 deletion allele.
6. Expand the Cnot3 heterozygous deletion clones from the stock 96-well plate (*see* Subheading 3.4).
7. Target the second allele for LoxP1 insertion by transfecting one Cnot3 heterozygous deletion cell line from **step 6** with pX330-sgRNA-LoxP1 and Cnot3-EcoRILoxP1-ssoligo ssODN (0.5 mg of each) using lipofectamine 2000.
8. Pick 48–96 clones and perform PCR genotyping as described in Subheadings 3.3 and 3.4. Set up one PCR reaction for each clone using the Cnot3 LoxP1 F-1 and Cnot3 LoxP1 R-1 primers. The expected PCR product for the wild-type Cnot3 allele is 362 bp, and the Cnot3 LoxP1 allele is 402 bp.
9. Identify targeted Cnot3-LoxP1 clones and expand the cells as described in Subheading 3.4.
10. PCR-amplify the LoxP1 site from the expanded cells using the Cnot3 LoxP1 F-1 and Cnot3 LoxP1 R-1 primers, and sequence the PCR product by Sanger sequencing to verify the LoxP sequence (*see* **Note 10**).
11. In one verified *Cnot3*-LoxP1 clone, insert LoxP2 by transfecting the cells with pX330-sgRNA-LoxP2 and Cnot3-BamHI-LoxP2-ssoligo ssODN.

12. Pick 48–96 clones and carry out PCR genotyping using the Cnot3 LoxP2 F-1 and Cnot3 LoxP2 R-1 primers, similarly as described above. The expected PCR products are as follows: WT = 235 bp, LoxP2 = 275 bp.
13. Expand the candidate *Cnot3*-LoxP2 clones, and verify the LoxP2 sequence by Sanger sequencing.
14. Expand and make frozen stocks (*see* Subheading 3.4) of the sequencing-verified *Cnot3*-LoxP2 clones as the *Cnot3* conditional knockout (*Cnot3*-cKO) ESC lines.

3.6 Assessment of Cnot3 Deletion

1. To assess gene deletion efficiency, culture E14-CreERT2 and *Cnot3*-cKO ESCs in the ESGRO medium and treat the cells with 4-OHT at 0.1 mM final concentration for 48 h in a 12-well plate (*see* **Note 11**).
2. Remove the medium and lyse the cells using lysis buffer from the RNA miniprep kit.
3. Purify total RNA using the RNA miniprep kit.
4. Use 1 μg of total RNA to set up reverse transcription (RT) reactions following the RT kit manufacturer's instructions. After RT is complete, dilute the reaction to 100 μl with TE.
5. Examine Cnot3 mRNA expression by real-time quantitative PCR (RT-qPCR): Mix 1 μl diluted RT reaction, 5 μl 2 × SYBR Green RT-qPCR master mix, 2 μl RT-qPCR primers (final concentration 0.4 μM each), 2 μl H_2O for each reaction. For each sample, set up 3 reactions as technical replicates using primer sets for the target gene (e.g., Cnot3-F and Cnot3-R). For normalization, set up another three reactions using primers for the housekeeping gene Gapdh (Gapdh-F and Gapdh-R).
6. Run the reactions on an RT-qPCR thermal cycler following the manufacturer's instructions.
7. Quantitate gene expression using software provided by the RT-qPCR thermal cycler manufacturer. For a representative result, please see Fig. 1b.
8. To assess the impact of *Cnot3* deletion on other Ccr4–Not complex components, culture and treat cells with 4-OHT as described in **step 1**.
9. 72 h later, remove medium and lyse cells in 100 μl/well NuPAGE LDS Sample Buffer supplemented with 1% (v/v) 2-Mercaptoethanol.
10. Transfer the lysate to microfuge tubes and briefly sonicate the lysate to reduce viscosity.
11. Separate the proteins by SDS-PAGE and examine the protein level of Ccr4–Not subunits by western blot. For representative results, please see Fig. 1c (*see* **Note 12**).

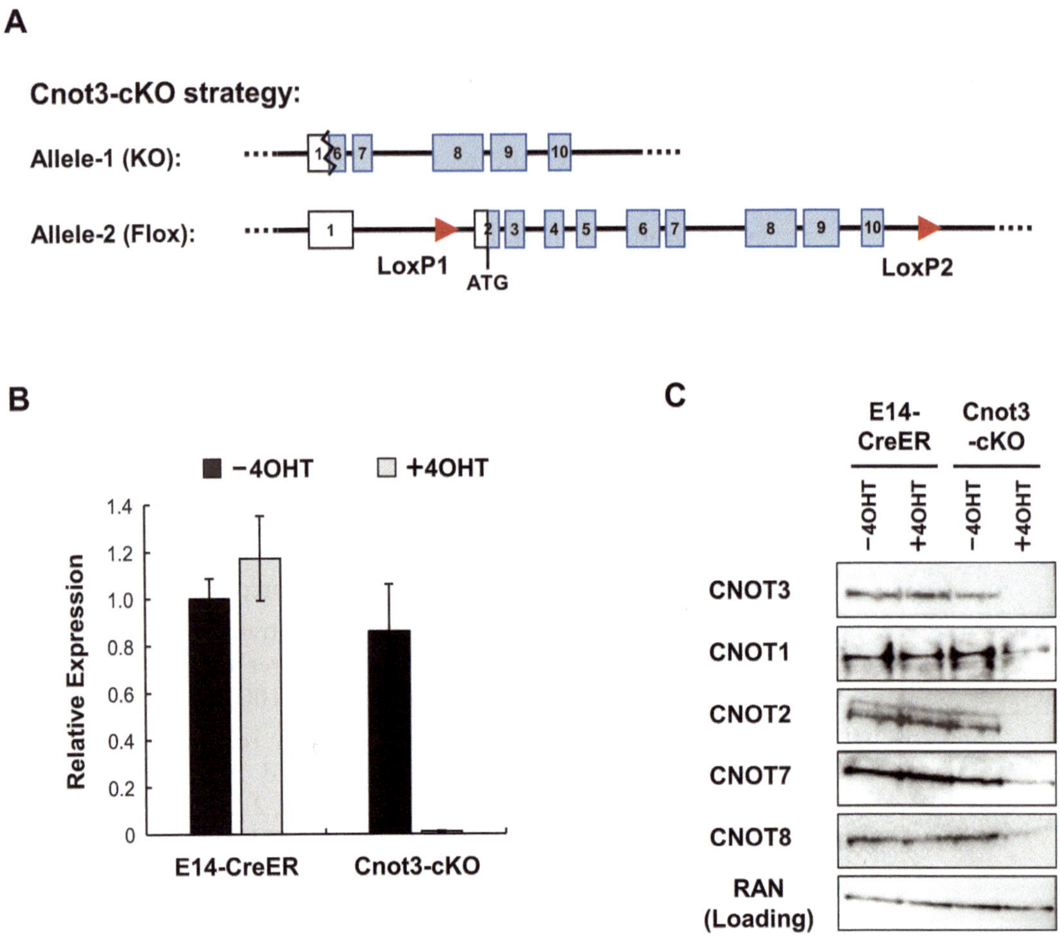

Fig. 1 Generation of conditional *Cnot3* deletion ESCs. (**a**) Illustration of the *Cnot3*-cKO strategy; (**b**) RT-qPCRs to verify Cnot3 deletion upon 4-OHT treatment; (**c**) Western blots to show the protein level of Ccr4–Not subunits after *Cnot3* deletion

3.7 Examination of the Impact of Cnot3 *Deletion on ESCs*

1. To assess the impact of *Cnot3* deletion on ESC maintenance, culture E14-CreERT2 and *Cnot3*-cKO ESCs in the ESGRO medium and treat the cells with 4-OHT at 0.1 mM final concentration for 48 h in a 12-well plate.
2. Passage 4-OHT-treated cells and plate them in the M10 medium. Culture the cells for an additional 3 days in M10 with daily medium refreshment (*see* **Note 13**).
3. Examine cell morphology using an inverted phase contrast microscope. Take bright-field images for documentation. For representative results, please see Fig. 2a (*see* **Note 14**).
4. Remove medium and lyse cells in lysis buffer from the RNA miniprep kit. Purify total RNA and carry out RT-qPCRs as described in Subheading 3.6 to examine lineage marker expression. For representative results, please see Fig. 2b.

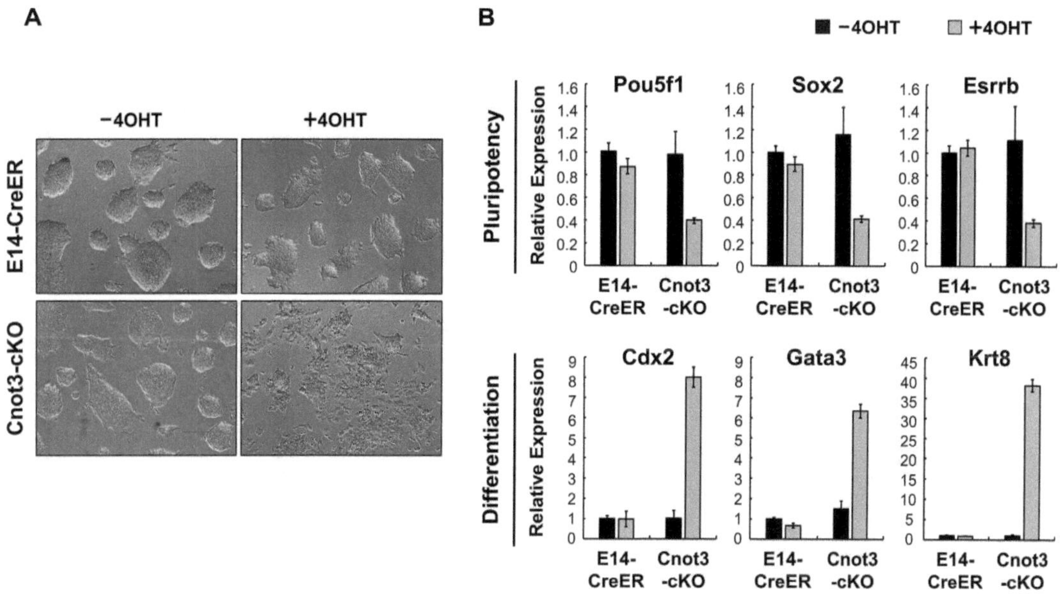

Fig. 2 Analysis of the impact of *Cnot3* deletion on mouse ESC maintenance. (**a**) Cellular morphology of E14-CreER and *Cnot3*-cKO cells treated with or without 4-OHT; (**b**) RT-qPCRs to examine the expression of lineage marker genes. Relative expression was normalized by *Gapdh* and plotted as mean ± SEM

5. To assess the impact of *Cnot3* deletion on ESC differentiation, dissociate 4-OHT-treated cells from **step 1** into single-cell suspension. Seed cells in the AggreWell-400 plates in M10 without LIF following the manufacturer's instructions for embryoid (EB) formation (*see* **Note 15**).
6. Examine EB morphology at different time points by an inverted phase contrast microscope. For representative results, please see Fig. 3a.
7. Collect the EBs after 4 days and examine the expression of lineage markers by RT-qPCR, similarly as described in **step 4**. For representative results, please see Fig. 3b.
8. To assess the impact of gene deletion on ESC differentiation into the germ cell lineage, coat 12-well plates with human plasma fibronectin (16.7 μg/ml) at room temperature for 1 h.
9. Plate 1×10^5 cells/well of the E14-CreERT2 and *Cnot3*-cKO cells in the fibronectin-coated plate in EpiLC medium [16]. Induce *Cnot3* deletion with 0.1 mM 4-OHT and culture the cells for 2 days in the EpiLC medium with daily medium change.
10. Dissociate the cells with trypsin and plate 1×10^3 cells/well in a low-cell-binding U-bottom 96-well plate in the PGC-LC medium for EB formation. Change the medium every 2 days.

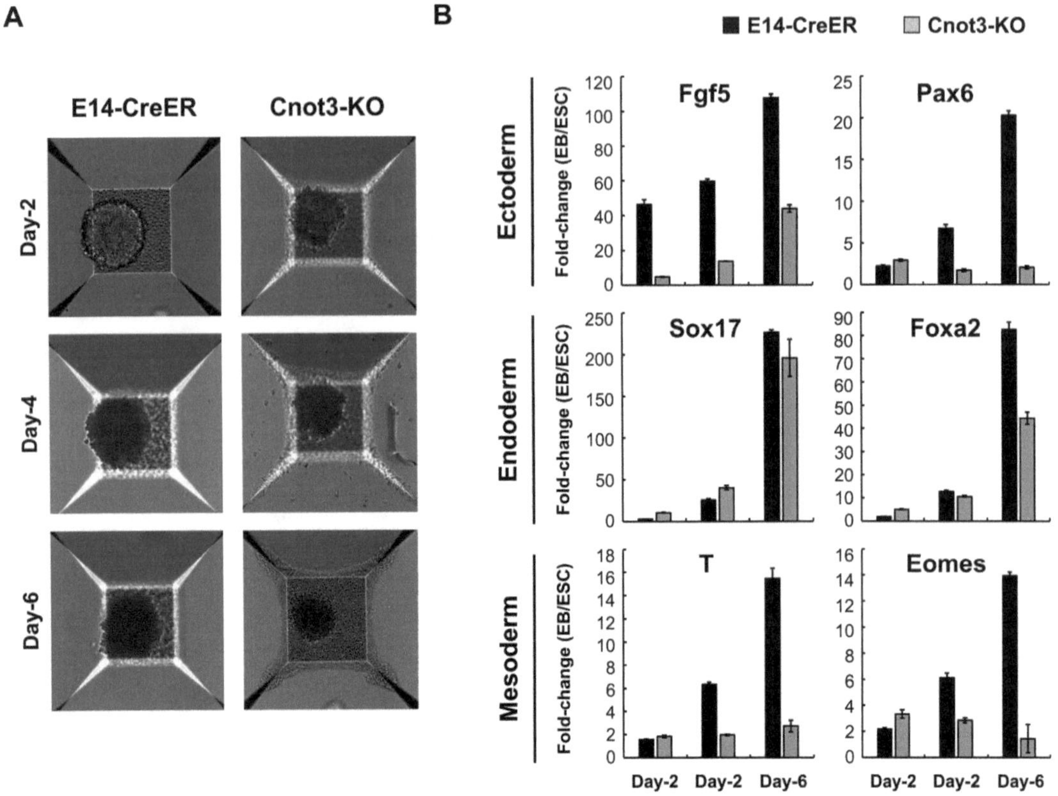

Fig. 3 Analysis of the impact of Cnot3 deletion on ESC differentiation. (**a**) Morphology of EBs formed at the indicated time points by E14-CreER or *Cnot3*-cKO cells after 4-OHT treatment; (**b**) RT-qPCRs to examine the expression of lineage marker genes in the EBs (in comparison to their expression in undifferentiated ESCs). Relative expression was normalized by *Gapdh* and plotted as mean ± SEM

11. To examine the expression of PGC surface markers CD61 and SSEA1, dissociate the EBs on day 4 with 0.05% Trypsin-EDTA. Pool EBs from 4–10 wells as needed to ensure sufficient cells for antibody staining and FACS analysis below.
12. Collect the cells by centrifugation at 300 × g for 5 min and resuspend cells in 100 μl DPBS with 0.1% BSA. Add PE-conjugated-CD61 and Fluor 647-conjugated-SSEA1 antibodies (1:100 dilution) to the cells and incubate at room temperature for 1 h.
13. Wash the cells in DPBS with 0.1% BSA three times.
14. Resuspend cells in 500 μl DPBS with 0.1% BSA and carry out fluorescence-activated cell sorting (FACS). For representative results, please see Fig. 4a.
15. To examine the expression of PGC markers by RT-qPCR, collect the EBs on day 4 and carry out RT-qPCRs as described in **step 4**. For representative results, please see Fig. 4b.

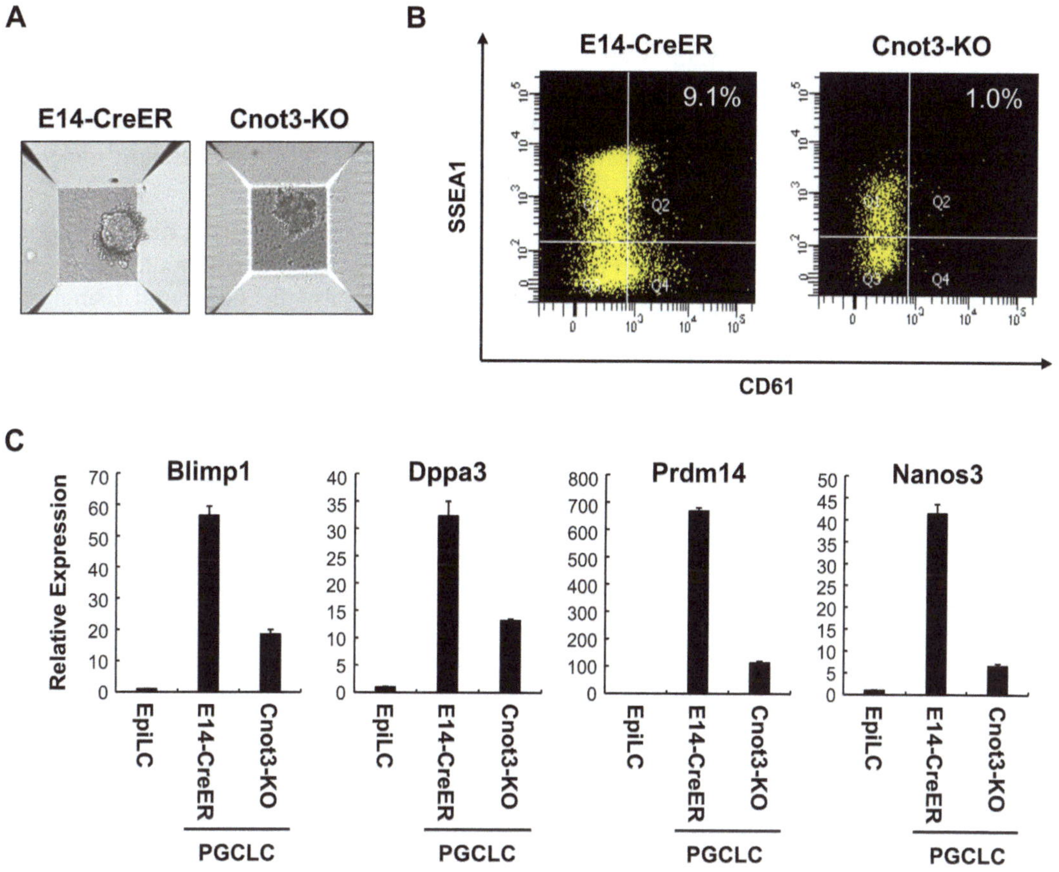

Fig. 4 Analysis of the impact of Cnot3 deletion on PGC-LC specification. (**a**) Morphology of EBs at day 4 formed by E14-CreER or *Cnot3*-cKO cells after 4-OHT treatment during PGC-LC differentiation; (**b**) FACS analysis of the EBs for the expression of PGC surface markers SSEA1 and CD61, with quantification of SSEA1 and CD1 double-positive cells; (**c**) RT-qPCRs to examine the expression of PGC markers in the EBs (in comparison to their expression in EpiLCs). Relative expression was normalized by *Gapdh* and plotted as mean ± SEM

4 Notes

1. For more details on sgRNA design, please refer to this review [17].
2. ESCs are often cultured in one well of a six-well plate for routine maintenance.
3. ESCs grow very fast and need to be split every 2–3 days.
4. Plating density can be adjusted to achieve optimal colony density for picking.
5. Do not leave picked colonies in trypsin for more than 30 min. If more time is needed to pick enough clones, dissociate and transfer the picked colonies to the gelatin-coated plate every 20–30 min.

6. Use multichannel pipettes to handle cells in 96-well plates.
7. Many Ccr4–Not subunits, including CNOT3, are essential in mammalian cells. Therefore, we generated a conditional deletion ESC line for *Cnot3* to study its role in pluripotency and differentiation. For nonessential subunits, straight deletion lines can be created instead.
8. We created the *Cnot3*-cKO line by introducing deletion in one allele and flanking loxP sites (flox) in the other allele. Alternatively, conditional deletion can also be generated using two floxed alleles.
9. For more details on ssODN design, please refer to this review [18].
10. It is important to verify the sequence of the inserted LoxP sites, as CRISPR-mediated targeting can sometimes cause mutations in the LoxP sequence.
11. Depending on the CreER ESC line used to create the cKO cells, the optimal concentration and duration of 4-OHT treatment to induce gene deletion may require titration. The goal is to produce efficient gene deletion in the conditional deletion cells without causing significant inhibition on cell growth in wild-type cells.
12. It has been shown that *Cnot3* deletion negatively affects the protein level of several other Ccr4–Not subunits.
13. For routine maintenance, ESCs are cultured in the ESGRO medium. To examine the impact of *Cnot3* deletion on self-renewal, cells are cultured in the M10 medium instead because M10 is more permissive for ESC differentiation.
14. ESC differentiation often leads to cell death and obvious changes in cell morphology. The extent of cellular differentiation can be quantitated as described in Fig. 3G in [9].
15. As an alternative to the AggreWell plates, EB differentiation can also be carried out using ultra-low attachment plates [19] or the hanging-drop method [20].

Acknowledgments

We wish to thank members of the Hu lab for their comments and discussions on this protocol. This work was supported in part by the Intramural Research Program of the NIH, National Institute of Environmental Health Sciences Z01ES102745 (to G.H.).

References

1. Passmore LA, Coller J (2022) Roles of mRNA poly(A) tails in regulation of eukaryotic gene expression. Nat Rev Mol Cell Biol 23:93–106
2. Yan YB (2014) Deadenylation: enzymes, regulation, and functional implications. Wiley Interdiscip Rev RNA 5:421–443
3. Temme C, Zaessinger S, Meyer S et al (2004) A complex containing the CCR4 and CAF1 proteins is involved in mRNA deadenylation in Drosophila. EMBO J 23:2862–2871
4. Nousch M, Techritz N, Hampel D et al (2013) The Ccr4-Not deadenylase complex constitutes the main poly(A) removal activity in C. elegans. J Cell Sci 126:4274–4285
5. Yi H, Park J, Ha M et al (2018) PABP cooperates with the CCR4-NOT complex to promote mRNA deadenylation and block precocious decay. Mol Cell 70:1081–1088.e1085
6. Collart MA (2016) The Ccr4-Not complex is a key regulator of eukaryotic gene expression. Wiley Interdiscip Rev RNA 7:438–454
7. Chalabi Hagkarim N, Grand RJ (2020) The regulatory properties of the Ccr4-not complex. Cell 9:2379
8. Zheng X, Dumitru R, Lackford BL et al (2012) Cnot1, Cnot2, and Cnot3 maintain mouse and human ESC identity and inhibit extraembryonic differentiation. Stem Cells 30:910–922
9. Zheng X, Yang P, Lackford B et al (2016) CNOT3-dependent mRNA deadenylation safeguards the pluripotent state. Stem Cell Rep 7:897–910
10. Joly W, Chartier A, Rojas-Rios P et al (2013) The CCR4 deadenylase acts with Nanos and Pumilio in the fine-tuning of Mei-P26 expression to promote germline stem cell self-renewal. Stem Cell Rep 1:411–424
11. Zhou B, Liu J, Ren Z et al (2017) Cnot3 enhances human embryonic cardiomyocyte proliferation by promoting cell cycle inhibitor mRNA degradation. Sci Rep 7:1500
12. Zukeran A, Takahashi A, Takaoka S et al (2016) The CCR4-NOT deadenylase activity contributes to generation of induced pluripotent stem cells. Biochem Biophys Res Commun 474:233–239
13. Raisch T, Valkov E (2022) Regulation of the multisubunit CCR4-NOT deadenylase in the initiation of mRNA degradation. Curr Opin Struct Biol 77:102460
14. Dovey OM, Foster CT, Cowley SM (2010) Histone deacetylase 1 (HDAC1), but not HDAC2, controls embryonic stem cell differentiation. Proc Natl Acad Sci U S A 107:8242–8247
15. Cong L, Ran FA, Cox D et al (2013) Multiplex genome engineering using CRISPR/Cas systems. Science 339:819–823
16. Hayashi K, Ohta H, Kurimoto K et al (2011) Reconstitution of the mouse germ cell specification pathway in culture by pluripotent stem cells. Cell 146:519–532
17. Li C, Chu W, Gill RA et al (2023) Computational tools and resources for CRISPR/Cas genome editing. Genomics Proteomics Bioinformatics 21(1):108–126
18. Schubert MS, Thommandru B, Woodley J et al (2021) Optimized design parameters for CRISPR Cas9 and Cas12a homology-directed repair. Sci Rep 11:19482
19. Stover AE, Schwartz PH (2011) The generation of embryoid bodies from feeder-based or feeder-free human pluripotent stem cell cultures. Methods Mol Biol 767:391–398
20. Wang X, Yang P (2008) In vitro differentiation of mouse embryonic stem (mES) cells using the hanging drop method. J Vis Exp 17:825

Chapter 9

Measuring Proximity-Mediated Function of mRNA Regulatory Proteins by Engineered Tethering

Breanne M. Hatfield, Chase A. Weidmann, and Kevin M. Weeks

Abstract

A powerful approach for studying the functional consequences of site-specific RNA–protein interactions is to artificially tether a protein to a messenger (or noncoding) RNA through a selective, high-affinity interaction. We share a strategy for evaluating the contribution of protein positioning within an mRNA on gene expression. We introduced an RNA hairpin recognition site for the MS2 coat protein into the untranslated regions or coding sequence of mRNAs expressing a luminescent reporter protein, NanoLuc. Effector proteins fused to the MS2 coat protein could thus be targeted to distinct regions across the mRNA. We illustrate this approach using ZFP36L2, which recruits the CCR4-NOT complex for poly(A) tail deadenylation. Tethering ZFP36L2 to the 3'-UTR decreased NanoLuc expression, as expected, given the known interaction of this adapter protein with adenine uridine-rich elements (AREs). Intriguingly, ZFP36L2 also decreased NanoLuc expression when bound within the coding sequence, revealing that ZFP36L2—and potentially many other mRNA regulatory proteins—can function when targeted to diverse locations within an mRNA. This multi-target tethering strategy enables exploration of the interplay between mRNA–protein proximity and gene expression.

Key words RNA-protein interactions, RNA tethering, mRNA proximity, RNA-binding proteins, Dual-luciferase assay, MS2 hairpin, MS2 coat protein

1 Introduction

Regulation of messenger RNA (mRNA) function, from transcription to degradation, involves hundreds of regulatory RNA-binding proteins [1, 2]. Many mRNA regulatory proteins are recruited to a specific location within an mRNA. For example, zinc finger proteins are recruited to 3′ untranslated regions (3′-UTRs) through binding to AU-rich elements (AREs) [3]. Cap-binding proteins bind the mRNA 5′-cap to promote translation and prevent decapping and thus 5′-to-3′ degradation [4]. However, the role that target site location plays in the function of proteins that regulate mRNA biology remains poorly understood. In this chapter, we describe a strategy for site-specific tethering of an mRNA

Eugene Valkov and Aaron C. Goldstrohm (eds.), *Deadenylation: Methods and Protocols*, Methods in Molecular Biology, vol. 2723,
https://doi.org/10.1007/978-1-0716-3481-3_9,

regulatory protein to a particular region within an mRNA to investigate how binding at a specific site on the mRNA influences the ability of a protein to effect gene expression.

To site specifically tether a protein to an RNA, we took advantage of the highly specific and high-affinity interaction between the bacteriophage MS2 coat protein and an RNA hairpin binding site [5]. An mRNA-regulating "Effector" protein is fused to the MS2 coat protein (the tethering protein, TP) and targeted to a specific site in the RNA of interest using the RNA target hairpin (TH) at desired sites. In the assay we describe here, the target mRNA expresses NanoLuc, a bioluminescent protein [6]. The NanoLuc signal is correlated with the expression of the target mRNA. It is normalized to the signal from firefly luciferase [7], a second bioluminescent protein expressed in the same cell. If binding by the Effector-TP fusion changes target mRNA expression levels, light output is altered relative to a control expressing the TP alone or a TP fusion with a non-functional control protein (we use superfolder green fluorescent protein, sfGFP) (Fig. 1).

We created a set of plasmids encoding target mRNAs containing a single target hairpin (TH) at one of four sites: the 5′-UTR, the coding region (CDS, two sites), or the 3′-UTR. An mRNA omitting the TH serves as a control (Fig. 2a). In the coding region, the TH sequence encodes amino acids with a neutral hydropathy index (Fig. 2b) and is positioned within beta-hairpin turns that face away from the NanoLuc protein active site (Fig. 2c) to support correct protein folding and retain NanoLuc enzyme activity.

Dual-luciferase assays require optimization and careful execution. First, all plasmids must be carefully validated for integrity and have a supercoiled topology (Fig. 3). Second, plasmids encoding the target mRNA (NanoLuc) and firefly luciferase control must be transfected at levels yielding a linear relationship between plasmid amount and measured luminescent signals (Fig. 4a). In particular, expression at high levels can lead to protein precipitation or loss of activity. The plasmids developed here yield broad linear expression ranges in HEK293 cells (Fig. 4b). Third, luminescence must be measured when the amount of expressed luciferase protein is proportional to light production. Waiting too long to measure luminescence after plasmid transfection can allow NanoLuc to accumulate to cytotoxic levels (Fig. 4a). Transfection conditions for the firefly luciferase control plasmid and for the target mRNA plasmid should be optimized before the introduction of the plasmid expressing the Effector-TP fusion protein.

To evaluate the effect of site-specific binding of an Effector-TP fusion on target mRNA expression, the three plasmids are co-transfected into cells and allowed to express their respective proteins (Fig. 1). Firefly control and target mRNA expression levels are subsequently quantified using sequential luminescence measurements for firefly luciferase and NanoLuc, respectively. Three

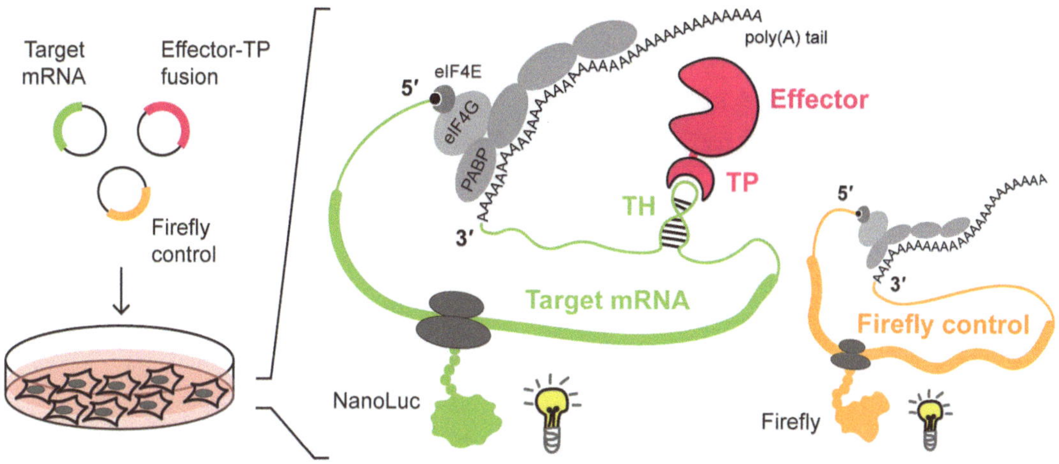

Fig. 1 Monitoring effects of site-specific binding of a protein to an mRNA via a luminescent reporter assay. Plasmids (3 total) expressing target mRNA with target hairpin (TH), Effector-tethering protein (TP) fusion, and firefly control are transfected into cells. The Effector-TP fusion protein binds the TH in the target mRNA. The signal due to NanoLuc, expressed from the target mRNA, is normalized to the signal due to firefly luciferase, expressed from the control plasmid. mRNAs are shown as lines; thicker regions indicate coding regions. Dark and light gray ovals indicate ribosomes and poly(A) binding proteins, respectively

calculations are then performed (Fig. 5): First, signals are corrected for background by subtracting the average signal from all wells containing cells but not transfected with plasmid. The NanoLuc signal, produced from the target mRNA, is divided by the firefly luciferase signal. This per-well normalization accounts for well-to-well differences in cell number, plasmid transfection efficiencies, and target mRNA expression. Second, data from triplicate wells are averaged. Third, the average normalized NanoLuc signal from Effector-TP fusion-expressing cells is divided by the signal from TP (only)-expressing cells to calculate a "fold change" value. This experiment ultimately reveals site-specific regulatory effects of the Effector protein, which may involve either a decrease or increase in protein production from the target mRNA.

2 Materials

2.1 Plasmids

1. Transfection carrier DNA (Promega).
2. pGL4.50 vector (Promega) encodes firefly control.
3. pNL3.2.CMV.NLuc-PEST.Hygro vector encodes target mRNA with no TH (no-TH control and parent vector for TH-containing mRNAs).
4. pNL3.2.CMV.5UTR-TH.NLuc-PEST.Hygro vector encodes target mRNA with TH in the 5'-UTR.

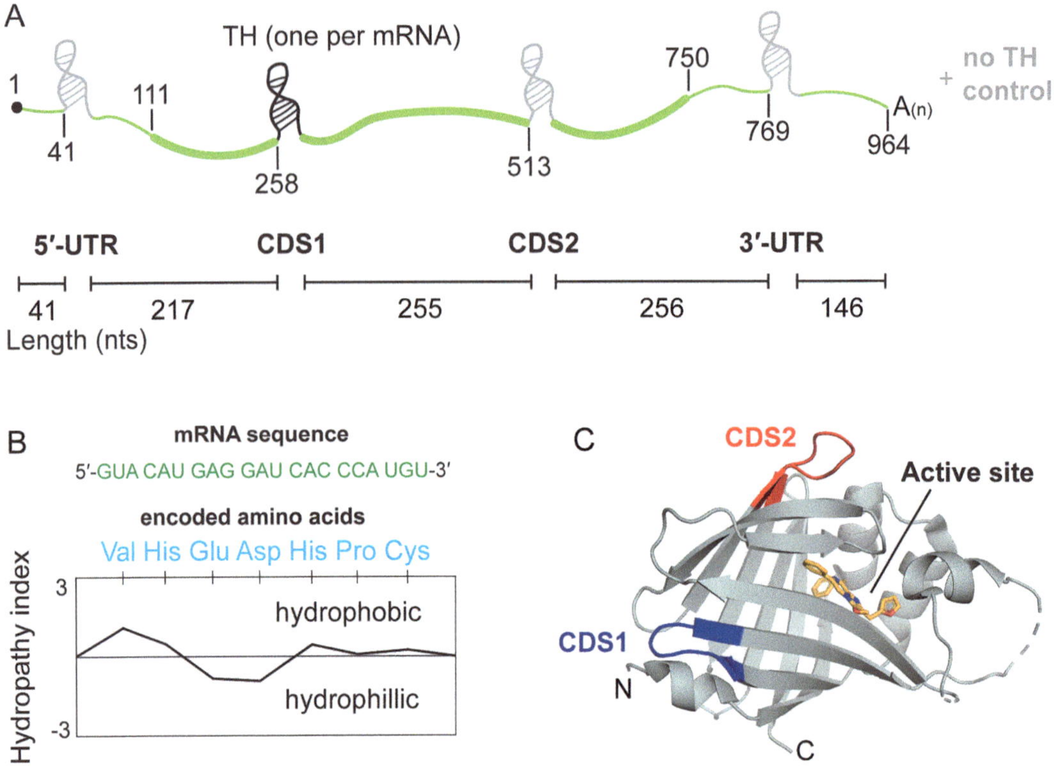

Fig. 2 Target mRNA design. (**a**) Four target mRNAs contain single target hairpins (TH) inserted at evenly spaced sites across the mRNA. An mRNA with no TH is used as a control. (**b**) Targeting hairpin sequence and hydropathy index for the encoded peptide. (**c**) Structure of NanoLuc with substrate analog in the active site (PDB 7SNT). Insertion sites at beta-hairpin turns, encoded by the CDS1 and CDS2 mRNAs, are indicated in blue and red, respectively. A single insertion occurs per mRNA

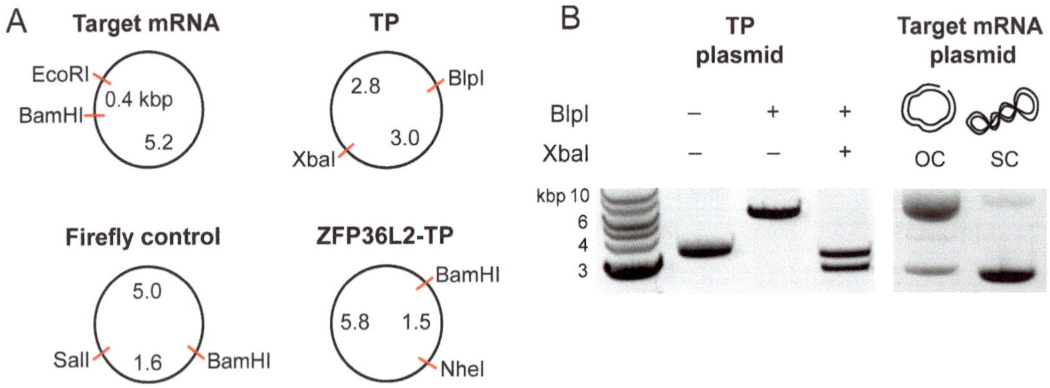

Fig. 3 Validation of plasmid size and topology. (**a**) Diagrams of useful restriction enzyme sites in assay plasmids. Numbers give expected band sizes in kbp. For single plasmid digestions, select one of the two indicated enzymes. (**b**) Plasmid quality control, visualized by agarose gel electrophoresis. Gels showing (*left*) DNA ladder, undigested TP plasmid, single and double digestions of the TP plasmid; and (*right*) no-TH target mRNA plasmid in nicked and supercoiled forms

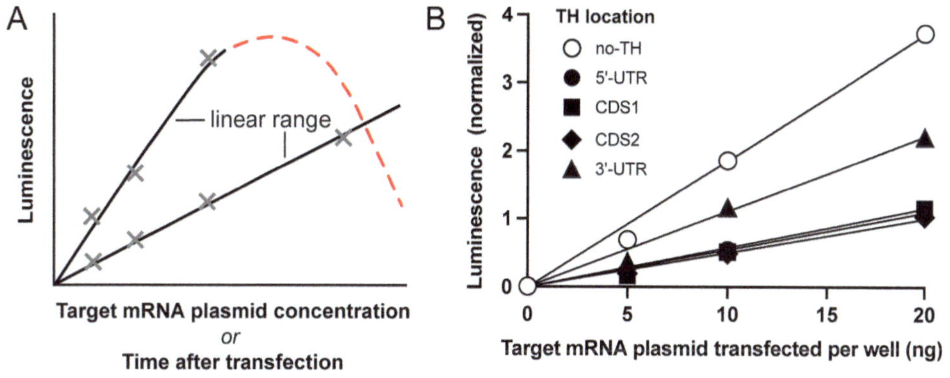

Fig. 4 Target mRNAs show broad linear expression ranges. (**a**) Schematic plot of luminescence as a function of target mRNA plasmid concentration (or time after transfection). Linear expression region shown in black; region of NanoLuc precipitation or cellular toxicity emphasized in red. (**b**) Representative experimental data showing linear expression for the five target mRNAs

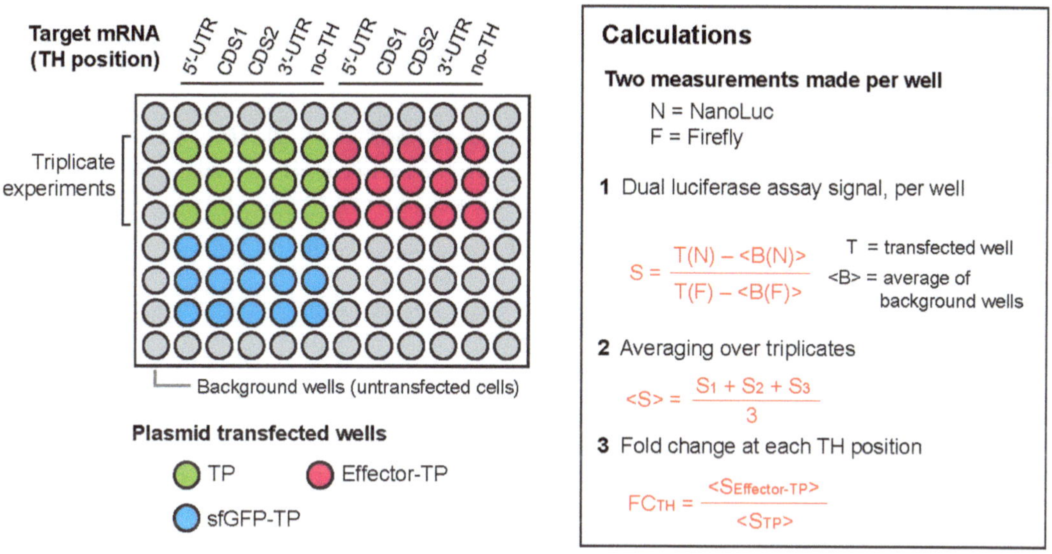

Fig. 5 Representative plate layout and calculations to determine effects of effector protein tethering. (*left*) Representative plate layout. Plate columns correspond to the target mRNA transfected. The well color indicates TP fusion protein transfected. All wells contain firefly plasmid. Samples are analyzed in triplicate. Wells containing untransfected cells are shown in gray. (*right*) Calculations to determine the effect of Effector-TP fusion protein on (NanoLuc expressing) target mRNAs

5. pNL3.2.CMV.CDS1-TH.NLuc-PEST.Hygro vector encodes target mRNA with TH in the CDS1 position.
6. pNL3.2.CMV.CDS2-TH.NLuc-PEST.Hygro vector encodes target mRNA with TH in the CDS2 position.
7. pNL3.2.CMV.3UTR-TH.NLuc-PEST.Hygro vector encodes target mRNA with TH in 3'-UTR.

8. pA.CBh.TP-V5H6.MCh.Puro encodes TP alone (negative control and parent vector for Effector-TP fusion proteins).
9. pA.CBh.TP-sfGFP-V5H6.MCh.Puro encodes sfGFP-TP fusion.
10. pA.CBh.TP-ZFP36L2-V5H6.MCh.Puro encodes ZFP36L2-TP fusion.

2.2 Plasmid Preparation

1. *E. coli* competent cells: any *E. coli* cloning strain with recA1 deficiency or mutation (*see* **Note 1**).
2. LB agar plates with 50 μg/ml ampicillin.
3. Sterilized (autoclaved) 500 ml cell culture flasks.
4. LB broth (Miller) with 50 μg/ml ampicillin.
5. Midiprep plasmid purification system that removes endotoxins.

2.3 Plasmid Quality Check

1. EcoRI-HF, SalI-HF, BamHI-HF, NheI-HF, XbaI, and BlpI enzymes.
2. Supplied restriction enzyme buffer(s).
3. Agarose.
4. Tris–borate–EDTA (TBE) buffer: weigh 10.8 g Tris base, 5.5 g boric acid, 4 ml of 500 mM EDTA (pH 8). Make up to 1 l with deionized water.
5. Ethidium bromide.
6. 1 kb DNA ladder.
7. DNA gel loading dye.
8. Nuclease-free water.

2.4 Mammalian Cell Preparation

1. HEK293 cells (ATCC CRL-1573).
2. Complete Dulbecco's modified Eagle's medium (DMEM), DMEM supplemented with 10% fetal bovine serum, streptomycin (100 μg/ml), and penicillin (100 U/ml).
3. 100 mm Tissue culture-treated dishes.
4. 1× Phosphate-buffered saline (PBS): 137 mM NaCl, 2.7 mM KCl, 8 mM Na_2HPO_4, and 2 mM KH_2PO_4. Adjust pH to 7.4.
5. Trypsin with EDTA: 0.05% trypsin with 0.02% (0.53 mM) EDTA.
6. 10 ml Serological pipettes.

2.5 Plasmid Transfection

1. 0.2 ml 8-strip PCR tubes.
2. 1.6 ml Microcentrifuge tubes, autoclaved.
3. Fugene 6 transfection reagent (Promega).
4. DMEM.

5. Hemocytometer.
6. Microscope coverslips.
7. Trypan blue.
8. 25 ml Reagent reservoirs.
9. 96-Well, white, clear-bottom tissue culture-treated plate with lid.

2.6 Dual-Luciferase Assay

1. 25 ml Reagent reservoirs.
2. Nano-Glo Dual-Luciferase Reporter Assay System (Promega).

2.7 Instruments

1. Tabletop centrifuge.
2. Vacuum manifold.
3. Gel imager.
4. Agarose gel electrophoresis system.
5. Mammalian cell culture CO_2 incubator.
6. Mammalian cell culture biological safety cabinet.
7. Microscope.
8. NanoDrop UV-Vis spectrophotometer.
9. Qubit 2.0 fluorometer.
10. Luminescence microplate reader.

3 Methods

3.1 Plasmid Preparation

1. Transform plasmid DNA into *E. coli* competent cells, according to manufacturer technical protocol.
2. Spread transformed cells on individual LB agar plates containing 50 μg/ml ampicillin and incubate at 37 °C overnight until individual colonies grow.
3. Using a sterile pipette tip, inoculate 100 ml sterilized LB broth (Miller) and 50 μg/ml ampicillin in a sterilized 500 ml cell culture flask with a single bacterial colony.
4. Incubate flask in a shaking incubator at 37 °C and 400 rpm for 12 h.
5. Measure optical density (OD) at 600 nm every 2 h until OD remains constant and *E. coli* reach stationary phase of growth (*see* **Note 2**).
6. Harvest plasmids with Midiprep plasmid purification system according to manufacturer protocol.

3.2 Plasmid Quality Control

1. Quantify plasmid DNA using a spectrophotometer (*see* **Note 3**).
2. Confirm integrity and topology of plasmids by performing single and double restriction enzyme digestions on 1 μg of the plasmid. Useful enzymes for each plasmid are shown in Fig. 3a. Optimal digestion conditions can be obtained from the NEBcloner restriction digest website.
3. Prepare 0.8% (w/v) agarose gel with TBE as the running buffer [*see* ref. 8].
4. Add loading dye to the DNA ladder and all DNA samples.
5. Load the DNA ladder in the leftmost and rightmost lanes on the gel. Load 1 μg of undigested plasmid and plasmids digested with one or two enzymes in adjacent lanes on the gel.
6. Run the gel at 120 V until tracking dye is 75% of the way through the gel (about 1–1.5 h).
7. Image gel under 300 nm UV light (Fig. 3b, left).
8. Analyze undigested plasmid and plasmid digested with one or two enzymes and confirm that all plasmids are predominantly in supercoiled topology (*see* **Note 2**). Nicked and linearized plasmids migrate more slowly than (the required) supercoiled plasmid (Fig. 3b, right).

3.3 Preparation of Mammalian Cells

1. Perform all mammalian cell work in a mammalian cell culture biological safety cabinet and incubate mammalian cells in dedicated mammalian CO_2 incubator at 37 °C, supplemented with 5% CO_2.
2. Pre-warm complete DMEM to 37 °C.
3. Obtain HEK293 cells (stored at −80 °C in aliquots) and thaw in a sterile 37 °C water bath for 2 min (*see* **Note 4**).
4. Spin cells down at 250 × g for 5 min, remove supernatant, and resuspend the cell pellet in 1 ml of complete DMEM.
5. Place resuspended HEK293 cells in a 100 mm tissue culture dish containing 9 ml complete DMEM and gently swirl to mix.
6. Place the cell culture dish in a CO_2 incubator and grow HEK293 cells to 80% confluency.
7. Passage cells twice (passaging cells involves performing **steps 8–12**). *See* **Note 5**.
8. Aspirate medium from cell culture dish and wash cells twice in 1× PBS to remove cell debris.
9. Add 2 ml trypsin with EDTA to the cell culture dish and incubate for 5 min at 37 °C to detach cells.

10. Quench trypsin with EDTA with 8 ml complete DMEM and suspend remaining adherent cells into solution by washing cells off the surface with a serological pipette.
11. Add 2 ml of detached HEK293 cells to a clean cell culture dish and dilute with 8 ml complete DMEM to complete cell passage.
12. Place the cell culture dish in a CO_2 incubator and grow HEK293 cells to 80% confluency.

3.4 Plating Mammalian Cells for Dual-Luciferase Assay

1. Pre-warm complete DMEM to 37 °C.
2. Wash cells twice in 1× PBS, detach cells with 2 ml of trypsin with EDTA, and dilute in 8 ml of complete DMEM (*see* **Note 6**).
3. To quantify cell number, add 100 μl of detached HEK293 cells and 100 μl of trypan blue to a sterile 1.6 ml microcentrifuge tube and pipette to mix (*see* **Note 7**).
4. Add 20 μl of mix to a clean hemocytometer with a cover slip.
5. Count cells using a microscope with a 10× objective and dilute cells to a density of 100,000 cells per ml with complete DMEM.
6. Place diluted HEK293 cells (10 ml per 96-well plate) in a 25 ml reagent reservoir and add 100 μl of cells to each well in a 96-well plate using a P200 multi-channel pipette. Two 96-well plates are required for identifying target mRNA (NanoLuc) linear expression ranges.
7. Place 96-well plate(s) in a CO_2 incubator for 24 h.
8. Passage the remaining HEK293 cells and place them in a CO_2 incubator for future experiments. Cells can be passaged up to ten times before discarding.

3.5 Identifying Linear Expression Ranges for Firefly Control and Target mRNA

1. First, complete this section for the firefly luciferase control plasmid (**step 1**). Second, repeat this section for each of the five target mRNA plasmids, using the firefly luciferase plasmid concentration optimized in the first step (**step 2**).
2. Allow serum-free DMEM and Fugene 6 to equilibrate to room temperature.
3. Individually dilute firefly control plasmid (or target mRNA plasmids) to 70 ng/μl and carrier DNA to 200 ng/μl with nuclease-free water to total volumes of 400 μl.
4. Quantify plasmid concentrations using the Qubit instrument following the manufacturer protocol (*see* **Note 8**).
5. Aliquot 40 μl of the firefly control plasmid (or target mRNA plasmids) into a fresh 0.2 ml strip PCR tube.

6. Perform a serial dilution of the firefly control plasmid (or target mRNA plasmids) in twofold steps with nuclease-free water to create 1:2 and 1:4 dilutions from stock, respectively.
7. (In **step 1**) Use serial dilutions to add 0 ng, 100 ng (1:4 dilution), 200 ng (1:2 dilution), and 400 ng (stock) of firefly control plasmid to 0.2 ml strip PCR tubes as samples 1–4, respectively. Add 1000 ng, 900 ng, 800 ng, and 600 ng of carrier DNA to samples 1–4, respectively. Each sample will contain 1000 ng of total plasmid DNA. (In **step 2**) For each target mRNA plasmid, add 0 ng, 100 ng (1:4 dilution), 200 ng (1:2 dilution), and 400 ng (stock) to 0.2 ml strip PCR tubes as samples 1–4, respectively. Then, add optimized firefly control plasmid concentration from **step 1** to samples 1–4. Finally, add carrier DNA to samples 1–4 so each sample contains 1000 ng of plasmid DNA.
8. Bring all samples to 20 μl with nuclease-free water and mix by pipetting.
9. Add 4 μl of each sample to a clean 0.2 ml strip PCR tube, sterilize the outsides of tubes with ethanol, and bring it into the mammalian cell culture hood.
10. Prepare 6× Fugene 6:DNA (0.6 μL of Fugene 6 per 100 ng of DNA) master mix. Each 4 μl sample requires 1.2 μl of Fugene 6 and 14.8 μl of serum-free DMEM. For identifying the firefly control plasmid linear expression range (**step 1**), add 7.2 μl of Fugene 6 to 88.8 μl of serum-free DMEM in a sterile 1.6 ml microcentrifuge tube (*see* **Note 9**). To identify each target mRNA plasmid linear expression range (**step 2**), add 30 μl of Fugene 6 to 370 μl of serum-free DMEM in a sterile 1.6 ml microcentrifuge tube.
11. Incubate 6× Fugene 6 master mix for 5 min at room temperature and mix gently by pipetting until a homogeneous mixture is obtained.
12. Add 16 μl of Fugene 6 master mix to samples and mix by pipetting.
13. Incubate samples for 15 min to form transfection particles.
14. Remove the 96-well plate with HEK293 cells from the CO_2 incubator and add 5 μl of each sample to the wells, in triplicate (*see* **Note 10**).
15. Mix by swirling the plate gently and place the plate in the CO_2 incubator for 40 h (*see* **Note 11**).

3.6 Dual-Luciferase Assay Readout

1. Allow ONE-Glo EX Luciferase substrate (firefly substrate) and ONE-Glo EX Luciferase buffer from the Nano-Glo Dual-Luciferase Reporter Assay System to reach room temperature. For the assays involving the target mRNA plasmids also allow

NanoDLR Stop and Glo buffer to reach room temperature. Keep NanoDLR substrate (NanoLuc substrate) in the freezer until needed.

2. Add all 10 ml of ONE-Glo EX Luciferase buffer to ONE-Glo EX Luciferase substrate bottle and vortex to mix.
3. Remove 96-well plate(s) with transfected HEK293 cells from the CO_2 incubator.
4. Remove 50 μl of media from each well with a P200 multi-channel pipette, not touching the bottom of the wells, and discard.
5. Add ONE-Glo EX Luciferase substrate mixture to a 25 ml reagent reservoir.
6. Add 50 μl of ONE-Glo EX Luciferase substrate mixture to each well and mix by gentle pipetting. Store remaining ONE-Glo EX Luciferase substrate mixture at −20 °C for future experiments.
7. Protect the 96-well plate from light and incubate it at room temperature for 30 min.
8. Measure luminescence (Fig. 5, Firefly measurement) on a microplate reader using a top reading of 9.5 mm for 0.5 s with a gain of 2500.
9. For firefly control linear expression range experiments, stop here and proceed to Subheading 3.7. For target mRNA (Nano-Luc) linear expression range experiments and Effector-TP fusion tethering experiments, proceed to the next steps.
10. Make sufficient NanoLuc substrate master mix for all samples by adding NanoDLR substrate to NanoDLR Stop and Glo buffer to a 15 ml conical tube in a 1:100 ratio. Vortex to mix and add to a reagent reservoir. The NanoLuc substrate mixture should be made fresh for each experiment.
11. Add 50 μl of NanoDLR mixture to each well with a P200 multi-channel pipette and mix by gently pipetting.
12. Protect the 96-well plate from light and incubate it at room temperature for 30 min.
13. Measure luminescence (Fig. 5, NanoLuc measurement) on a microplate reader using a top reading of 9.5 mm for 0.5 s with a gain of 2500.
14. For target mRNA linear expression range experiments, proceed to Subheading 3.8.
15. For Effector-TP tethering experiments, proceed to Subheading 3.10.

3.7 Firefly Control Linear Expression Range Calculations

1. Export firefly luminescence data from the microplate reader.
2. Correct for firefly background signal by subtracting the average signal from all wells with cells but not transfected with plasmids.
3. Average triplicate wells.
4. Plot average firefly signal as a function of firefly control plasmid transfection amount.
5. Select the firefly control transfection amount within the middle of the linear response range. 10 ng of firefly control plasmid was used for the experiments described below.

3.8 Target mRNA (NanoLuc) Linear Expression Range Calculations

1. Export firefly and NanoLuc luminescence data from the microplate reader.
2. Correct for firefly and NanoLuc background signal by subtracting the average signal from all wells with cells but not transfected with plasmids from firefly and NanoLuc measurements, respectively.
3. Normalize background-subtracted NanoLuc signal to background-subtracted firefly signal for each well (Fig. 5, calculation 1).
4. Average triplicate wells (Fig. 5, calculation 2).
5. Plot normalized luminescence as a function of target mRNA plasmid transfection amounts (Fig. 4b).
6. Select the target mRNA plasmid transfection amount within the middle of the linear response range. For experiments described here, 7 ng of target mRNA plasmid was used per well.

3.9 Dual-Luciferase Assay to Evaluate Effector Protein Function as a Function of mRNA Location

1. Plate one 96-well plate with HEK293 cells. Place 100 μl of HEK293 cells at 100,000 cells/ml or 10,000 cells/well (Subheading 3.4).
2. Perform the following steps for all target mRNA plasmids.
3. Prepare master mix with optimized amounts of firefly control, target mRNA, and carrier DNA plasmids for four 20 μl (20 wells) samples. Optimized carrier DNA plasmid concentration is the sum of firefly control, target mRNA, and Effector-TP plasmid concentrations subtracted from 50 ng. For example, here, carrier DNA plasmid transfection concentration per well is:

50 ng (1 well) − 10 ng (firefly control) − 7 ng (target mRNA) − 5 ng (TP fusions) = 28 ng.

A single 20 μl sample (20 wells) will contain 200 ng of firefly control, 140 ng of a target mRNA, and 560 ng of carrier

DNA. Master mix for four 20 μl reactions contains 800 ng of firefly control, 560 ng of a target mRNA, and 2240 ng of carrier DNA.

4. Add appropriate amounts of master mix to three separate 0.2 ml strip PCR tubes, samples 1–3.
5. Add 100 ng of TP, sfGFP, and ZFP36L2-TP plasmids to samples 1–3, respectively (*see* **Note 12**).
6. Bring the total volume of each sample to 20 μl with nuclease-free water.
7. Add 4 μl of each sample to clean 0.2 ml strip PCR tubes, sterilize the outside of tubes with ethanol, and bring samples into the cell culture hood.
8. Prepare 6× Fugene 6:DNA master mix by adding 24 μl of Fugene 6 to 296 μl of serum-free DMEM.
9. Transfect plasmids into a 96-well plate with HEK293 cells (Subheading 3.5, **steps 11–15**). *See* Fig. 5 for the suggested plate layout.
10. Place cells in a CO_2 incubator for 40 h.
11. Perform dual-luciferase assay and measure firefly and NanoLuc luminescence (Subheading 3.6).

3.10 Dual-Luciferase Assay Calculations to Evaluate Effector Protein Function

1. Export firefly and NanoLuc luminescence readings, background-subtract firefly and NanoLuc signals, and normalize all NanoLuc signals by firefly signals per well (Fig. 5, calculation 1).
2. Average normalized NanoLuc signals for triplicate wells (Fig. 5, calculation 2).
3. Calculate fold change of Effector-TP (ZFP36L2-TP was used here) (and sfGFP) sample to TP control sample at each TH position within target mRNA. The calculation involves dividing the (normalized and averaged) NanoLuc signal for the Effector-TP experiments (and sfGFP-TP) by the NanoLuc signal for TP-alone experiments (Fig. 5, calculation 3).
4. Examine the fold change for the no-TH target mRNA wells. Effector-TP transfected samples should be within 10% of TP-alone control samples when performed with the no-TH target mRNA. If not, repeat Subheadings 3.9 and 3.10 with a decreased amount of Effector-TP (and TP) plasmid. If Effector-TP transfected sample is within 10% of the TP-alone sample with the no-TH target mRNA, consider repeating Subheadings 3.9 and 3.10 with an increase in Effector-TP (and TP alone) expression to maximize the effect of the tethered Effector protein.

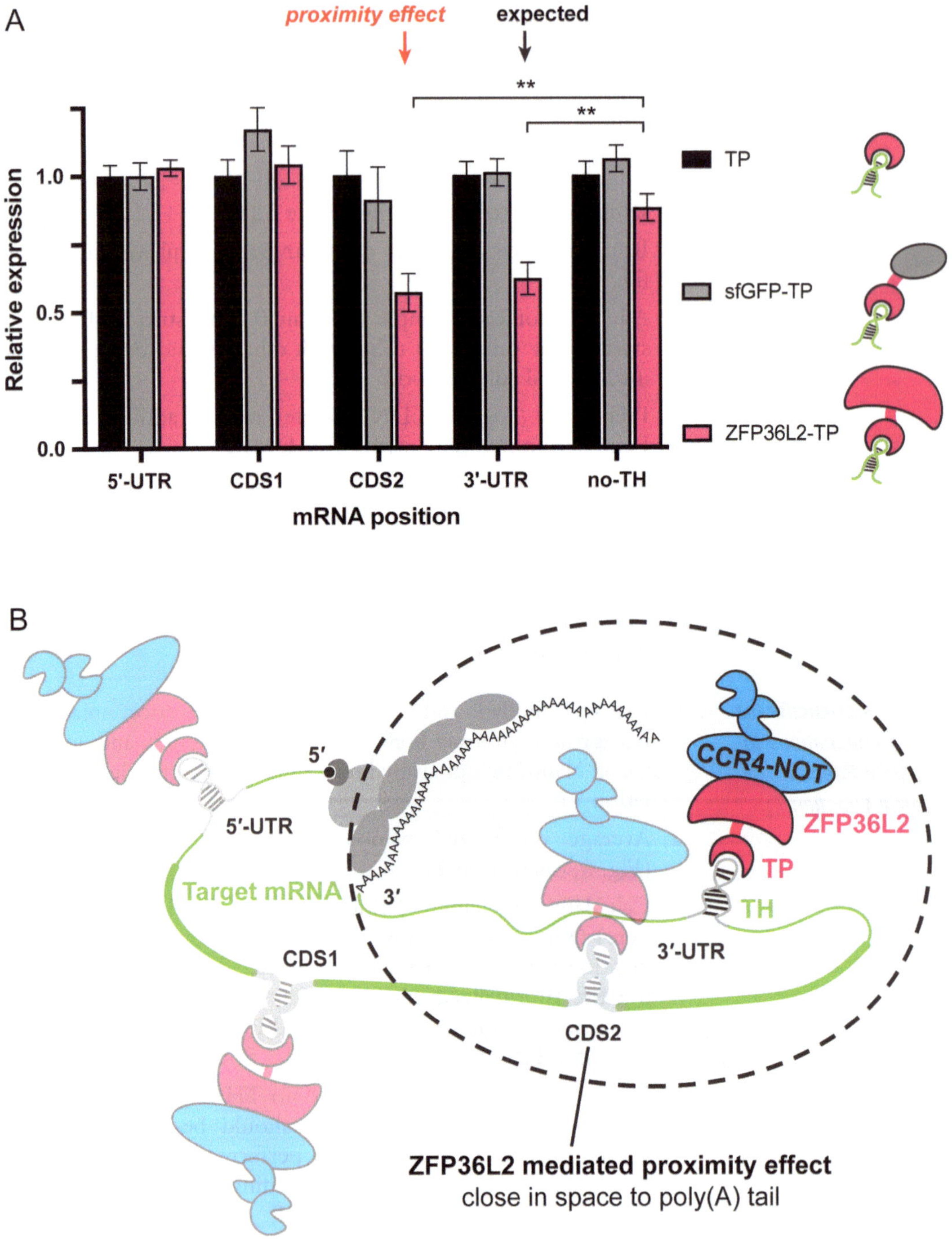

Fig. 6 Proximity effects on ZFP36L2 function. (**a**) Relative expression of target mRNAs as a function of target hairpin (TH) position for fusion proteins linked to the tethering protein (TP) (black), sfGFP-TP (gray), and ZFP36L2-TP (magenta). Means ± SEM are plotted. $^{**}p \leq 0.005$ (two-tailed t-test). (**b**) Scheme illustrating ZFP36L2-mediated proximity effect on target mRNA expression

We used our tethering strategy (Fig. 1) to explore proximity-mediated effects of zinc finger protein 36-like 2 (ZFP36L2), a zinc finger RNA-binding protein that recognizes AU-rich elements (AREs) within the 3'-UTRs of mRNAs. ZFP36L2 acts as an adapter protein that recruits CCR4-NOT, the major deadenylase in mammalian cells [9]. When a ZFP36L2-TP fusion was directed to the TH located in the 3'-UTR of the target mRNA, reporter expression decreased (Fig. 6a). This reduction in expression, based on targeting the 3'-UTR, is consistent with (and expected from) both the known localization of AU-rich elements in 3'-UTRs and with previous studies on zinc finger proteins [10]. Intriguingly, we also observed decreased expression when ZFP36L2 was tethered in the 3′ half of the coding sequence of our target mRNA (CDS2) but not when the protein was tethered in the 5'-UTR or the 5′ half of the coding sequence (CDS1) (Figs. 2a and 6a). ZFP36L2 thus needs only to bind near the poly(A) tail and can act from a site distant from the 3'-UTR via a proximity effect, enabling CCR4-NOT mediated degradation (Fig. 6b).

mRNA regulatory proteins broadly regulate mRNA stability and translation efficiency [1]. The mechanisms by which protein positioning on a given mRNA influence function are poorly understood. Here, we devised an engineered tethering system to address this knowledge gap. We developed a suite of plasmids that allow testing the effect of localizing any Effector protein across an mRNA. The strategy shared here will allow the exploration of proximity-mediated effects for diverse mRNA-binding regulatory proteins on mRNA expression, stability, and translation.

4 Notes

1. Growing plasmids in recombinase deficient strains (recA- or recA mutant) is recommended as using these strains decreases the frequency of multimeric plasmids. Multimerized plasmids have reduced transfection efficiencies but higher expression levels, leading to artifacts in the dual-luciferase assays [11].
2. *E. coli* in the stationary phase is not actively replicating and yields a high proportion of plasmids with supercoiled topology, which have high transfection efficiency and nuclear uptake. Transfection of predominantly supercoiled plasmid DNA is required for optimal dual-luciferase assay results [12].
3. DNA concentration can be calculated from the absorbance at 260 nm. An absorbance ratio (260 nm/280 nm) can be used to assess DNA quality. High absorbance at 280 nm indicates the presence of proteins or phenol. An acceptable 260/280 ratio is in the range of 1.7–2.0, with 1.8 ideal. High absorbance at 230 nm is indicative of the presence of organic compounds. An

acceptable 260/230 range is 2.0–2.2. Using a plasmid preparation with absorbance ratios outside these acceptable ranges can result in poor cellular expression.

4. Dual-luciferase assays can be performed in many different cell lines. Here, we use HEK293 cells as they are transfected with high efficiency and have broad linear expression ranges for both reporters. Note that these differ from HEK293T cells, often used to overexpress recombinant proteins. Plasmids used here contain the SV40 polyadenylation signal, which yields (often excessively) high levels of expression in HEK293T cells.
5. HEK293 cells must be passaged 2 or 3 times after being thawed from −80 °C to yield efficient transfection.
6. HEK293 cells from an over-confluent (>80%) cell culture dish have slower growth rates and lower transfection efficiencies than those grown to (the optimal) 80% confluency. If cells are over-confluent, passage an additional time and then plate cells for the dual-luciferase assay.
7. Trypan blue staining is indicative of a dead cell. Total cell viability above 90% is recommended before plating HEK293 cells for a dual-luciferase assay.
8. Quantification of DNA concentrations using a Qubit instrument often yields lower concentrations than NanoDrop quantification. If DNA concentrations are quantified using a Nanodrop instrument, the total amount of DNA transfected into HEK293 cells may require heuristic correction. We find doubling all individual plasmid amounts and using a Fugene 6: DNA ratio of 3× to work well.
9. Fugene 6 is "sticky" and readily adheres to plastic. Invert the tube several times before preparing the 6× Fugene 6 master mix. Be cautious when adding Fugene 6 to the media. Do not touch the side of the microcentrifuge tube with your pipette. Also, when Fugene 6 and DMEM are combined, the operable volume will be significantly less than expected based on summing the volumes of the two reagents individually. Prepare Fugene 6 and DMEM master mix for 3–5 "extra" samples.
10. Do not use wells on the perimeter of a 96-well plate for samples to be quantified for luminescence (*see* Fig. 5). Corner and edge wells are most prone to evaporation, affecting luciferase assay results [13]. These wells may be used for background luminescence calculations.
11. Steady-state expression occurs 24–48 h after transfection for this assay. Testing a range of time points is important to ensure that NanoLuc does not precipitate or show toxicity during target mRNA expression (*see* Fig. 4a).
12. The optimized transfection concentration for the ZFP36L2-TP fusion in the experiments reported here was 5 ng of plasmid

per well. This low transfection amount for the ZFP36L2-TP plasmid reflects that HEK293 cells are sensitive to exogenous ZFP36L2 expression. A range of transfected DNA concentrations should be tested for other Effector-TP fusions.

Acknowledgments

We are indebted to Christina McCutchin for assistance with plasmid construction and to Brittany Bowman, Matt Smola, and Devon Blake for many helpful discussions. This work was supported by a sponsored research agreement with Ribometrix and the NIH (R35 GM122532 to K.M.W.)

Disclosure K.M.W. is an advisor to and holds equity in Ribometrix.

References

1. Hentze MW, Castello A, Schwarzl T, Preiss T (2018) A brave new world of RNA-binding proteins. Nat Rev Mol Cell Biol 19:327–341. https://doi.org/10.1038/nrm.2017.130
2. Van Nostrand EL, Freese P, Pratt GA et al (2020) A large-scale binding and functional map of human RNA-binding proteins. Nature 583:711–719. https://doi.org/10.1038/s41586-020-2077-3
3. Otsuka H, Fukao A, Funakami Y et al (2019) Emerging evidence of translational control by AU-rich element-binding proteins. Front Genet 10:332. https://doi.org/10.3389/fgene.2019.00332
4. Topisirovic I, Svitkin YV, Sonenberg N, Shatkin AJ (2011) Cap and cap-binding proteins in the control of gene expression. WIREs RNA 2: 277–298. https://doi.org/10.1002/wrna.52
5. Talbot SJ, Goodman S, Bates SR et al (1990) Use of synthetic oligoribonucleotides to probe RNA-protein interactions in the MS2 translational operator complex. Nucleic Acids Res 18: 3521–3528. https://doi.org/10.1093/nar/18.12.3521
6. England CG, Ehlerding EB, Cai W (2016) NanoLuc: a small luciferase is brightening up the field of bioluminescence. Bioconjug Chem 27:1175–1187. https://doi.org/10.1021/acs.bioconjchem.6b00112
7. Gibbons AE, Luker KE, Luker GD (2018) Dual reporter bioluminescence imaging with NanoLuc and firefly luciferase. Methods Mol Biol, Clifton NJ 1790:41–50. https://doi.org/10.1007/978-1-4939-7860-1_4
8. Lee PY, Costumbrado J, Hsu C-Y, Kim YH (2012) Agarose gel electrophoresis for the separation of DNA fragments. J Vis Exp:3923. https://doi.org/10.3791/3923
9. Raisch T, Valkov E (2022) Regulation of the multisubunit CCR4-NOT deadenylase in the initiation of mRNA degradation. Curr Opin Struct Biol 77:102460. https://doi.org/10.1016/j.sbi.2022.102460
10. Redmon IC, Ardizzone M, Hekimoğlu H et al (2022) Sequence and tissue targeting specificity of ZFP36L2 reveals Elavl2 as a novel target with co-regulation potential. Nucleic Acids Res 50:4068–4082. https://doi.org/10.1093/nar/gkac209
11. Maucksch C, Bohla A, Hoffmann F et al (2009) Transgene expression of transfected supercoiled plasmid DNA concatemers in mammalian cells. J Gene Med 11:444–453. https://doi.org/10.1002/jgm.1310
12. Tudini E, Burke LJ, Whiley PJ et al (2019) Caution: plasmid DNA topology affects luciferase assay reproducibility and outcomes. BioTechniques 67:94–96. https://doi.org/10.2144/btn-2019-0060
13. Mansoury M, Hamed M, Karmustaji R et al (2021) The edge effect: a global problem. The trouble with culturing cells in 96-well plates. Biochem Biophys Rep 26:100987. https://doi.org/10.1016/j.bbrep.2021.100987

Chapter 10

Tethered mRNA Amplifier: A Novel Approach to Increase Protein Expression

Bahareh Torkzaban, Austin Sponaugle, and Jeff Coller

Abstract

Posttranscriptional RNA modification has become a revolutionary clinical tool to improve the underlying condition in genetic disorders. The cell achieves translational regulation through sequence and/or structural elements that recruit specific positive- or negative-acting factors to mRNAs. Targeting mRNA expression offers a less invasive therapeutic approach than other well-known gene therapy approaches. We have utilized our understanding of mRNA translational regulation to develop a novel disease-modifying treatment called the "Tethered mRNA Amplifier." Specifically, our approach forces a key positive-acting mRNA regulator, polyadenosine binding protein (PABPC1), to bind and remain resident on the target mRNA. This enhances the target mRNA's expression precisely and restores deficient protein levels to normal. This approach effectively increases the steady-state expression level of several transcripts associated with haploinsufficiency disorders in cell culture.

Key words mRNA, Expression, Translation, RNA therapeutics, Genetic disorders

1 Introduction

Not only do mRNAs transmit genetic information accurately, but they also confer this information at the correct level. The amount of protein that comes from an mRNA is intrinsically timed. And the key feature that times how long an mRNA makes its protein is its 3′ polyadenosine tail. Though changes in gene expression are commonly considered to reflect programmed transcriptional variability, it is a lesser-known fact that extensive regulation of mRNA expression also occurs during their translation [1]. mRNA translation rate is a key feature defining the posttranscriptional regulation [2]. The cell achieves translational regulation through sequence and/or structural elements that recruit specific positive- or negative-acting factors to mRNAs and control their transcription rate, which can dramatically impact protein output per mRNA molecule.

Eugene Valkov and Aaron C. Goldstrohm (eds.), *Deadenylation: Methods and Protocols*, Methods in Molecular Biology, vol. 2723,
https://doi.org/10.1007/978-1-0716-3481-3_10,

The polyadenosine (poly(A)) tail has been greatly attributed to mRNA's translational status and overall stability. Providing a docking site that favors poly(A) binding protein (PABP), the poly(A) tail serves to prevent the destruction of a transcript by exonucleases. Additionally, the poly(A) tail, in conjunction with PABPC1, works to prevent decapping enzymes from reaching the 7-methylguanosine (m7G) cap [2, 3]. The poly(A) tail is necessary for recruiting PABPC1 and its well-known stimulatory function. The translational stimulation by PABPC1 is a necessary component of the design and basic principle of the mRNA tether. PABPC1 is a highly conserved RNA-binding protein in eukaryotes. This protein has four N-terminal RNA recognition motif (RRM) domains, which bind poly(A) RNA with a nanomolar affinity [4, 5]. A proline-rich linker and a C-terminal MLLE domain follow the RRMs. The MLLE domain recognizes a peptide motif called poly (A)-interacting motif 2 (PAM2), found in several PABPC1-interacting proteins that regulate mRNA metabolism (stability and translation). PABPC1 interactions with mRNA's 3′ poly (A) tail significantly increase translation efficacy. A poly(A)-bound PABPC1 is a platform for binding eukaryotic initiation factors eIF4G and eIF4E. This PABPC1-eIF4G-eIF4E complex facilitates binding to the 5′ cap of mRNA [6]. This closed-loop system brings constituent factors at the 3′ UTR and the poly(A) tail in contact with the cap. This closed-loop system further facilitates the recycling of terminating ribosomes with the translation termination factor eRF3. Additional models suggest that PABPC1 stimulates the formation of the 40S ribosomal subunit complex and 60S subunit joining [7, 8].

We have utilized our understanding of mRNA translational regulation to develop a novel disease-modifying technology that targets mRNA stability and translation to improve the level of the protein product. This technology uses the tethering approach to bind PABPC1, a positive mRNA regulator, to the 3′UTR of the target mRNA. This enhances stability and, ultimately, the wild-type target mRNA translation, which can boost protein production, potentially compensating for any deficiencies in levels. This technology holds considerable promise as a therapeutic strategy in treating haploinsufficiency disorders. PABPC1 tethering to target mRNA is achieved using a novel RNA-binding protein (dCas13) and a guide RNA (Fig. 1).

CRISPR-based research tools established a system capable of selective recognition of RNA molecules. This novel RNA-targeting system has enabled the detecting and manipulating of specific RNA using different approaches, such as RNA knock-down, site-specific RNA editing, RNA localization, and destruction of toxic RNAs that lead to human neurodegenerative disorders [9, 10]. mRNA regulon therapy utilizing the CRISPR-based RNA-binding approach can target genetic disorders at the transcript level. The

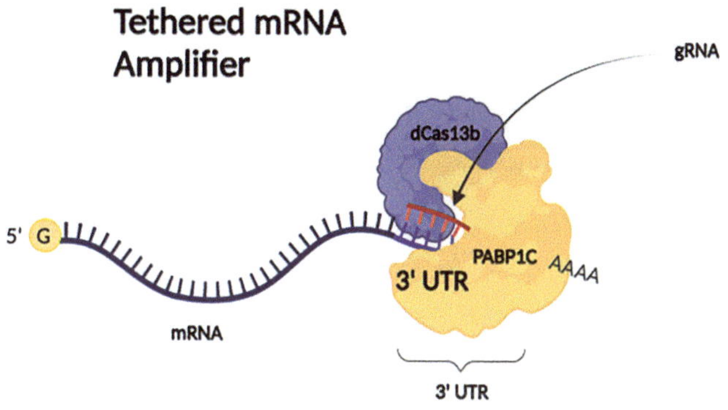

Fig. 1 The mRNA amplifier approach enhances mRNA expression by binding a "fused PABPC1 to dCas13b" to mRNAs of interest via a gRNA targeted to the 3′ UTR. (Created with BioRender.com)

RNA-targeting CRISPR-Cas13 exhibits RNA strand specificity and binds with high affinity, enabling efficient and precise delivery to relevant transcripts [11]. The mRNA-specific targeting capacity of Cas13 makes it the preeminent tool to be fused to the regulon of the interest and target a specific mRNA without affecting global cellular transcription.

In vivo studies aimed to investigate Cas13b interactions with CRISPR RNA (crRNA) revealed valuable insight into the structural interactions of Cas13b with crRNA. The structure of Cas13b assumes a triangular domain distribution around a central L-shaped crRNA. One side of this structure comprises Helical-1, HEPN-1, and HEPN-2 domains, while the other consists of RRI-1, RRI-2 domains, and the linker region [12]. Notably, the HEPN domains of Cas13b function as RNases, indicative of Cas13b's interference with RNA [13]. These HEPN domains are necessary for the cleavage activity of Cas13b. Modifications made in the HEPN domains show a loss of cleavage activity by Cas13b while maintaining RNA-binding capacity [14].

Tethered mRNA amplifier is a novel molecular therapeutic strategy using posttranscriptional regulation of the target mRNA to overcome haploinsufficiency in genetic disorders and a wide range of diseases, including cancer. Our technology is based on what is termed a "Tethered Function Approach." In brief, this technology tethers a specific regulatory factor to the 3′ UTR of a wild-type allele of the target mRNA, thereby changing its stability and/or translational rate (Fig. 2) [15–17]. The tethering protein is an RNA-binding protein that recognizes a short guide RNA tag sequence with high specificity and affinity. The fusion protein's effects on mRNA metabolism are then assayed by conventional means (i.e., Western blot, northern blot, and reverse transcriptase polymerase chain reaction (RT–PCR)).

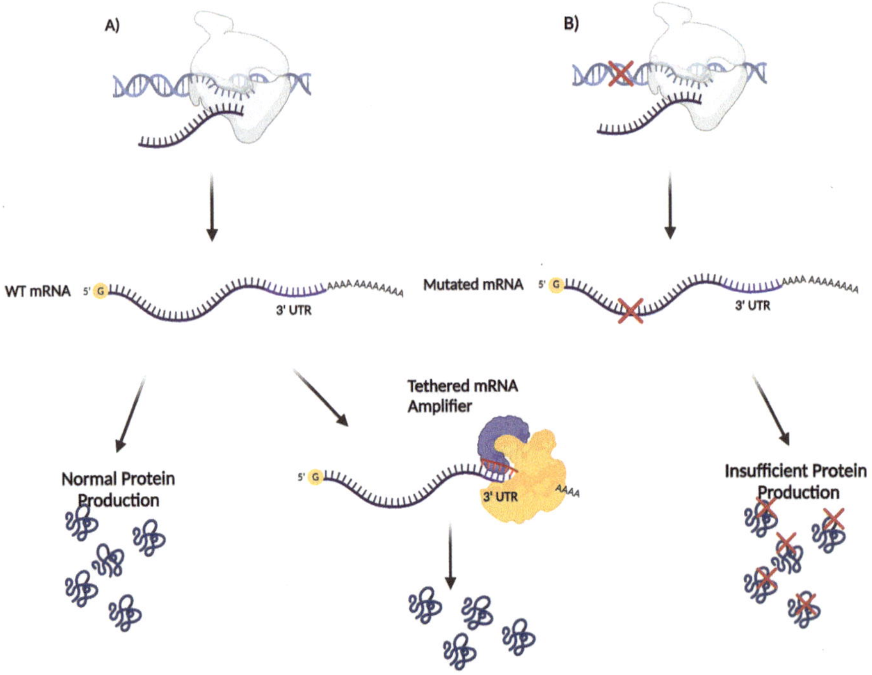

Fig. 2 The mRNA amplifier therapeutic approach for treating haploinsufficiency disorders. (**a**) The normal copy of the gene produces an intact protein on time, so the amount of the protein for the normal copy depends on its mRNA half-life. (**b**) In haploinsufficiency disorders, one of the DNA copies is mutated, decreasing the amount of functional protein by 1/2. The mRNA Amplifier targets the normal copy of messenger RNA, restoring protein expression to normal levels. The amplifier is a unique fusion protein of PABPC1 and dCas13. mRNA specificity is achieved using guide RNAs that bind dCas13 and the mRNA. (Created with BioRender.com)

The tethered mRNA amplifier has proven to be a promising therapeutic for the treatment and study of human diseases associated with an imbalance of protein products. The implementation of PABPC1 allows this technology to be used in a broad range of disease studies as it regulates most human mRNAs [2, 18].

2 Materials

All components and constructs must be prepared and stored in sterile conditions using ultrapure, nuclease-free distilled, deionized water (ddH_2O), and ribonuclease (RNase) free reagents and plasticware.

2.1 Tethered mRNA Amplifier Constructs and Primers

All constructs are available according to their specific pJC number. The backbone plasmids were purchased from Addgene and are denoted with their original references number in the Subheading 2.

1. pJC 1210 pC0054-CMV-dPspCas13b-longlinker-ADAR2DD (*see* **Note 1**).

2. pJC 1211 dCas13-PABPC1 construct (*see* **Note 2**).
3. pJC1208 PC0043-Cas3b-crRNA backbone (Addgene #103854) (*see* **Note 3**).
4. PABPC1 F primer: 5′-AATGGGATCCTGGCAAGCCCATC.
5. PABPC1 R primer: 5′-CTAAAGGGAAGCGGCCGCTTAAA.
6. The sequences of forward oligos designed across 3′UTR of the human MeCP2 gene for the purpose of this study are listed below.

   Mecp2-G1 (oJC5051): 5′-CACCGAAACACTTAGAGTTTC GGAGCTT.

   Mecp2-G2 (oJC5053): 5′-CACCGTGCCATTCAAGAAGAA CTCCGAG.

   Mecp2-G3 (oJC5055): 5′-CACCGATTACACCATAGTAAG GAAGCCAGG.

   Mecp2-G4 (oJC5057): 5′-CACCGAAATTCACTGAAAGAG GCCCCCA.

   Mecp2-G5 (oJC5059): 5′- CACCGACAGAAGATTGTCTG CACCCGCA.

   Mecp2-G6 (oJC5061): 5′- CACCGGCTAAAAATGTATATG CCCAAAG.

   Mecp2-G7 (oJC5063): 5′-CACCGCCAAAAGACAGGTTCA GAAGCCA.
7. Maxi-prep DNA purification kit from commercial source.
8. BamHI restriction enzyme with reaction buffer.
9. NotI restriction enzyme with reaction buffer.
10. BbsI restriction enzyme with reaction buffer.
11. 1× NEB buffer 3.1 (New England Biolabs).

2.2 Mammalian Cell Culture and Transfection

1. HEK293 cell line (ATTC CRL-3216) (*see* **Note 4**).
2. Dulbecco's modified Eagle's medium (DMEM).
3. Opti-MEM reduced serum medium (Gibco).
4. Fetal bovine serum (FBS).
5. 0.05% (w/v) Trypsin, 0.53 mM EDTA solution.
6. Dulbecco's phosphate-buffered saline (D-PBS).
7. 25 cm^2 cell culture flasks.
8. 75 cm^2 cell culture flasks.
9. 6-well cell culture plates.
10. Cell culture biosafety cabinet.
11. 37 °C, 5% CO_2 air-humidified cell culture incubator.
12. GeneXPlus Transfection Reagent (ATCC ACS-4004).

2.3 Electrophoresis and Western Blotting

1. Phosphate-buffered saline (1× PBS)(pH 7.4): 137 mM NaCl, 2.7 mM KCl, 8 mM Na_2HPO_4, and 2 mM KH_2PO_4.
2. Tween-20.
3. Cell scraper.
4. RIPA buffer: 50 mM Tris–HCl (pH 7.5), 150 mM NaCl, 2.5% (v/v) deoxycholic acid, 0.5% (v/v) NP-40 with protease inhibitor cocktail.
5. Mini-protean TGX 4–15% gradient gels (Bio-Rad).
6. Protein molecular weight marker.
7. 5× SDS dye: 10% SDS, 500 mM DTT, 50% (v/v) glycerol, 500 mM Tris–HCl, 0.05% (w/v) bromophenol blue dye.
8. 1× cold transfer buffer: 25 mM Tris–HCl (pH 8.3), 192 mM glycine, 20% (v/v) methanol.
9. Vertical electrophoresis apparatus, electroblotting apparatus, and power supply.
10. Odyssey nitrocellulose membrane (Li-Cor, Lincoln, NE).
11. Enhanced chemiluminescence detection reagent.
12. Powdered dry milk in 1× phosphate-buffered saline (PBS) with 0.5% (v/v) Tween 20 (PBST) or blocking buffer (*see* **Note 5**).
13. Appropriate primary and secondary antibodies. The primary antibodies used in our study are listed below:

    Anti-MeCP2 (D4F3) (Cell Signaling Technology).

    Anti-GAPDH (6C5) (Santa Cruz Biotechnology).

2.4 RNA Isolation and Quantitative PCR

1. Direct-zol RNA Kit (Zymo Research).
2. SuperScript III First-Strand Synthesis System (Invitrogen).
3. Real-time PCR instrument.
4. PowerUp SYBR Green master mix (Applied Biosystems).
5. We used the following oligos in this study for qPCR.

    MeCP2-F (oJC5065): 5′- GTGTATTTGATCAATCCCCAG GGAA.

    MeCP2-R (oJC5066): 5′- CAGTTCCTGGAGCTTTGGGA GATT.

    Actb-F (oJC5067): 5′-CACCATTGGCAATGAGCGGTTC.

    Actb-R (oJC5068): 5′-AGGTCTTTGCGGATGTCCACGT.

3 Methods

Tethered function assays dissect the function of an RNA-binding fusion protein in the posttranscriptional control of mRNA. A limitation of this approach is that a corresponding RNA-binding element has to be engineered into a reporter mRNA. Tethered mRNA amplifier approach described here surpasses that step be creating a novel protein: RNA interaction based on dCas13b that tethers the activity of PABPC1 to a target mRNA.

3.1 dCas13b In Vitro Optimization

To amplify the expression of the protein of interest in vitro, PABC1, an RNA binding protein fused to the tethering protein (Cas13b) that recognizes a short guide RNA (sgRNA) sequence with high specificity and affinity. The fusion protein's regulate mRNA translation are assayed by conventional means [i.e., Western blot, reverse transcriptase polymerase chain reaction (RT-PCR), etc.]. Only positive results are meaningful: lack of effects cannot be interpreted. Some RNA-binding proteins may require other proteins or their cognate RNA-binding sites to function, or be inactive as chimeras, or require appropriate positioning on the mRNA.

3.2 Plasmid and Construct Preparation

1. To generate the dCas13-PABPC1 construct (pJC1211), PCR amplify human PABPC1 from PC1-MS2V5-P ABPC1 plasmid using PABPC1 F and PABPC1 R primers which were designed to add BamHI restriction site to the 5′ end and NotI to the 3′ end of the sequence.
2. Digest plasmid pJC1210 and PCR product separately with BamHI and NotI to remove ADAR2DD sequence from the plasmid and make sticky ends for the insert, compatible with plasmid.
3. Insert PCR-amplified PABPC1 from **step 1** by relegating the BamHI and NotI compatible ends into the mutated and modified pC0054-CMV-dPspCas13b-longlinker-ADAR2DD (pJC 1210).

3.3 Designing sgRNAs to Target the 3′ UTR of the Desired Gene

1. Using the nygenome online tool (Cas13design [nygenome.org]), predict dCas13b guides that target the 3′UTR of your specific target genes.
2. The predicted guides with the highest score depend on their position on the 3′UTR selected to be synthesized and cloned into the guide RNA plasmid backbone. Selected sequences need to be ordered in both 5′ to 3′ (Forward) and 3′ to 5′ (Reverse) directions.
3. Create a reaction mixture including forward and reverse oligos in a 1× NEB buffer 3.1.
4. Incubate for 5–10 min at boiling temperature (100 °C).

5. Incubate at room temperature to cool down gradually for 2 h.
6. Clone each individual guide into PC0043-Cas3b-crRNA backbone (pJC1208) at BbsI sites using the GreenGate protocol [19].
7. Amplify the generated plasmids in *E. coli* and then purify using a commercially available maxi-prep DNA purification kit to obtain pure plasmid DNA suitable for transfection.

3.4 Mammalian Cell Culture and Transfection

3.4.1 HEK 293 Cell Culture

1. Culture HEK293 cells in Dulbecco's modified Eagle's medium (DMEM) with 10% FBS.
2. Renew growth medium every 2–3 days and subculture them when cells reach 70–80% confluency.
3. Briefly rinse the cell layer with Ca^{2+} and Mg^{2+} free Dulbecco's phosphate-buffered saline (D-PBS) to remove possible trypsin inhibitors.
4. Add an appropriate amount of Trypsin–EDTA solution to the culture plate and incubate at 37 °C for 5 min.
5. Add an appropriate amount of complete growth media, collect cells into 15 ml conical tubes, and spin at approximately 125 × *g* for 5 min.
6. Aspirate the supernatant and resuspend cells in a warm fresh media, count, and add 500×10^5–700×10^5 cells per well of the six-well plates.
7. Cells are ready to transfect when they reach 50% confluency, change the media every two days.

3.4.2 Mammalian Cell Transfection

1. Prepare transfection reagents using GeneXPlus Transfection Reagent in Opti-MEM reduced serum medium following the manufacturer's protocol by adding 1 μg of dCas13b-PABPC1 + 1.5 μg of sgRNA constructs (*see* **Note 6**).
2. Co-transfect cells with individual sgRNAs + dCas13b-PABPC1 fusion plasmid by adding the prepared transfection reagents throughout the well and incubating cells for 48 h at 37 °C, 5% CO_2 air-humidified incubator.

3.5 Analysis of RNA Amplifier Technology

3.5.1 RNA Analysis

1. 48 h post-transfection, isolate RNA using the direct-zol RNA kit, following the manufacturer's protocol.
2. Use equal amount of the purified RNA (between 100 and 500 ng) to synthesize cDNA using random hexamers and SuperScript III First-Strand Synthesis System following the manufacture protocol.
3. Prepare qPCR using PowerUp SYBR Green master mix and your designed PCR primers.
4. Perform qPCR with an appropriate real-time PCR instrument.

5. Analyze data using cycle threshold (Ct) value (double delta Ct analysis method) [20] and compare to mock-transfected control cells (i.e., no sgRNA transfection) while normalizing to a housekeeping gene internal control, such as ACTB.

3.5.2 Protein Analysis

1. Rinse cells with 1 ml cold D-PBS, aspirate liquid, and then add 120 μl of RIPA buffer to each well.
2. To isolate protein from transfected cells using RIPA buffer, scrape cells with RIPA buffer on ice, collect it in the 1.5 μl tube and invert them for 30 min in 4 °C by to homogenize and lysate cells.
3. Centrifuge the cell lysate at 10,000 × g for 10 min at 4 °C and collect the supernatant.
4. The isolated protein can be run for Western blotting immediately or be stored at −80 °C.
5. Determine the protein concentration using the Bradford method and then analyze equal amounts of each cell lysate by Western blot.
6. Separate equal amounts of proteins on SDS–PAGE using the Mini-protean TGX 4–15% gels, after diluting protein in 1× SDS dye and boiling at 95 °C for 5 min.
7. Load protein marker ladder alongside the samples to determine the protein bands molecular weight.
8. Transfer protein in gel to the nitrocellulose membrane by wet transfer using cold 1× transfer buffer with 20% methanol. Transfer should be performed in a cold room at 250 mA for 2 h.
9. Block the membrane in 10% (w/v) powdered dry milk in PBST for 1 h.
10. Probe the membrane with the appropriate antibody following the manufacturer's suggested concentrations. We use 1:1000 concentration for anti-MeCP2, with overnight probing. For GAPDH detection, we used 1:5000 dilution for anti-GAPDH, with 1 h probing at room temperature. The secondary antibodies were diluted 1:10000 and incubated in the blocking buffer for 1 h at room temperature.
11. Detect the proteins using enhanced chemiluminescence detection reagent and a Western blot imager instrument according to the manufacturer's instructions.
12. Use software compatible with your Western blot imager or ImageJ tools to quantify Western blotting bands.

4 Notes

1. To generate pJC1210, the original plasmid pC0054-CMV-dPspCas13b-longlinker-ADAR2DD (Addgene 103870)- site mutated and modified at nucleotide 5606 to create a unique BamHI site using site mutation oligo A (TCAAGCTGCC CAAGGACCGCATCCACAGCGAGAAGTCCAAC) (oJC4996) and site mutation oligo B (GTTGGACTTCTCG CTGTGGATGCGGTCCTTGGGCAGCTTGA) (oJC4996).
2. To generate pJC1211, The ADAR2DD sequence was removed from pJC1210 using BamHI + NotI and replaced with PCR amplified, which the PC1-MS2V5-PABPC1 plasmid (Addgene#65807) was used as a template for PABPC1 amplification using PABPC1 F (AATGGGATCCTGGCAAGCC CATC) and PABPC1 R (CTAAAGGGAAGCGGCCGCTTAAA). The generated construct included the full human PABPC1 sequence.
3. In the Cas13 plasmid backbone, multiple guide RNAs (gRNA) targeting the 3′UTR of the genes of interest were designed using an online tool for the prediction dCas13b guide. They were individually inserted into Cas13b-crRNA compatible backbone. The gRNA is critical because it has anti-sense homology to specific mRNAs and a short hairpin required for dCas13b binding (Fig. 1). The generated plasmids must be maxi-prep purified using commercial kits to have high-quality plasmids for transfection in mammalian cells.
4. We tested our RNA amplifier approach in HEK293, HepG2, and SH-SY5Y cell lines targeting MeCP2 (methyl CpG binding protein 2) gene. We demonstrated that this gRNA-targeted tethering enhances targeted endogenous mRNAs in both RNA and protein level in a gRNA-dependent manner, and no stimulation occurs when the dCas13b-PABPC1 fusion is expressed alone (Fig. 3).
5. To detect MeCP2 protein, we blocked our membrane with 10% milk in 1× PBS with 0.5% Tween-20. Commercial blocking buffers (Every blot blocking buffer, Bio-Rad) have been used for other antibodies, such as anti-GAPDH.
6. It may be necessary to optimize the co-transfection of the different plasmids. For example, we tested ratios of dCas13-PABPC1 and guide expressing plasmids (1:1, 1:1.5, 1:2) that 1:1.5 selected as the optimal working ratio.

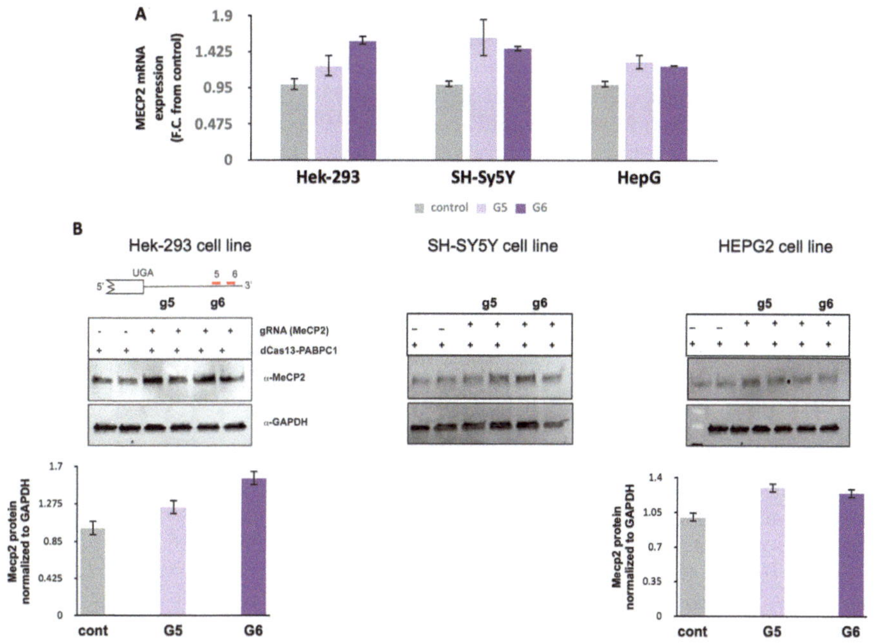

Fig. 3 The tethered mRNA amplifier enhances gene expression in a gRNA-dependent manner at both mRNA and protein levels. (**a**) qPCR analysis for the MeCP2 expression in multiple cell lines (Hek-293, HEPG2, and SH-SY5Y cells) 48 h after transfection with fused PABPC1 to dCas13b alongside specific gRNAs designed to target MeCP2 mRNA 3′UTR. (**b**) Immunoblotting detected a 20–80% increase in the level of MeCP2 protein in the three cell lines

Acknowledgments

This research was supported by grants to J.C. from the National Institutes of Health and Bloomberg Philanthropies (J.C.).

References

1. Wang Z-Y, Leushkin E, Liechti A et al (2020) Transcriptome and translatome co-evolution in mammals. Nature 588:642–647
2. Passmore LA, Coller J (2022) Roles of mRNA poly (A) tails in regulation of eukaryotic gene expression. Nat Rev Mol Cell Biol 23:93–106
3. Nicholson AL, Pasquinelli AE (2019) Tales of detailed poly (A) tails. Trends Cell Biol 29: 191–200
4. Nietfeld W, Mentzel H, Pieler T (1990) The Xenopus laevis poly(A) binding protein is composed of multiple functionally independent RNA-binding domains. EMBO J 9:3699–3705
5. Deardorff JA, Sachs AB (1997) Differential effects of aromatic and charged residue substitutions in the RNA-binding domains of the yeast Poly(A)-binding protein. J Mol Biol 269:67–81. 11 Edited by K. Nagai
6. Eliseeva I, Lyabin D, Ovchinnikov L (2013) Poly (A)-binding proteins: structure, domain organization, and activity regulation. Biochem Mosc 78:1377–1391
7. Kahvejian A, Svitkin YV, Sukarieh R et al (2005) Mammalian poly (A)-binding protein is a eukaryotic translation initiation factor, which acts via multiple mechanisms. Genes Dev 19:104–113
8. Derry M, Yanagiya A, Martineau Y et al (2006) Regulation of poly (A)-binding protein through PABP-interacting proteins. Cold Spring Harb Symp Quant Biol, Cold Spring Harbor Laboratory Press 71:537–543

9. Cox DB, Gootenberg JS, Abudayyeh OO et al (2017) RNA editing with CRISPR-Cas13. Science 358:1019–1027
10. Gootenberg JS, Abudayyeh OO, Lee JW et al (2017) Nucleic acid detection with CRISPR-Cas13a/C2c2. Science 356:438–442
11. Terns MP (2018) CRISPR-based technologies: impact of RNA-targeting systems. Mol Cell 72: 404–412
12. Zhang B, Ye W, Ye Y et al (2018) Structural insights into Cas13b-guided CRISPR RNA maturation and recognition. Cell Res 28: 1198–1201
13. Abudayyeh OO, Gootenberg JS, Essletzbichler P et al (2017) RNA targeting with CRISPR–Cas13. Nature 550:280–284
14. Smargon AA, Cox DB, Pyzocha NK et al (2017) Cas13b is a type VI-B CRISPR-associated RNA-guided RNase differentially regulated by accessory proteins Csx27 and Csx28. Mol Cell 65:618–630.e617
15. Coller J, Wickens M (2007) Chapter fourteen – tethered function assays: an adaptable approach to study RNA regulatory proteins. In: Lorsch J (ed) Methods in enzymology. Academic Press, pp 299–321. https://doi.org/10.1016/S0076-6879(07)29014-7
16. Coller JM, Gray NK, Wickens MP (1998) mRNA stabilization by poly(A) binding protein is independent of poly(A) and requires translation. Genes Dev 12:3226–3235
17. Keryer-Bibens C, Barreau C, Osborne HB (2008) Tethering of proteins to RNAs by bacteriophage proteins. Biol Cell 100:125–138
18. Kajjo S, Sharma S, Chen S et al (2022) PABP prevents the untimely decay of select mRNA populations in human cells. EMBO J 41: e108650
19. Lampropoulos A, Sutikovic Z, Wenzl C et al (2013) GreenGate-a novel, versatile, and efficient cloning system for plant transgenesis. PLoS One 8:e83043
20. Schmittgen TD, Livak KJ (2008) Analyzing real-time PCR data by the comparative CT method. Nat Protoc 3:1101–1108

Chapter 11

RNA-Binding Protein-Mediated mRNA Deadenylation in Mammalian Cell Extracts

Wi S. Lai, Stephanie N. Hicks, and Perry J. Blackshear

Abstract

Removal of the poly(A) tail, or deadenylation, is a crucial step in destabilizing mRNAs in eukaryotes. In this chapter, we describe a cell-free deadenylation assay that uses cytoplasmic cell extracts from human HEK293 cells transiently transfected with DNA encoding RNA-binding proteins (RBP), and in vitro-transcribed, radiolabeled, RNA probes. We include methods to evaluate the effects of RBPs or deadenylases on various in vitro-transcribed probes, with or without poly(A) tails. Finally, we also demonstrate the adaptability of these assays to test purified protein components in our cell-free deadenylation assay. In our experience, these methods are well suited for the initial assessment of the effects of RBPs on the deadenylation of mRNAs.

Key words mRNA-binding proteins (RBPs), Cytosolic extracts, Capped RNA probes, Poly(A), Deadenylation

1 Introduction

Destabilization and ultimate decay of eukaryotic mRNAs involve the initial activity of 3′-exonucleases, or deadenylases, and these are often modulated by the binding of RNA-binding proteins (RBPs) to target sequences within the 3′-untranslated regions (3′UTR) [1, 2]. Once the poly(A) tail is removed, the mRNA body is destroyed by other exonucleases and endonucleases [3]. The mRNA poly(A) tail-binding protein, PABP, has been shown to play multifaceted roles in protecting the mRNA poly(A) tail, and also in binding to components of the translation initiation apparatus to facilitate translation [4]. Multiple molecules of PABP are associated with a given poly(A) tail [5]. It has also been shown that PABP, when in complex with a given poly(A) tail, can participate in protein-protein interactions with other RBPs to carry out a variety of cellular activities [6]. However, the interplay between

Eugene Valkov and Aaron C. Goldstrohm (eds.), *Deadenylation: Methods and Protocols*, Methods in Molecular Biology, vol. 2723, https://doi.org/10.1007/978-1-0716-3481-3_11,

RBPs and PABP in the destabilization of mRNAs is not well defined and is an area of ongoing investigation.

We describe here a cell-free deadenylation method that uses in vitro-transcribed 5′-capped, radiolabeled RNA probes, with or without poly(A)-tails, together with cytosolic extracts prepared from HEK293 cells transiently transfected with various protein expression constructs. With this cell-free deadenylation method, we address whether an RNA-binding protein can promote deadenylation, using the RBP human tristetraprolin (TTP or ZFP36) to illustrate the proof of principle of this assay. TTP is the prototype member of the TTP family of tandem zinc finger (TZF) domain proteins. Members of the TTP family of proteins bind, through their TZF domain, to AU-rich elements (ARE) in single-stranded RNA, such as UUAUUUAUU, usually located within the 3′UTRs of mRNA [7, 8]. Some of the in vitro-transcribed RNA probes we used in the in vitro deadenylation method described here include multiple repeating units of UUAUUUAUU (Fig. 1), the optimal TTP target sequence. In some of these experiments, along with transiently expressed TTP in HEK293 cells, we also include the human cytosolic poly(A)-binding protein 1 (PABPC1) and the human poly(A)-specific ribonuclease (PARN) [9]. We also demonstrate the applicability of the cell-free deadenylation assay described here to assess the deadenylation activity of purified proteins, such as PARN, in the presence and absence of TTP.

In an initial cell-free deadenylation experiment, with TTP protein expressed in HEK293 cells together with the native human cell deadenylation machinery, TTP promoted the decay of the ARE-containing polyadenylated probe (ARE.A50), but not the A50 (without a TTP-binding site), or the ARE (without a poly (A) tail) probes (Fig. 2a, reprinted from [10] by permission of Taylor & Francis Ltd). For these assays, 10 μg of total cytosolic protein was used, as this amount of HEK293 cell extract contained sufficient TTP protein to allow ready detection of the final deadenylated product, but still showed appreciable fully adenylated probe (Fig. 2a, lanes 11–12). After demonstrating that a TTP TZF domain mutant, C124R, expressed in the same way, could not bind the ARE-containing probe (Fig. 2b, compare lane 4 to lane 3, reprinted from [10]), we showed that it also could not promote deadenylation in this assay (Fig. 2c, compare lane 6 to lane 4, reprinted from [10]). To ensure that roughly equivalent amounts of intact protein are used in the assay, we usually first perform western blots with epitope-tagged versions of the proteins after each batch of transfected HEK293 cells is harvested.

One parameter that should be optimized in the cell-free deadenylation assay is the concentration of $MgCl_2$. As shown in Fig. 3a (reprinted from [10]), at a low concentration of $MgCl_2$ (lane 7), or with none added (lane 8), TTP had little or no ability to promote deadenylation. However, the trace amount of $MgCl_2$ in the cell

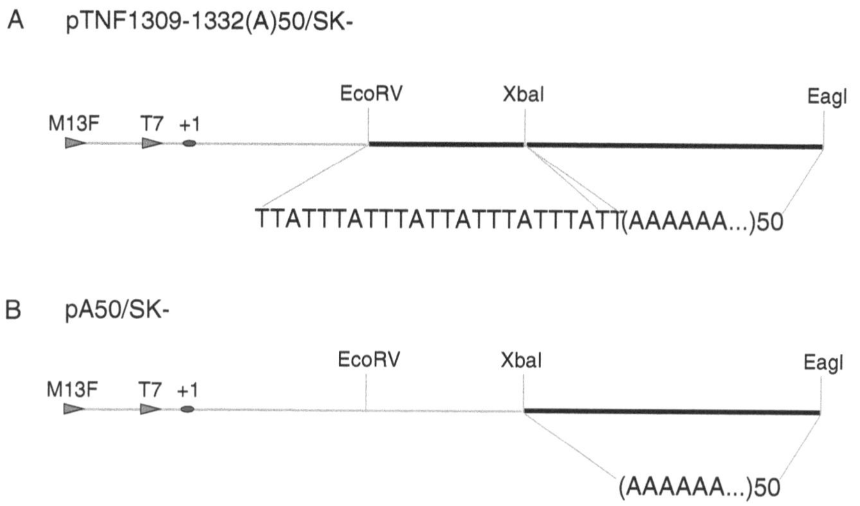

Fig. 1 Plasmids for generating RNA probes. The thin gray lines in the schematics **a** and **b** represent the vector pBluescript (SK-) sequence, with indicated M13 Forward priming site (M13F), T7 promoter (T7), and the transcription start site (+1). In **a**, the thick black line represents the inserted *TNF* mRNA ARE (bp 1309 to 1332 of GenBank accession no. NM_13693. 3), and the 50As, respectively, with the corresponding sequences displayed below the schematic. In **b**, the thick black line represents the inserted 50 As shown below the schematic. Plasmid A is for transcribing RNA probes V, ARE, or ARE.A50. Plasmid B is for transcribing RNA probe A50

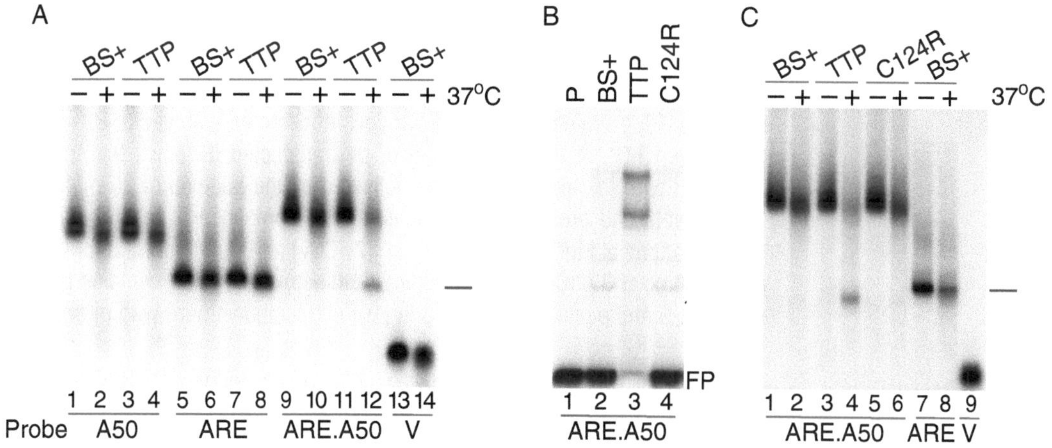

Fig. 2 Cell-free deadenylation of a polyadenylated ARE-containing RNA probe by TTP. HEK293 cell extracts were prepared from cells transfected with vector BS+ alone or CMV-driven TTP and its TZF domain mutant C124R. In **a** and **c**, HEK293 extracts (10 μg protein) were incubated with the indicated probes in the presence of 3 mM $MgCl_2$ on ice (−) or at 37 °C (+) for 60 min in the presence of 3 mM $MgCl_2$, followed by the addition of 20 mM EDTA. The migration position of the ARE probe is indicated on the right. In **b**, a gel-shift mobility assay is shown. Cell extracts (10 μg of protein per binding reaction) were incubated with the probe ARE.A50 in the presence of 3 mM $MgCl_2$ at room temperature for 60 min, followed by RNase T1 digestion. The products were separated using an 8% nondenaturing polyacrylamide gel. Gel lane 1 labeled P was loaded with RNase T1 digested probe alone. The position of the free probe (FP) is indicated on the right. (This figure contains panels A, B, and part of panel D from Fig. 2 in [10] and is shown here with permission)

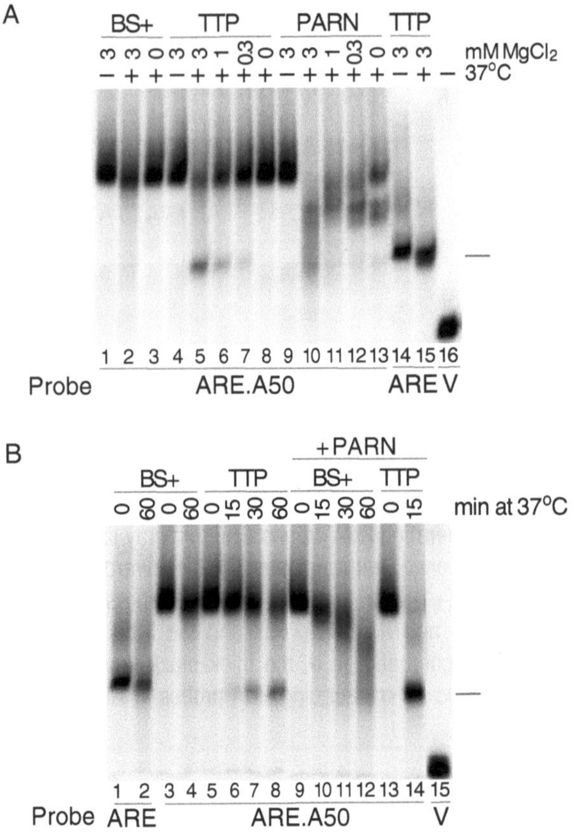

Fig. 3 Characterization of the TTP-induced deadenylation activity. HEK293 cell extracts (10 μg protein) prepared from cells transfected with vector BS+ alone, CMV-driven TTP, or PARN, singly (**a**) or co-transfected (B, lanes 9–14)), were incubated with the indicated probes, followed by the addition of 20 mM EDTA. Cell extracts and probes were incubated in the presence of (**a**) the indicated concentrations of $MgCl_2$ on ice (−) or at 37 °C (+) for 60 min, or (**b**) 3 mM $MgCl_2$ on ice (0) or at 37 °C (+) for the indicated times. (This figure contains panel A from Fig. 5 and panel A from Fig. 6 in [10] and is shown here with permission)

extract was perhaps sufficient for the deadenylase PARN to promote the reaction (Fig. 3a, lane 13).

Another parameter to be considered is the incubation time of the cell-free deadenylation reactions. We demonstrated that the deadenylation activity changed with increased incubation time, using extracts from HEK293 cells transfected with the TTP expression plasmid, as shown in Fig. 3b (lanes 5–8, reprinted from [10]). Another consideration includes the effects of other deadenylases and RBPs on the ability to promote or inhibit deadenylation activity. When we co-transfected two protein expression constructs in HEK293 cells, as shown in Fig. 3b with TTP and PARN (lanes 13 and 14), we found that TTP and PARN could work together in the deadenylation reaction. Similarly, when we co-expressed additional proteins in HEK293 cells, as illustrated in Fig. 4a, we found

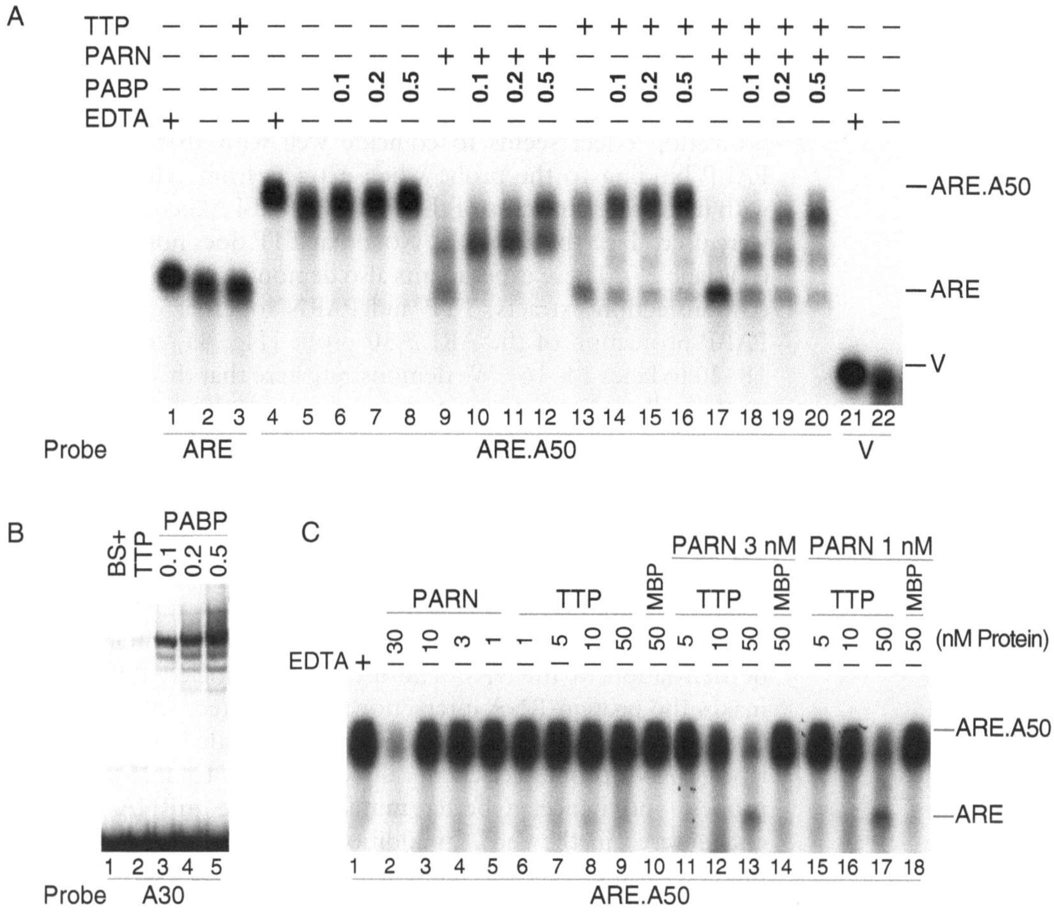

Fig. 4 Stability of RNA probes. (**a**) Destabilization of the probe ARE.A50 in the presence of PARN and the ARE-binding protein TTP. Extracts (10 μg total protein) were used from HEK293 cells transfected with vector alone (BS+, lanes 1, 2, 4, 5, 21, and 22), or with various expressed plasmids, singly or in combination (as indicated by the + or – signs, or numbers, above the graph), including CMV.TTP.HA (TTP, 0.2 μg/plate), CMV.PARN.Flag (PARN, 0.5 μg/plate), and CMV.PABP.Flag (PABP, 0.1, or 0.2, or 0.5 μg/plate). These were incubated with the designated RNA probes (ARE, lanes 1–3; ARE.A50, lanes 4–20; V, lanes 21–22) at 37 °C for 60 min in the presence of 3 mM $MgCl_2$, with (+) or without (–) 20 mM EDTA. The migration positions of the RNA probes are indicated on the right. (**b**) An electrophoretic gel-shift assay of binding to the 30A probe by PABP. Extracts from HEK293 cells used in A lanes 1, 3, and 6–8 were incubated for 60 min at room temperature in the presence of 3 mM $MgCl_2$ with a synthetic 30A RNA oligonucleotide 3′-end-labeled with [^{32}P]pCp and T4 RNA ligase. The products were separated using an 8% nondenaturing polyacrylamide gel. Note that TTP did not bind to the 30A probe (compare lane 2 to lane 1). (**c**) Probe ARE.A50 stability in the presence of purified recombinant MBP.TTP and His.PARN. Various concentrations of purified proteins, including the control, MBP alone, were incubated with probe ARE.A50 at 37 °C for 60 min in the presence of 3 mM $MgCl_2$ (lanes 2–18). As a control for the probe stability, 20 mM EDTA (+) was added to the sample containing ARE.50 probe alone (lane 1). TTP alone (lanes 6–9) did not exhibit deadenylating activity

that PABP provided relative protection of the ARE.A50 probe against the native HEK293 cell deadenylation machinery, in both the absence or presence of TTP (lanes 6–8 compared to lane 5, and lanes 14–16 compared to lane 13). The increase of the PABP protection effect seems to coincide well with that of increased PABP binding to the probe when extracts from cells transfected with increasing amounts of the CMV.PABP.FLAG construct DNA were used (Fig. 4b, lanes 3–5. Note that TTP does not bind to poly (A) (lane 2)). These experiments also demonstrate that in the triple co-transfection extracts, TTP and PARN together can overcome PABP protection of the ARE.A50 probe (Fig. 4a, compare lanes 18–20 to lanes 14–16). We demonstrate here that this in vitro assay works well as a preliminary test to investigate the activities of two recombinant proteins together, in this case showing that TTP and PARN worked in concert to promote deadenylation of the ARE. A50 probe (Fig. 4c).

In our experience, the cell-free deadenylation system described here using HEK293 cell extracts, and in vitro-transcribed RNA probes incorporating radioactive nucleotides, is suitable for the initial assessment of the effects of RNA-binding proteins on the deadenylation of mRNAs. The use of radiolabeled RNA probes makes the protein-RNA interaction easy to detect with simple lab equipment, such as a film developer, with the added benefit that the labeled RNA probes are chemically very similar to unlabeled RNA probes (and hence, a better mimic of native mRNA 3′-UTRs encountered in the cell). In addition, the transfection-expression method ensures that the relevant proteins are appropriately post-translationally modified; this is particularly relevant in the case of mammalian TTP, which is very heavily phosphorylated [11].

2 Materials

2.1 Cell Cytosolic Extracts

1. HEK293 cells (ATCC CRL-1573).
2. Culture Medium: Gibco Minimum Essential Media (MEM) supplemented with 10% fetal bovine serum, 2 mM L-glutamine, and 100 U/ml penicillin-streptomycin.
3. Protein expression plasmids for transfection: CMV.TTP.HA (human TTP, NP_003398.3), CMV.TTP(C124R).HA, CMV.PARN.FLAG (human poly(A)-specific ribonuclease (PARN; NP_002573.1), and CMV.PABP.FLAG (human PABP or PABPC1, NP_002559.2). These constructs are all CMV-driven and have the pBluescribe (BS+, *see* **Note 1**) backbone.
4. 2× HN: 50 mM HEPES-NaOH, 280 mM NaCl (pH 7.0), do not go below pH 6.95 or above pH 7.05. Filter the solution through a 0.2 μM filter, and store it at 4 °C.

5. Monobasic 70 mM NaH_2PO_4 filtered through a 0.2 μM filter, stored at 4 °C.
6. Dibasic 70 mM Na_2HPO_4 filtered through a 0.2 μM filter, stored at room temperature.
7. 0.5 M $CaCl_2$ filtered through a 0.2 μM filter, stored at room temperature.
8. Phosphate-buffered saline (PBS): 137 mM NaCl, 2.7 mM KCl, 10 mM Na_2HPO_4, 2 mM KH_2PO_4 (pH 7.4), stored at 4 °C.
9. Diethyl pyrocarbonate (DEPC)-treated water.
10. Cell wash water: DEPC-treated water, to which is added (final concentrations) 0.1 mM phenylmethylsulfonyl fluoride (PMSF) and 1 μg/ml pepstatin A, immediately before use.
11. HEK293 lysis buffer: 10 mM HEPES-NaOH (pH 7.6), 40 mM KCl, 5% (v/v) glycerol, 0.2% (v/v) NP-40, to which is added (final concentrations) 0.1 mM PMSF, 1 μg/ml pepstatin A, 5 μg/ml leupeptin, and 10 mM dithiothreitol (DTT), immediately before use.

2.2 Protein Purification Solutions

All solutions should be filtered through a 0.22 mM filter.

1. Lysis buffer: 20 mM Na_2HPO_4 (pH 8), 100 mM KCl, 5% (v/v) glycerol, 10 mM imidazole, to which is added one EDTA-free protease inhibitor tablet immediately before use.
2. Lysis buffer 2: 20 mM HEPES-NaOH (pH 7.5), 100 mM KCl, and 10% (v/v) glycerol.
3. Elution buffer1: Lysis buffer1 with 40 mM imidazole.
4. Elution buffer2: Lysis buffer2 with 10 mM maltose.
5. S1 buffer: 20 mM Na_2HPO_4 (pH 7.5), 30 mM KCl, 5% (v/v) glycerol, 2 mM β-mercaptoethanol, 25 mM $ZnSO_4$.
6. S2 buffer: S1 buffer with 1 M KCl.
7. S200 buffer: 20 mM Na_2HPO_4 (pH 7.5), 100 mM KCl, 25 mM $ZnSO_4$, 2 mM β-mercaptoethanol, 5% (v/v) glycerol.
8. S200 buffer 2: 20 mM HEPES-NaOH (pH 7.5), 100 mM KCl, 10 mM maltose, and 10% (v/v) glycerol.
9. 10× SDS-PAGE buffer: 1% (w/v) sodium dodecyl sulfate (SDS) powder, 0.25 M Tris–HCl, and 1.92 M glycine (pH 8.3).

2.3 Reagents for Protein Purification

1. One Shot BL21 Star (DE3) competent cells.
2. 5 ml HisTrap HP column.
3. 5 ml HiTrap Q HP column.
4. Superdex S200 10/300 GL column.
5. Amylose resin.

6. 10,000 MWCO Vivaspin 20 centrifugal device.
7. Isopropyl b-D-1-thiogalactopyranoside (IPTG).
8. Rosetta 2(DE3) competent cells.
9. 2.5 × 20 cm Econo-column.
10. 2× Laemmli buffer.
11. 10% polyacrylamide gels.
12. Human full-length TTP (GenBank accession NP_003398.3) is expressed in a plasmid encoding an N-terminal MBP affinity tag and a C-terminal six-residue histidine tag (a kind gift from Marc R. Fabian, McGill University [12]).
13. MBP (GenBank accession WP_000695387.1; amino acids 27–392) protein is expressed from a modified pMAL-c5X vector (a kind gift from Lars Pederson, NIEHS [13]).
14. Human PARN (GenBank accession NP_002573.1) is expressed from the pET33b(+) vector encoding a six-residue histidine tag located at the amino terminus.

2.4 PCR-Generated Template for In Vitro Transcription

1. Plasmid pTNF1309–1332(A)50/SK-: for template AER.A50.
2. Plasmid pA50/SK-: for template A50.
3. 20 mM dNTP mix.
4. Taq DNA Polymerase High Fidelity.
5. M13 forward primer: 5′-GTAAAACGACGGCCAGT-3′.
6. T50Xba primer: 5′-$T_{(n = 50)}$CTAGA-3′.
7. Agarose gel extraction kit.

2.5 In Vitro Transcription and Deadenylation

1. PCR-generated template for probe ARE.A50.
2. PCR-generated template for probe A50.
3. Restriction enzyme EcoRV linearized Plasmid pTNF1309–1332(A)50/SK- for probe V.
4. Restriction enzyme XbaI linearized Plasmid pTNF1309–1332 (A)50/SK- for probe ARE.
5. T7 RNA polymerase (10 U/μl) with supplied reaction Buffer.
6. RNasin ribonuclease inhibitor (20 U/μl, Promega).
7. 40 mM m^7G cap analog (m^7G(5′)ppp(5′)G).
8. NTPs: 10 mM ATP, 10 mM CTP, 1 mM GTP, 0.3 mM UTP.
9. 0.1 M DTT.
10. [α ^{32}P]UTP (800 Ci/mmol).
11. 1 U/μl RQ1 RNase-free DNase.
12. 20 mg/ml glycogen.
13. Absolute ethanol.

14. 2× Formamide Stop solution: each ml contains 0.92 ml formamide, 30 μl 0.5 M EDTA, 25 μl 1% (w/v) bromophenol blue, 25 μl 4% (w/v) xylene cyanol.
15. TBE-Urea gels, 6%: 10-well, 15-well.
16. 5× TBE solution: 0.45 M Tris, 0.45 M boric acid, 10 mM EDTA disodium salt dihydrate.
17. 3MM chr Blotting paper.
18. X-ray film.
19. Gel elution buffer: 0.5 M ammonium acetate, 1 mM EDTA-NaOH (pH 8) in DEPC water.
20. Acidic phenol/chloroform.
21. 10× RNA-protein-binding buffer: 0.1 M HEPES-NaOH (pH 7.6), 0.4 M KCl, 50% glycerol.
22. Gel fixation solution: 10% (v/v) methanol/10% (v/v) acetic acid in water.
23. Nylon mesh.

2.6 Equipment

1. PCR machine.
2. NanoDrop spectrophotometer.
3. Fast-performance liquid chromatography (FPLC) system.
4. Incubator shaker (with a capacity to heat and cool).
5. Platform shaker.
6. Gel dryer.
7. Film developer.

3 Methods

3.1 Transfection of HEK293 Cells

1. Maintain HEK293 cells in Culture Medium at 37 °C. One day before transfection, plate cells onto 100 mM dishes at approximately 0.6–0.7 million per dish in 10 ml of Culture Medium. On the day of transfection, the cells should be 40–50% confluent.
2. In a tissue culture hood, prepare CMV-driven protein expression constructs for calcium-phosphate transfection (*see* **Note 2**) by using the following amounts of DNA of the appropriate expression plasmids per dish in 0.5 ml of 0.125 M $CaCl_2$: 0.2 μg of DNA encoding human TTP (TTP); 0.1, 0.2, or 0.5 μg of PABP; 0.5 μg of PARN; alone or in various combinations with the carrier vector Bluescribe (BS+), for a total of 5 μg of plasmid DNA per dish. For extracts that serve as controls, use 5 μg of BS+ per dish of cells. We routinely transfect three dishes pooled together at harvest time for each condition.

Therefore, for each transfection condition, prepare the DNA needed for four dishes in the same tube (DNA in 2 ml of 0.125 mM $CaCl_2$).

3. In a tissue culture hood, prepare 2× HNP by mixing 39.2 ml of 2× HN and 0.8 ml of pH 7.0 70 mM NaH_2PO_4/70 mM Na_2HPO_4 (*see* **Note 3**). Dispense 2 ml to each 15 ml tube (0.5 ml for each dish).
4. While vortexing the tube containing the 2 ml 2× HNP, dropwise, add the 2 ml DNA in 0.125 M $CaCl_2$ to the 2× HNP. After 2–5 min, the finished mixture should look barely cloudy. If precipitation is visible at this point, indicating a higher pH, prepare the 70 mM NaH_2PO_4/70 mM Na_2HPO_4 (pH 7.0) solution again and check the pH carefully.
5. Add 1 ml mixture prepared at **step 4** to each dish (*see* **Note 4**). Incubate the dishes at 37 °C.
6. After 16–24 h, about 50–60% area at the bottom of the dish should be covered by a fine pepper and salt-like precipitate. Replace the transfection medium with fresh Culture Medium and allow the cells to recover for 24 h before harvesting.

3.2 Preparation of HEK293 Extracts

The HEK293 cells are ready to harvest 48 h after transfection (*see* **Note 5**). Before harvesting, prepare a tube of ice-cold DEPC-treated water (for Cell Wash water preparation, 3–4 ml per 3 dishes of cells), and a basin of ice that is slightly sloped so that the dishes rest on a slight incline.

1. Decant the culture medium from the dish into a vessel and set the dish on ice. Aspirate the remaining culture medium (*see* **Note 6**).
2. Add 4 ml of ice-cold PBS to each dish. Scrape the cells off the dish with a cell scraper and pool the cells from the same transfection condition (three dishes) into a 15 ml tube (*see* **Note 7**).
3. Centrifuge the cells for 5 min at 600 g at 4 °C. Aspirate the PBS.
4. While turning the tube slowly on ice, carefully add 3–4 ml of ice-cold Cell Wash Water along the tube wall to not disturb the cell pellet. Aspirate the wash water. Rinsing the tube's wall and the cell pellet's surface aids in the cell lysis process by removing any residual isotonic buffer, especially when the cells are lysed in a low-detergent buffer.
5. Carefully remove residual water using a pipette fitted with a 200 μl pipet tip. Add 0.7–1 ml (about 10× the pellet volume) of HEK293 lysis buffer to the cell pellet using a pipette fitted with a 1000 μl pipet tip, and pipet the cells up and down three times

before transferring the suspension to a 1.5 ml microfuge tube. Set the tube on ice (*see* **Note 8**).

6. Prepare the remaining samples, handling the cell pellets in the same manner as described above (**step 5**). Next, centrifuge the samples in microfuge tubes for 15 min at 14,000 g at 4 °C.
7. While the samples are centrifuging, prepare a new set of microfuge tubes. Add glycerol to each tube at a volume equivalent to 10% of the sample's supernatant. Transfer the supernatant to the new tube, and then rotate the tubes at 4 °C for 15–30 min to mix with the glycerol.
8. Check the protein concentration of each sample using the Bradford Protein assay. Snap-freeze the samples in liquid nitrogen and store them at −80 °C.

3.3 Purification of Recombinant Human TTP, PARN, and MBP Using the BL21 (DE3) or Rosetta 2(DE3) Bacterial Expression Systems

To purify his-tagged human PARN (His.PARN), follow the steps below. All chromatography steps should be performed at 4 °C.

1. Pellet 5 l of bacterial cells (BL21 Star (DE3)) expressing His. PARN at 3600 g for 15 min at 4 °C, after inducing expression with 0.1 mM IPTG for 16 h at 22 °C.
2. Resuspend the cell pellet in 50 ml of lysis bufferl.
3. Lyse the cell suspension by sonication using four 30 s intervals while on ice.
4. Centrifuge lysate at 4 °C at 35,000 g for 30 min.
5. Apply supernatant to a 5 ml HisTrap HP column (Cytiva), wash with 5 column volumes (CV) of lysis bufferl, and elute with 5 CV of elution bufferl containing 40 mM imidazole.
6. Prepare samples for SDS-PAGE by mixing 20 ml of each elution fraction with 20 ml of 2× Laemmli buffer.
7. Perform SDS-PAGE using a 10% polyacrylamide gel in 1× SDS-PAGE running buffer.
8. Pool fractions containing His.PARN and dilute with S1 buffer to reach a final KCl concentration of 30 mM.
9. Load protein onto a 5 ml HiTrap Q HP column. Wash with 10 CV of S1 buffer and elute protein in a linear gradient (0–80%) in S2 buffer.
10. Repeat **steps 6** and 7 above to verify the fractions containing His.PARN.
11. Concentrate eluate at 4 °C using a 10 kDa MWCO Vivaspin 20 centrifugal device at 5500 g.
12. Apply the concentrated sample to a Superdex 200 10/300 GL column equilibrated in S200 bufferl.
13. Repeat **steps 6** and 7 above to perform SDS-PAGE to analyze the purity of His.PARN.

14. Pool fractions containing purified His.PARN and concentrate in a Vivaspin 20 10 kDa MWCO centrifugal device at 5500 g at 4 °C.
15. Measure the UV absorbance at 280 nm to determine the final protein concentration.
16. Store the purified protein at −80 °C in a buffer with 10% (v/v) glycerol (final) after flash freezing in liquid nitrogen.

To purify full-length human TTP:

1. Express full-length TTP containing an N-terminal MBP-tag and a C-terminal six-residue histidine tag in Rosetta 2(DE3) competent cells after induction with 0.1 mM IPTG for 16 h at 20 °C.
2. Proceed with the purification of full-length TTP, using **steps 2–16** from the above procedure describing the purification of His.PARN.

To purify MBP protein:

1. When 5, 1000 ml cultures of BL21 Star (DE3) cells transformed with the bacterial expression vector encoding MBP reach a density between 0.6 and 0.8, as measured by OD_{600} nm, induce expression of the MBP protein by adding 0.3 mM IPTG for 16 h at 20 °C.
2. Pellet cells by centrifugation at 3600 g for 15 min at 4 °C.
3. Resuspend the cell pellet in 50 ml of lysis buffer 2.
4. Lyse cell suspension by sonication using four 30 s intervals while on ice.
5. Centrifuge lysate at 4 °C at 35,000 g for 25 min.
6. Incubate clarified lysate with 8 ml of amylose resin after equilibrating resin with 10 column volumes (CV) of lysis buffer 2, and transfer to a 2.5 × 20 cm gravity flow Econo-column (Bio-Rad).
7. Wash with 5 CV of lysis buffer 2.
8. Elute MBP protein with 5 CV of elution buffer 2 containing 10 mM maltose, collecting 3 ml per fraction.
9. Prepare samples for SDS-PAGE by mixing 20 ml of each fraction with 20 ml of 2× Laemmli buffer.
10. Perform SDS-PAGE to evaluate the purity of MBP, using a 10% polyacrylamide gel in 1× SDS-PAGE running buffer.
11. Concentrate eluate using a 10 kDa MWCO Vivaspin 20 centrifugal device at 5500 g at 4 °C.
12. Apply concentrated eluate to a Superdex S200 10/300 GL column equilibrated in S200 buffer2.

13. Perform SDS-PAGE to evaluate the purity of MBP using a 10% polyacrylamide gel in 1× SDS-PAGE running buffer, preparing samples as described in **step 9**.
14. Concentrate fractions containing purified MBP using a 10,000 MWCO Vivaspin 20 centrifugal device at 5500 g at 4 °C.
15. Measure the UV absorbance at 280 nm to determine the final concentration.
16. Flash freeze the purified protein in liquid nitrogen and store aliquots at −80 °C in a final concentration of 10% (v/v) glycerol.

3.4 RNA Probe Transcription Templates

By linearizing plasmid pTNF1309–1332(A)50/SK- (Fig. 1a) using restriction enzymes, or by amplifying a fragment from the same construct by PCR, T7 RNA polymerase can be used to transcribe the three RNA probes used in the deadenylation assays described in this chapter.

(a) Probe V: a 55 nt transcript of vector sequence alone (ends with 3 nt of EcoRV sequence).

(b) Probe ARE: an 80 nt transcript with the 24 nt ARE sequence flanked by 55 nt vector sequence at the 5′ end and 1 nt XbaI sequence at the 3′ end.

(c) Probe ARE.A50: a 141 nt transcript with the 24 nt ARE sequence flanked by 55 nt vector sequence at the 5′ end and 50 As at the 3′ end, with a 6 nt XbaI sequence in between the ARE and 50 As.

1. To prepare templates for Probe V, or for Probe ARE, digest the plasmid (*see* **Note 9**) with EcoRV, or XbaI, respectively.
2. To make the ARE.A50 probe transcription template (or the A50 probe transcription template from plasmid pA50/SK- (Fig. 1b), *see* **Note 10**), set up 5 PCR reactions as follows, per reaction:
   - 2 μl of 20 mM dNTP mix.
   - 2 μl of 50 mM $MgSO_4$.
   - 30 pmol each of the M13 Forward and T50Xba primer.
   - 1 U of high-fidelity Taq DNA polymerase.
   - 5 μl of 10× buffer.
   - 50 ng plasmid template.
   - Water (to a final volume of 50 ml).
3. Run PCR using this Program: (1) 95 °C 3 min, (2) 95 °C 1 min, (3) 55 °C 1 min, (4) 72 °C 1 min, (5) repeat **steps 2–4** 30 times, (6) 72 °C 3 min, (7) 4 °C hold.

4. Separate the PCR products in a 2% agarose gel. The templates for probe ARE.A50 and probe A50 should migrate at approximately 174 bp and 150 bp, respectively.
5. Extract the bands from the agarose gel, use an agarose gel extraction kit following the manufacturer's protocol.
6. Measure the concentration of the elution product using a NanoDrop spectrophotometer.
7. Sequence the gel-purified PCR products using the M13 Forward primer. The T7 promoter sequence and the +1 transcription start site are within the resulting PCR fragments. Store the gel-purified PCR products at −20 °C.

3.5 In Vitro Transcription of Radiolabeled RNA Probes

We use a final 100 μl volume for each transcription reaction and add a m^7G cap analog to make capped transcripts. We routinely purify the in vitro-transcribed probes from Urea-TBE gels because there are higher molecular weight aggregates that may interfere with the interpretation of the deadenylation results. This method uses radioactivity and thus requires proper licensing, training, protective equipment, and waste disposal for safety and regulatory compliance.

1. Set up each transcription reaction as follows:
   - 20 μl of 5× T3/T7 Reaction Buffer.
   - 10 μl of 0.1 M DTT.
   - 5 μl of 10 mM ATP.
   - 5 μl of 10 mM CTP.
   - 5 μl of 1 mM GTP.
   - 4 μl of 0.3 mM UTP.
   - 1.25 μl of 40 mM $m^7G(5')ppp(5')G$.
   - 4 μl of Rnasin.
   - 50 U of T7 RNA polymerase.
   - 100 μCi of [α-^{32}P]UTP (800 Ci/mmol).
   - Probe transcription template (2 pmol of gel-purified PCR product, or 2 μg of linearized and gel-purified plasmid).
   - Water (to a final volume of 100 μl).
2. Incubate the reaction samples at 30 °C for 3 h.
3. To digest the DNA template, add 4 U of RQ1 DNase and incubate at 37 °C for 15 min, then add 100 μl of 2× Formamide Stop Solution. The mixtures can be stored at −20 °C overnight or taken to the next step, gel purification of the probe.
4. Boil the transcribed probe in Formamide Stop solution for 10 min.

5. Perform electrophoresis in a 6% Urea-TBE denaturing acrylamide gel, at 300 V in 1× TBE, until the bromophenol blue dye is close to the bottom of the gel.
6. Lift one of the plates from the gel and wrap the gel while it is still on one of the plates in plastic wrap.
7. Tape the wrapped gel to a transparent spent film and mark with several fluorescent marks.
8. In a dark room, place a piece of X-ray film on top of the gel.
9. Put a piece of plexiglass on top of the film and expose the film for 3–5 min. Develop the film in a film developer.
10. Under normal lab light, place the developed film underneath the spent transparent film (that has the wrapped gel plate taped onto it), and position it in reference to the fluorescence marks (*see* **Note 11**). Through the plastic wrap, cut out the probe band (avoiding higher molecular weight aggregates).
11. Slice the probe band into small pieces (don't crush them) and transfer to a 2 ml tube. Add 1.8 ml of Gel Elution Buffer. Incubate at 37 °C for 30–45 min, or at 4 °C overnight.
12. From the tube in **step 11**, transfer 0.3 ml aliquots of elution buffer into 1.5 ml tubes (6 tubes total). Add 40 μg glycogen and 1 ml absolute ethanol to each tube and incubate on dry ice for 30 min, or at −20 °C overnight.
13. Centrifuge the probes in ethanol at 4 °C for 15 min at 14,000 g.
14. Rinse the pellet in 70% (v/v) ethanol.
15. Resuspend each pellet in 15 μl of H_2O by vortexing.
16. Briefly centrifuge the tubes, transfer the sample to a new tube, mix well, and dispense into several tubes in small aliquots. Store the aliquots at −20 °C. Use the probes within 10 days.

3.6 In Vitro Deadenylation Assays

Use 10 μg of HEK293 total cytosolic protein in each reaction (*see* **Note 12**). When purified recombinant proteins are used, use nanomolar amounts.

1. Add protein samples to 1.5 ml tubes on ice in 50 μl of 1× RNA-protein-binding buffer. For each probe, also set up two control tubes of 50 μl of 1× RNA-protein-binding buffer with protein prepared from extracts from cells transfected with vector BS+ alone. Add 10 μl of 0.2 M EDTA into one of these tubes.
2. For each RNA probe that is used, prepare a Probe Mix on ice. For every ten reactions in the experiment, prepare one extra sample of Probe Mix in excess in a 1.5 ml tube. For each reaction sample, 50 μl of 1× RNA-protein-binding buffer contains 50,000 counts per minute of the probe and 0.3 μl of 1 M $MgCl_2$ (3 mM in the final 100 μl reaction volume).

3. Add 50 μl of Probe Mix into each tube containing the protein samples for a final volume of 100 μl.
4. Incubate the reaction mixtures at 37 °C for the designated times (e.g., 15, 30, or 60 min). While the reactions run, prepare a set of 1.5 ml tubes containing 50 μl of 2× Formamide Stop Solution at room temperature.
5. At the end of the incubation period, transfer the reaction tubes to an ice bath, and add 10 μl of 0.2 M EDTA into each tube to terminate the reaction.
6. Add 100 μl of acidic phenol/chloroform to each sample and vortex well.
7. Centrifuge for 2 min at 14,000 g to separate aqueous (upper) phase from the organic phase (lower).
8. Transfer an aliquot of 50 μl of aqueous phase to a tube that contains 50 μl of 2× Formamide stop solution, and lightly vortex to mix well.
9. Heat the mixtures at 70 °C for 5 min. Store the mixture at −20 °C overnight or proceed to the next step to analyze the reaction products.
10. To resolve the products, load 15 μl of each reaction (from **step 9**) into a gel well on a 6% Urea-TBE acrylamide gel.
11. Perform electrophoresis in 0.5 × TBE at 220 V for 45–55 min (or until the bromophenol blue dye is near the bottom of the gel).
12. Lay the gel cassette on a glass plate, lift one plate off the gel (the plate facing you when loading the gel, so that the samples are oriented in the same order as when they were loaded into the gel), and cut off a small piece at a corner of the gel to use as an orientation marker. Wet a piece of nylon mesh, about 3–4 times the size of the gel, in the gel fixation solution, and lay it on top of the gel. Invert it so that the nylon mesh is on the glass plate and the remaining gel plate is on the top. Slightly lift the gel plate, using a small spatula to ease the gel off the plate gradually, and drop the gel onto the nylon mesh.
13. Lower the nylon mesh (now with the gel on top) into a small tray containing 100 ml of gel fixation solution and shake at 30 rpm for 30 min on a platform shaker.
14. Return the nylon mesh with the gel on top to the glass plate. Wet a piece of 3MM chr blotting paper in the gel-fixing solution and lay it on top of the fixed gel. Invert so that the nylon mesh is now on top. Lift off the nylon mesh. Now, the gel samples are in exactly the same orientation as when you loaded the gel, from left to right.

15. Cover the gel with plastic wrap and lay it on a piece of backing paper to dry in a gel dryer for 45 min at 80 °C.
16. Expose the dried gel to X-ray film overnight.

4 Notes

1. pBluescribe (BS+, or BS−) and pBluescript (SK+, or SK−) are different plasmids.
2. We are satisfied with the calcium-phosphate transfection method. Some may find the lipofectamine method useful or easier to use.
3. It is important to use pH 7.0 to form a fine calcium-phosphate precipitate. To reach this pH, mix appropriate amounts of 70 mM NaH_2PO_4 (monobasic) and 70 mM Na_2HPO_4 (dibasic), usually in a range of 38:62 to 45:55 of monobasic:dibasic. Mix the two buffers fresh for each transfection and check the pH.
4. The precipitation mixture should not sit in the tube for more than 15 min before adding it to the dishes. Longer times make the precipitates coarser, which tends to damage the cells.
5. Handle with care: most of the cells after transfection tend to look "rounder" than before they are transfected and can easily be detached from the dish. An additional day in culture will not aid in the recovery of the transfected cells; instead, the un-transfected cells generally grow faster than the transfected cells and propagate to a high density.
6. Do not rinse the cells on the dish; otherwise, many will float and be lost in the rinsing process.
7. The cell suspension in PBS can sit on ice for up to 30 min.
8. At this point, almost all of the cells should be lysed, and the vast majority of the cell nuclei should be intact. To verify that the cells are lysed, add 5 μl of the cell lysate to a glass slide, add a cover slip, and then view under a light microscope under phase contrast. Checking the first and the last samples is sufficient to confirm lysis.
9. After the restriction enzyme digest, the linearized plasmids need to be gel-purified to avoid contamination with nonlinearized plasmids.
10. Probes ARE.A50 and A50 need to end with a string of adenosines to allow the deadenylases to work. A transcription template made by linearizing plasmids using a restriction enzyme (EagI in these two constructs) would frequently end with nucleotides other than As. Therefore, to ensure the RNA probes end with As, prepare transcription templates for the

ARE.A50 probe, or for the A50 probe, using a pair of PCR primers and plasmid pTNF1309–1332(A)50/SK-, or plasmid A50/SK-, as the PCR templates, respectively. Next, perform agarose gel purification and sequencing of the PCR product (primer: M13 Forward) to verify the presence of 50 As at the 3′ end.

11. To aid in the visualization of the bands and fluorescence marks, place films, with the gel on top, on a light box.
12. In most of our deadenylation experiments, we use cytosolic extracts prepared from HEK293 cells transiently transfected with plasmid constructs that express RNA-binding proteins (Figs. 2, 3, and 4a), using 10 mg of HEK293 total cytosolic protein in each reaction. The presence of this relatively high amount of protein often interferes with the sample migrating through the denaturing gel. Therefore, after the deadenylation reaction is complete, a phenol/chloroform extraction is often necessary prior to loading the samples into the gel. For each sample, use a 100 μl reaction volume and take out 50 μl of the aqueous phase of the phenol/chloroform extraction, avoiding the interface. In experiments using purified proteins, such as PARN and TTP (Fig. 4c), omit the phenol/chloroform extraction step, since the amount of protein used in these experiments is relatively small. Regardless of the protein source (either purified protein or from cell extracts), use 7.5% of the initial reaction per lane in the gel.

Acknowledgments

We thank Guang Hu and Mack Sobhany for constructive comments on the manuscript. This work was supported by the Intramural Research Program of the National Institute of Environmental Health Sciences, NIH. Panels shown in Figs. 2 and 3 were originally published in "Lai et al. [10]", reprinted by permission of Taylor & Francis Ltd, http://www.tandfonline.com. T&F Ref: la/tmcb/02894182

References

1. Sachs AB (1993) Messenger RNA degradation in eukaryotes. Cell 74:413–421
2. Hasan A, Cotobal C, Duncan CD et al (2014) Systematic analysis of the role of RNA-binding proteins in the regulation of RNA stability. PLoS Genet 10:e1004684
3. Goldstrohm AC, Wickens M (2008) Multifunctional deadenylase complexes diversify mRNA control. Nat Rev Mol Cell Biol 9: 337–344
4. Yi H, Park J, Ha M et al (2018) PABP cooperates with the CCR4-NOT complex to promote mRNA deadenylation and block precocious decay. Mol Cell 70:1081–1088
5. Smith BL, Gallie DR, Le H et al (1997) Visualization of poly(A)-binding protein complex

formation with poly(A) RNA using atomic force microscopy. J Struct Biol 119:109–117

6. Mangus DA, Evans MC, Jacobson A (2003) Poly(A)-binding proteins: multifunctional scaffolds for the post-transcriptional control of gene expression. Genome Biol 4:223
7. Carballo E, Lai WS, Blackshear PJ (1998) Feedback inhibition of macrophage tumor necrosis factor-alpha production by tristetraprolin. Science 281:1001–1005
8. Stumpo DJ, Lai WS, Blackshear PJ (2010) Inflammation: cytokines and RNA-based regulation. Wiley Interdiscip Rev RNA 1:60–80
9. Virtanen A, Henriksson N, Nilsson P et al (2013) Poly(A)-specific ribonuclease (PARN): an allosterically regulated, processive and mRNA cap-interacting deadenylase. Crit Rev Biochem Mol Biol 48:192–209
10. Lai WS, Kennington EA, Blackshear PJ (2003) Tristetraprolin and its family members can promote the cell-free deadenylation of AU-rich element-containing mRNAs by poly (A) ribonuclease. Mol Cell Biol 23(11): 3798–3812
11. Carballo E, Cao H, Lai WS et al (2001) Decreased sensitivity of tristetraprolin-deficient cells to p38 inhibitors suggests the involvement of tristetraprolin in the p38 signaling pathway. J Biol Chem 276:42580–42587
12. Fabian MR, Frank F, Rouya C et al (2013) Structural basis for the recruitment of the human CCR4-NOT deadenylase complex by tristetraprolin. Nat Struct Mol Biol 20:735–739
13. Moon AF, Mueller GA, Zhong XJ et al (2010) A synergistic approach to protein crystallization: combination of a fixed-arm carrier with surface entropy reduction. Protein Sci 19:901–913

Chapter 12

Transcriptome-Wide Analysis of mRNA Adenylation Status in Yeast Using Nanopore Sequencing

Pawel S. Krawczyk, Agnieszka Tudek, Seweryn Mroczek, and Andrzej Dziembowski

Abstract

There are multiple methods for studying deadenylation, either in vitro or in vivo, which allow for observation of mRNA abundance or poly(A) tail dynamics. However, direct RNA sequencing using the Oxford Nanopore Technologies (ONT) platform makes it possible to conduct transcriptome-wide analyses at the single-molecule level without the PCR bias introduced by other methods. In this chapter, we provide a protocol to measure both RNA levels and poly(A)-tail lengths in the yeast *Saccharomyces cerevisiae* using ONT.

Key words Poly(A) tail, Adenylation, Deadenylation, Nanopore sequencing, Direct RNA sequencing

1 Introduction

Eukaryotic RNA 3′ ends are ubiquitously polyadenylated, and poly (A) tails are crucial to defining transcript metabolism in the nucleus and cytoplasm. In budding yeast (*Saccharomyces cerevisiae*), there are two main sources of polyadenylation: canonical Pap1 and the Trf4 and Trf5 non-canonical poly(A)-polymerases. Pap1 enzyme is part of the conserved Cleavage and Polyadenylation Factor complex (CPF), which terminates the transcription of protein-coding transcripts (messenger RNAs—mRNAs). The other sources of polyadenylation are the Trf4 and Trf5 proteins, which produce 4–40 adenosine-long tails on noncoding RNAs and form two isoforms of the nuclear TRAMP complex (Trf4/5-Mtr4-Air1/2). While Pap1-produced poly(A) tails ensure mRNA stability in the nucleus and export [1], Trf4/5-mediated adenylation directs the

Pawel S. Krawczyk and Agnieszka Tudek contributed equally to this work.

Eugene Valkov and Aaron C. Goldstrohm (eds.), *Deadenylation: Methods and Protocols*, Methods in Molecular Biology, vol. 2723,
https://doi.org/10.1007/978-1-0716-3481-3_12,

RNA to decay or processing by the major nuclear 3′-5′ exonuclease complex, the exosome [2, 3].

The classical mRNA decay model in the cytoplasm postulates that degradation of the transcript depends on prior poly(A) tail shortening, termed deadenylation. Two conserved deadenylase complexes exist: Pan2/3, which targets mRNAs of high abundance most efficiently, and Ccr4-NOT, which is preferentially recruited to mRNAs of low abundance [4]. Both complexes bind the cytoplasmic poly(A)-binding protein Pab1, but only for Pan2/3 this interaction is the main means of recruitment to the mRNA, whereas Ccr4-NOT interacts with a collection of sequence-specific cofactors, Puf1-5, likely explaining the difference in substrate specificity [5, 6].

Deadenylases and their cofactors play an important role in adjusting the transcriptome to changing growth conditions. Many of those processes are still poorly understood, partly because, over the years, few techniques have been developed to simultaneously measure transcriptomic RNA levels and poly(A) tail lengths [7]. Most of the published transcriptomic data only addressed RNA levels and were produced using either, now obsolete, microarrays or Illumina short-read RNA sequencing. Although there are Illumina-based protocols that allow for poly(A) tail length determination [8–10], they are limited by sequencing read length. In addition, PCR amplification steps during library synthesis will inevitably introduce a bias preferentially amplifying short poly(A) tails, thus further limiting the length of the poly(A)-tail that can be observed. To this end, single-molecule technologies like PacBio or nanopore sequencing by Oxford Nanopore Technologies (ONT) can be used to overcome some of these issues.

In this chapter, we describe the application of nanopore sequencing to study poly(A) tail lengths. Direct RNA sequenceng (DRS) it is the only currently available technique to read native RNA molecules, without reverse transcription or PCR steps. This provides a clear technical advantage over Illumina, which is, however, lessened by the higher cost of the protocol, lower read coverage, and, as of today, the lack of official multiplexing. It was successfully applied for studying poly(A) lengths in various organisms [11–14], including yeast [4, 15]. Herein, we focus on the protocol employing nanopore sequencing, which we developed to study poly(A) tail lengths transcriptome-wide in *Saccharomyces cerevisiae*.

2 Materials

2.1 An Example Dataset

Steady-state and dynamic measurement of RNA levels and poly(A)-tail length are the two possible types of experiments in which DRS data can be used. In this protocol, we focus on dynamic profiles of mRNA abundance and poly(A)-tail lengths during the response to heat stress in a wild-type strain (follow up of [4], unpublished dataset) (*see* **Note 1**).

Dataset required to proceed with the final part of the protocol is available for download from GitHub (http://github.com/LRB-IIMCB/yeast_deadenylation_methods_chapter/). It was processed using *S. cerevisiae* and *Schizosaccharomyces pombe* transcriptomes available from Ensembl (https://ftp.ensemblgenomes.org/pub/fungi/release-55/fasta/saccharomyces_cerevisiae/cds/ and https://ftp.ensemblgenomes.org/pub/fungi/release-55/fasta/schizosaccharomyces_pombe/cds/, respectively). The GitHub repository also contains the RMarkdown document showing all the data analysis steps regarding poly(A) length profiles, the code used, and the expected output.

2.2 Yeast Strains and Media

1. *Saccharomyces cerevisiae* strains of interest (W303, BY, or other genetic backgrounds).
2. Optional: *S. pombe* wild-type strain can produce a spike-in to normalize the sequencing data (*see* **Note 2**).
3. Growth medium choice will depend on the experiment planned; the standard rich medium is YPDA, which can be modified to change the sugar level or identity (glucose, galactose, and raffinose). For certain applications, such as metabolic labeling, a minimal synthetic medium will be preferred. However, as the medium composition can greatly impact mRNA abundance and poly(A) tail length distribution [4], we recommend planning any additional experiments to allow comparison with other datasets. All media are produced according to standard protocols.
4. Depending on the application, various drugs will be used (like those inducing transcriptional arrest: thiolutin—CAS 87-11-6 or 1,10-phenanthroline—CAS 66-71-7). The concentration should always be consulted with standard protocols or independently tested.

2.3 RNA Extraction and Preparation for Library Synthesis

1. Phenol solution saturated with 0.1 M citrate (pH 4.3).
2. TES buffer: 10 mM Tris–HCl (pH 7.5), 5 mM EDTA, 1% w/v SDS (sodium dodecylsulfate).
3. Chloroform.
4. 2 M LiCl
5. Ethanol, methanol.
6. Nuclease-free water.
7. (optional) *S. pombe* total RNA (*see* **Note 2**)

2.4 Nanopore Direct RNA Sequencing

1. oligo-dT$_{(25)}$ beads.
2. KAPA Pure Beads (Roche).
3. Direct RNA sequencing kit (SQK-RNA002, ONT) (*see* **Note 3**).
4. SuperScript™ III reverse transcriptase (Invitrogen).
5. T4 DNA ligase (highly concentrated, 2 k U/ul).
6. Nanopore sequencing device. Either MinION, GridION, or PromethION (ONT).
7. Nanopore sequencing flow cell (*see* **Note 4**).

2.5 Computational Software

It is assumed that all analysis steps are run in the Linux environment. Installation instructions can be found on the documentation websites for each program.

1. Guppy base-caller, proprietary software available from ONT.
2. Rstudio (Posit, https://posit.co/download/rstudio-desktop/).
3. nanopolish (https://github.com/jts/nanopolish).
4. (optional) tailfindr (https://github.com/adnaniazi/tailfindr).
5. minimap2 (https://github.com/lh3/minimap2).
6. samtools (https://github.com/samtools/samtools).
7. nanotail (https://github.com/LRB-IIMCB/nanotail).

3 Methods

3.1 Cell Culture and Sample Collection

Regardless if steady-state and dynamic experiments are performed, one can be interested in the total RNA fraction or a subfraction of the transcriptome produced or degraded within a given time frame. The latter is typically achieved either during the response to stimuli or by metabolic labeling using 4-thiouracil or 4-thiouridine. Various metabolic labeling protocols were described [16–19].

When preparing samples, special care should be taken to maintain equal conditions, especially regarding temperature and media composition, since both significantly impact poly(A) tail lengths [4]. The protocol below specifies how to perform a heat stress chase and matches the way the DRS samples of the test datasets related to this protocol were prepared.

1. Grow cells at 25 °C overnight.
2. Measure OD in the morning, dilute strains to OD 0.075–0.12, and allow to produce a minimum of 3 generations at 25 °C. Ensure a sufficient culture volume to collect all the samples in the desired dataset with a 15% surplus. A minimum of 10 ml of culture at OD between 0.3 and 0.6 should be collected for one

sample. Remember not to overgrow the culture, especially past OD 1, which will initiate entry of the cells into the stationary phase, changing the entire cell metabolism.

3. 1 h before the experiment, preheat an equal volume (regarding the culture) of YPDA to 51 °C in a water bath. At least 1 h before the experiment, preheat a separate water bath to 37 °C.
4. 30 min before, prepare several 50 ml tubes that match the number of samples that will be collected. Fill them with 20 ml of ethanol or methanol and place them on ice (*see* **Note 5**).
5. When the culture obtains the desired OD, transfer a multiple of 10 ml volume per sample with one half to one sample in surplus to a fresh Erlenmeyer flask preheated to 25 °C.
6. Collect a control sample of wild-type cells grown at 25 °C.
7. Add an equal volume of YPDA preheated to 51 °C to the culture, start the timer, mix, and place the culture in a water bath at 37 °C.
8. Collect 20 ml of culture at the desired time and transfer it to 20 ml of precooled alcohol. Mix and place on ice. (*see* **Note 5**).
9. Centrifuge samples for 2 min at 3220 rcf.
10. Discard the supernatant, resuspend the cell pellet in 1 ml of sterile deionized water, and transfer to a 1.5 ml tube.
11. Centrifuge the samples using a bench centrifuge until the cell pellet settles.
12. Wash the pellet again in 1 ml of sterile deionized water.
13. Discard the supernatant and flash freeze the pellet in liquid nitrogen or dry ice.
14. Extract RNA or store cell pellets at −80 °C until RNA extraction can be performed.

The protocol can be modified to measure response to certain drugs that induce transcriptional arrest or result in protein depletion (*see* **Note 6**).

3.2 RNA Extraction

Special care should be taken while handling RNA to avoid degradation. Samples should be kept on ice unless otherwise stated. Use ribonuclease-free plastics, filter tips, and nuclease-free water, and wear gloves.

The RNA samples from *S. cerevisiae* and *S. pombe* for experiments cited in this protocol were prepared using the hot acid phenol method, which is given below. However, other methods (such as GTC/phenol) can be considered in some cases (e.g., cell starvation experiments [20]).

1. Resuspend cell pellet in 400 μl phenol solution saturated with 0.1 M citrate (pH 4.3).
2. Add 400 μl of TES buffer.
3. Vortex samples for 45 min at 65 °C and then centrifuge at 4 °C for 10 min.
4. Collect the supernatant into a fresh tube and add 400 μl phenol solution saturated with 0.1 M citrate (pH 4.3). While collecting the supernatant, do not touch the interphase, as this might contaminate the RNA sample with DNA, proteins, or phenol.
5. Vortex samples for 20 min at 65 °C and then centrifuge at 4 °C for 10 min.
6. Collect the supernatant into a fresh tube and add 400 μl of chloroform.
7. Vortex the samples briefly at room temperature and centrifuge them at 4 °C for 10 min.
8. Collect the supernatant into a fresh tube and add 45 μl of 2 M LiCl and 1 ml of 95% ethanol.
9. Precipitate samples for at least 30 min or overnight at −80 °C (RNA samples can be safely stored indefinitely at this stage).
10. Centrifuge for 25 min at 4 °C.
11. Discard the supernatant and wash the pellet with 400 μl of 80% ethanol.
12. After removing the supernatant, dry pellets at 37 °C.
13. Pellets should be resuspended in nuclease-free water.
14. Measure the RNA concentration using Nanodrop. Set the software to RNA detection. First, use 1–1.5 μl of deionized water as a blank control. Then, separately measure RNA concentration using 1–1.5 μl of each sample. Generally, start with the samples expected to have the lowest RNA concentration and proceed incrementally. Between consecutive measurements, thoroughly clean the pedestal and the cap with a soft tissue soaked with deionized water. The RNA quality and purity can be assessed from the 260/280 and 260/230 ratios (*see* **Note 7**).

RNA prepared from *S. pombe* can be used as a spike-in for *S. cerevisiae* samples. Alternatively one can also synthesize a collection of artificial spike-ins (*see* **Note 2**), requiring additional validation. External spike-ins are added to the total RNA before further purification steps (see below) or DRS library synthesis.

3.3 Library Preparation and Sequencing

A DRS library can be synthesized using total RNA (*see* **Note 8**). However, we recommend enriching the sample for polyadenylated transcripts on oligo-$dT_{(25)}$ magnetic resin using one of the many commercially available kits or purifying the capped RNA fraction (*see* **Note 9**). For studies involving rapid changes to transcription or decay, sometimes metabolic labeling with 4-thiouridine (4tU/4sU) is performed (*see* **Note 10**). There is no formal possibility to barcode mRNA samples and analyze them in a single run (*see* **Note 11**). However, it is possible to mix mRNA from different evolutionary distant organisms and synthetic IVT mRNA (*see* **Note 12**).

In general, we use up to 5 μg of total RNA or 0.5 μg cap-enriched mRNA from higher eukaryotes (murine or human) mixed with 50–200 ng of poly(A)+ isolated from yeast and 50 ng of spike-ins, which in aggregate results in a higher input than that recommended by ONT (50 ng of poly(A)+ or 500 ng of total RNA) for library preparation. Please note that the sequencing kit does not contain all the required reagents (e.g., T4 DNA ligase, SuperScript™ III reverse transcriptase), which should be ordered separately. Sequencing libraries should be prepared according to protocols available from ONT, with minor modifications. We recommend extending all ligation steps to 30 min and two rounds of the final cleaning of the library on magnetic beads (KAPA Pure Beads), to get rid of short RNA and DNA species (including adapters).

As nanopore sequencing is still being dynamically developed, and there are frequent modifications of the protocols and updates of software (MinKNOW) controlling the sequencing process, we do not provide the exact protocol for loading the prepared library on the nanopore flow cell and running the sequencing. All up-to-date information can be found on the Community Website provided by ONT (https://nanoporetech.com/community).

However, some good practices help to obtain satisfactory sequencing results:

1. When loading the library, try to avoid air bubbles, as they will destroy protein pores on a flow cell, consequently leading to reduced sequencing throughput.
2. Always check the flow cell before sequencing to ensure there are enough pores to obtain satisfactory data. Flow cells with >1400 active pores at the beginning of the run should provide sufficient data for subsequent analyses (but the ONT-guaranteed number of pores is 800). The number of active pores slowly decreases during sequencing. The typical output from a single continuous DRS run (72 h of sequencing) is one to three million reads.
3. Monitor the run using provided software (MinKNOW) to check the flow cell health status. A substantial number of sequencing pores characterizes a good sequencing run, slowly decreasing with time. Rapid (within first 2 h) decrease in the number of pores in the "sequencing" (or "strand") state suggests there are problems with the library.

4. Running real-time base-calling (getting the sequence out of raw nanopore signal) is not necessary, as off-site base-calling is required for subsequent analyses.

3.4 Base-Calling

Please note that we do not provide raw nanopore output files (fast5 files) in the example dataset. Such data can be downloaded from public repositories. Accession numbers for multiple yeast samples can be found in [4, 15].

Currently, for Direct RNA Sequencing, the only available base-caller-producing files compatible with subsequent poly(A) length analysis steps is Guppy (provided by ONT). However, other algorithms exist, developed mostly for base-calling of DNA sequencing runs. It should be noted that direct RNA sequencing and base-calling are still characterized by low accuracy (~92%).

Base-calling is the most computationally demanding part of the analysis and should be run on machines equipped with Graphics Processing Units (GPU). The example command for running base-calling of DRS data with Guppy version 6.0.0 is:

```
guppy_basecaller -i input_folder -s basecalling_folder--recursive --records_per_fastq 0 --fast5_out --flowcell FLO-MIN106 --kit SQK-RNA002 -x "cuda:all"
# -i input_folder - folder containing raw fast5 files (option--recursive will ensure that all sub-folders are processed as well)
# -s base-calling_folder - folder where base-called fastq and fast5 files will be stored
# --fast5_out - ensures that base-called sequences, in addition to fastq, will be stored in fast5 files (required for subsequent analyses)
# -x "cuda:all" - base-calling will be run on all available GPU devices. Change to -x "cuda:n" where n={0,1,2,..} to select single device
# --flowcell FLO-MIN106 and --kit SQK-RNA002 - specify the types of flowcell and sequencing kit which were used, respectively. All possible types can be listed with guppy_basecaller --print_workflows
```

Upon completion, the basecalling_folder will contain:

- sequencing_summary.txt—a text file containing basic information about each sequencing read (with its location in fast5 files).
- pass folder—containing fastq sequence files with reads passing quality filters (fail folder, analogously, contains reads below the quality threshold, *see* **Note 13**).
- workspace—folder containing base-called fast5 files (*see* **Note 14**).

3.5 Mapping

Reads must be mapped to a reference to allow their identification. Choosing a proper transcriptome (*see* **Note 15**) reference is an important prerequisite for obtaining good-quality results.

It is advised to use Minimap2 [21] for mapping DRS reads to a reference transcriptome. As after base-calling multiple fastq files are produced in both pass and fail folders, they must be merged into a single file before further analysis steps. This can be done using standard Linux commands, e.g.,

```
cat *.fastq > merged.fastq
```

Example minimap2 mapping command:

```
minimap2 -ax map-ont -k 14 –secondary=no -t 30 index
reads | samtools view -b -F 2320 -@ 30 | samtools
sort -o mapping.bam -@ 30
# index – reference sequences in fasta
(or previously created minimap2 index file)
# reads – single (merged) fastq file with the base-
called reads
# --secondary=no – output only best hit for each
read
# -k 14 is the kmer size, allowing better mapping
accuracy of noisy DRS reads
# samtools view -b -F 2320 – produce a bam file, with
all reads mapped to a reverse strand filtered out
```

This will produce a bam file (mapping.bam) with alignments sorted by position in the reference file. As DRS is by default strand-specific, reads mapping to the reverse strand can be safely removed as a possible artifact (unless possible antisense transcripts are considered).

Please note that in the case of libraries containing yeast RNA mixed with material from other species (*see* **Note 12**), the output will contain a large fraction of unmapped reads.

After mapping the output file should be indexed:

```
samtools index mapping.bam
```

3.6 Determination of Poly(A) Lengths

There are currently two principal algorithms for the determination of poly(A) lengths from nanopore sequencing data: nanopolish polya [16] and tailfindr [17]. They use different underlying algorithms to find poly(A)-specific regions in raw signal from nanopore sequencing and also differ in terms of computational efficiency (*see* **Note 16**), but results obtained with both methods correlate well [22]. The protocol for using tailfindr is provided elsewhere [23]. For subsequent analyses, we will use nanopolish polya.

1. Index reads using the command nanopolish index.

```
nanopolish index -d folder containing fast5 files
-s sequencing_summary_file reads_fastq
# -d Folder containing fast5 files is usually "work-
space" located inside the folder containing base-
called data
# -s sequencing_summary_file - sequencing_summary.
txt can be found inside the folder containing base-
called data
# reads.fastq - the same file containing base-
called reads which was used for mapping
```

2. Run poly(A) estimator:

```
nanopolish polya -b bam_file -r reads.fastq -t 20 -g
genome_used_for_mapping > output_file
# -b bam_file mapping.bam, produced in section
3.4.2
# -r reads.fastq the same file containing base-
called reads which were used for mapping
# -g genome_used_for_mapping - fasta file with re-
ference sequences which were used for mapping in
section 3.4.2
# -t 20 - number of parallel processing threads
(adjust number to capabilities of your hardware)
```

As a result, a tab-separated (tsv) file is produced, containing information about all sequencing reads. The most important columns included in the file are as follows: read_id (unique identification of sequencing read), contig (assigned transcript id, derived from reference), polya_length (calculated length of poly(A) tail), and qc_tag (category summarizing reliability of poly(A) length calculation, *see* **Note 17**).

3.7 Poly(A)-Tail Length Data Analysis

Next, it is possible to continue with statistical data analysis, as all sequencing reads already have poly(A) lengths called with nanopolish. As this is direct RNA sequencing, each read represents a single transcript present in the library. At this point, there are multiple routes of analysis possible, depending on the experimental setup chosen at the beginning. Here we will cover only the basic analysis possible with obtained data. All subsequent steps are performed in the R environment and require the installation of the nanotail package, which we developed to ease the analysis of polyadenylation data coming from nanopore sequencing.

1. Please clone the GitHub repository containing the example dataset.

```
git clone https://github.com/lrb-iimcb/yeast_dea-
denylation_methods_chapter
```

2. Open Rstudio and set the working directory to the one with cloned data.

```
setwd("/path/to/yeast_deadenylation_methods_chap-
ter")
```

3. Load the required libraries (ensure they are installed).

```
library(nanotail)
library(tidyverse)
library(ggplot2)
```

4. Load the example metadata tables (*see* **Note 18**) for both *S. cerevisiae* and *S. pombe* spike-in data into the R session using:

```
metadata <- read.table("metadata_heat.csv",
sep=",",header=T)
metadata_pombe <- read.table("metadata_heat_pombe.
csv",sep=",",header=T)
```

5. Data are loaded into single data.frame using the read_polya_-multiple() function from the nanotail package (*see* **Note 19**), taking the metadata table as an argument.

```
polya_data_table <- read_polya_multiple(metadata)
# set a proper order of samples (groups) for plot-
ting polya_data_table$group <-
factor(polya_data_table$group,levels=c
("t0","t2","t6","t10","t18"))
polya_data_table_pombe <- read_polya_multiple(me-
tadata_pombe)
```

6. Basic QC of analyses can be done using.

```
nanopolish_qc <-
get_nanopolish_processing_info(polya_data=polya_-
data_table,grouping_factor = "sample_name")
# polya_data is the table loaded in point 5.
# grouping_factor can be any of the metadata columns
in the metadata table. Setting to NA will produce a
summary of the whole dataset
plot_nanopolish_qc(nanopolish_processing_info=na-
nopolish_qc,frequency=F)
```

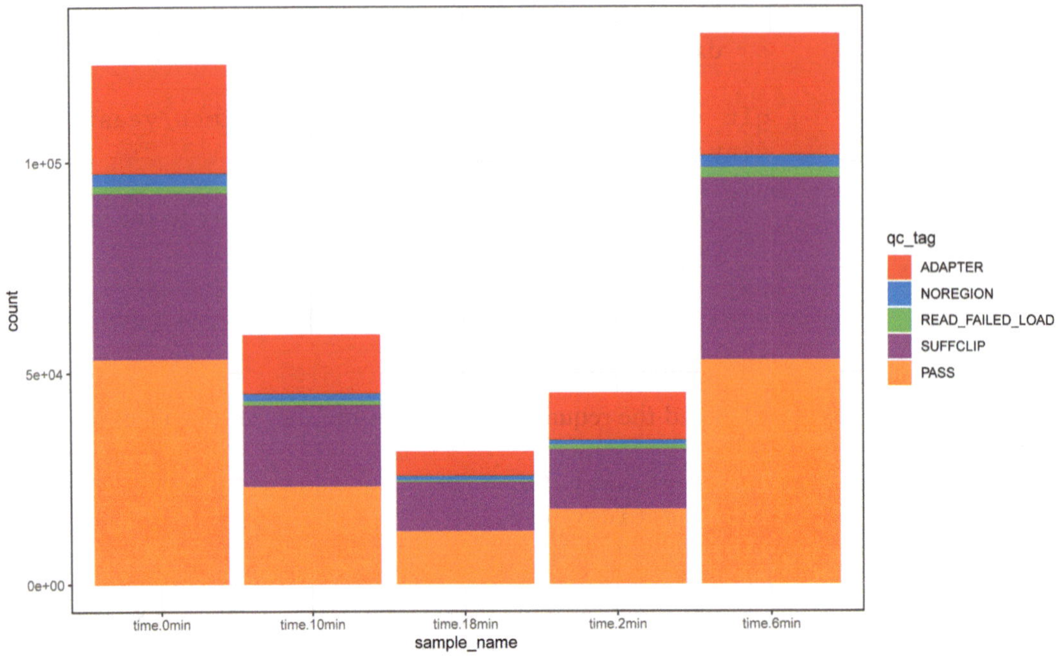

Fig. 1 QC plot of nanopolish polya analysis. All samples are shown as separate bars, with nanopolish polya qc_tag represented as colors. The height of each bar indicates the number of reads in each sample and category

This will produce a plot showing the number of reads in each dataset and QC categories assigned by nanopolish (Fig. 1; please *see* Subheading 3.6 and **Notes 17** and **20** for details).

As we can see, a large proportion of reads are assigned to the SUFFCLIP category. This is due to the reference lacking UTR sequences (*see* **Note 15**). As this should not bias poly(A) length estimations, reads with such a category can be safely included in the further processed dataset. However, caution should be taken every time such a QC plot shows a large proportion of categories other than PASS (*see* **Note 20**).

7. In the next step, reads that failed internal nanopolish QC (assigned to categories NOREGION, READ_FAILED_PASS, or ADAPTER) can be removed. It can be done using dplyr package:

```
polya_data_filtered <- polya_data_table %>%
dplyr::filter(qc_tag %in% c("PASS","SUFFCLIP"))
```

8. Next, we can show global distribution of poly(A) tail lengths on a density plot using the plot_polya_distribution() function (*see* **Note 21**):

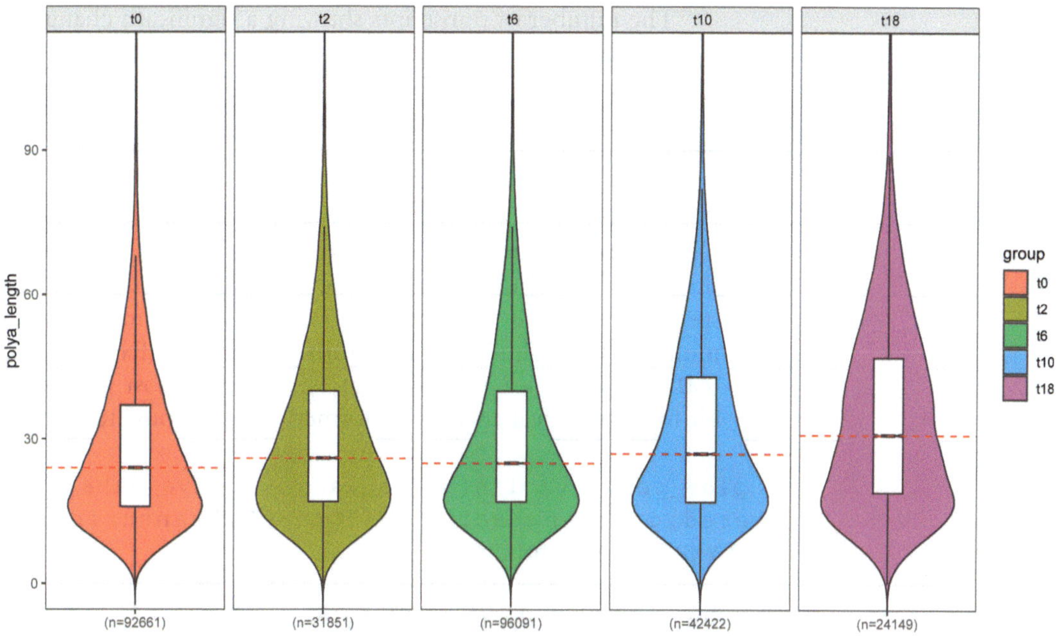

Fig. 2 Distribution of poly(A) lengths in each experimental condition represented as a violin plot. Please note increased poly(A) lengths across time points following heat-shock

```
plot_polya_distribution(polya_data=polya_data_-
filtered, groupingFactor="group",show_center_va-
lue="median")
```

For multiple samples, it is better to use a violin plot instead (Fig. 2):

```
plot_polya_violin(polya_data = polya_data_fil-
tered, groupingFactor = "group",add_boxplot=T)
```

9. Next, find transcripts with significant changes in poly (A) lengths between conditions using the kruskal_polya() function. It uses the Kruskall-Wallis test for comparison between multiple groups:

```
polya_data_stats<-kruskal_polya(input_data=po-
lya_data_filtered, grouping_factor="group", tran-
script_id="transcript")
```

The result table can be reviewed with the command:

```
View(polya_data_stats %>% dplyr::filter
(padj<0.05))
```

The number of transcripts showing a significant change in poly(A) lengths can be assessed with:

```
polya_data_stats %>% dplyr::filter(padj<0.05) %>%
dplyr::count()
```

10. Poly(A) lengths for individual transcripts can be viewed using plot_polya_distribution() or plot_polya_violin() functions, similar to global distribution. The transcript identifier can be specified with the "transcript_id" argument and may contain multiple transcript IDs in a form of vector. Example for *YMR251W-A_mRNA* transcript, which is the most significant hit from statistical analysis performed in a previous step:

```
plot_polya_violin(polya_data = polya_data_fil-
tered,groupingFactor = "group",transcript_i-
d="YMR251W-A_mRNA")
```

11. Next, calculate the per-transcript summary of poly(A) length distributions. This will produce mean, median, and quantile (0.03, 0.05, 0.95, 0.97) values for each transcript at each condition. Such numeric values can nicely show changes in poly(A) length between analyzed conditions as they represent poly(A) length dynamics with better sensitivity than just looking at mean or median values (*see* **Note 22**).

```
polya_summary <- summarize_polya_per_transcript
(polya_data = polya_data_filtered,group-
By="group",transcript_id="transcript",quantiles=c
(0.03,0.05,0.95,0.97),summary_functions=c
("mean","median")
```

Obtained values can be plotted to show changes in selected transcripts. Quantiles distribution can be plotted with the function plot_quantiles() (Fig. 3).

```
plot_quantiles(summarized_data=polya_summary,
transcript_id="YMR251W-A_mRNA",groupBy="group")
```

To look at transcripts' abundance, it is necessary to normalize counts to sequencing depth (or to *S. pombe* spike-in or other spike-in if used). For this purpose, please use:

```
polya_summary_normalized <-
normalize_counts_to_depth(summarized_data=polya_-
summary,raw_data=polya_data_table,spike_in_data=-
polya_data_table_pombe,groupBy="group")
```

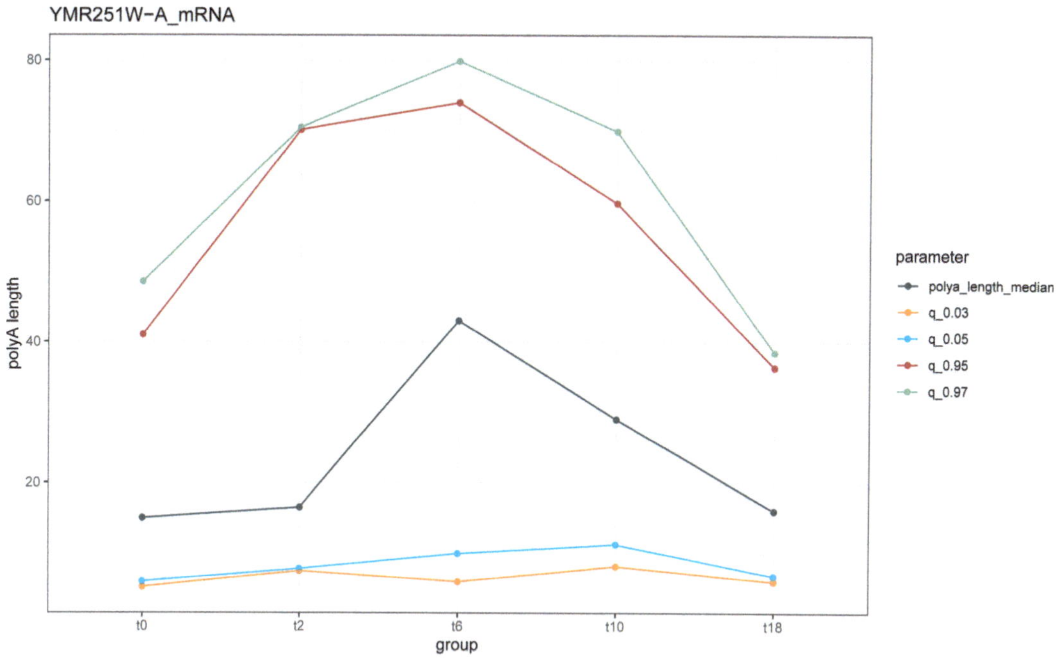

Fig. 3 Poly(A) lengths dynamics of YMR251W-A_mRNA transcript. Poly(A) lengths dynamics of YMR251W-A_mRNA transcript represented as changes in quantile values

This will produce a separate table containing the norm_counts column with normalized expression values (*see* **Note 23**). Obtained data can be used to analyze the relationship between transcript abundance, poly (A) lengths, and their dynamics.

All obtained values can be useful for data interpretation, but only if the appropriate number of reads is provided for each transcript. We propose that such a cut-off should be at least 10 reads for mean and median values and 20 reads for mode and quantile values. Many solid conclusions can be obtained when analyzing mean and median value changes. The mean will be more sensitive to scarce but extremely long poly(A) tails and can show more differences between strains or conditions, but one must bear in mind that those differences are more solid if also seen in median values. Budding yeast is an exceptional species that does not contain cytoplasmic adenylases. This means the poly(A) tail can only be shortened or remain stable after de novo synthesis. In principle, this allows for modeling deadenylation rates by comparing changes to either the mean/medians or the quantile values. As in the case of decay rates, this can be done by fitting function parameters into experimental data, as proposed for decay previously [19, 24].

The advantage of DRS is that this method simultaneously provides information about mRNA levels and poly(A) tail lengths. As standard normalization of the per-transcript read counts to sequencing depth in some cases (e.g., metabolic labeling or global change in mRNA abundance) can lead to bias in abundance estimates (*see* **Note 24**), we suggest providing a proper spike-in (e.g., *S. pombe* RNA) which can be used at the step of RNA enrichment by oligo-dT to provide help for normalization of the DRS datasets. The spike-in will also be useful to control for substantial down- or up-regulation of certain transcripts. Simultaneous comparison of changes in mean/median poly(A) tail lengths and fold-change in RNA abundance can help to interpret the data. For example, a strong decrease in poly(A) tail length and a decrease in mRNA abundance indicate transcript decay. Conversely, increasing both factors could suggest increased RNA stability or de novo transcription.

4 Notes

1. Another dataset with steady-state profiles of mRNA abundance and poly(A)-tail lengths in *ccr4Δ* and *pan2Δ* mutant strains was already published [4]. The basic processing protocol is essentially the same as presented here.
2. Each sequencing run will have a different depth, and this technical bias will prevent comparing samples directly. One workaround is the normalization of transcript levels by the size of the sequencing library, which can still be insufficient when significant transcriptome reconstruction occurs. An alterative approach is to include an external RNA spike-in, which can be introduced using two methods. First, when the sample of interest derives from budding yeast, at the step before sample collection, the culture can be mixed with an equal amount of *S. pombe* cells [19, 25]. One must, however, bear in mind that the media used for the growth of both species has to be similar not to induce any response to changing growth conditions. The second method is to introduce the spike-in at the level of purified RNA. This RNA will be added in equal amounts to each sample in the experiment to allow cross-comparison of transcript abundance. The spike-in can be an in vitro transcribed RNA [20, 26] or total RNA isolated from a different organism. In our experiments, we used *S. pombe* total RNA mixed with the sample of interest at a ratio of 1:100 w/w (1 μg total spike-in RNA: 100 μg total RNA of interest).
3. Both direct RNA and cDNA sequencing kits are available from ONT. cDNA kits (like SQK-PCS111) require significantly less amount of input material, but PCR steps can lead to the shortening of poly(A) tails. Additionally, only one software package can currently be used for poly(A) length determination from standard cDNA sequencing protocol (tailfindr). There are

also alternative protocols allowing for simultaneous analysis of polyadenylated transcripts and those lacking poly(A) tails [27, 28], which require additional pre-processing. T4 DNA ligase and reverse transcriptase are not included in the ONT kits and should be purchased separately.

4. Currently, only pore version 9 flowcells (FLO-MIN106 or FLO-FLG001 for MinION and GridION devices, FLO-PRO002 for PromethION) are compatible with direct RNA sequencing and some cDNA sequencing kits (like SQK-PCS111). However, nanopore technology undergoes constant development and novel sequencing chemistries are being developed. Thus, it is advised to always compare with the official ONT website.

5. Mixing the cell culture with high concentrations of cold alcohol, methanol or ethanol, will instantaneously freeze cellular metabolism, ensuring that the RNAs are preserved intact in the state at which they were at the time or growth state of interest. This also allows the collection of multiple samples in a relatively short time frame, with reduced technical bias. The alcohol can be pre-cooled on ice or dry ice, and the fixed sample can be kept on ice before cell pellet collection, but no longer than 20 min. This is especially important for samples produced in YPDA since prolonged incubation in alcohol causes precipitation of the media, which later impairs RNA concentration measurements.

6. Instead of shifting cells to a higher temperature, one adds the desired drug and collects the samples in the time course. It is important to ensure sufficient drug concentration or test it independently before producing samples for sequencing. Two potent protocols exist to rapidly deplete the protein of interest from a given cellular compartment, usually the nucleus (the Anchor Away [AA] system [29]), or to remove it altogether by inducing its decay by the proteasome (the Auxin-Inducible Degron [AID] system [30]). Both require a specific genetic background, deriving from the W303 wild-type strain. The AID system is closer to a wild-type situation as it bears only an insertion of the *Oryza sativa TIR1* gene. At the same time, the AA background provides a mutation in the *TOR1* gene (*tor1-1*), which is key to ensuring the background strain survival on rapamycin. The AA system should not be used when studying the mTOR pathway not only due to the *tor1-1* mutation, but also for using rapamycin (CAS 53123-88-9). To deplete the factor of interest, older protocols would also suggest placing the gene of interest under a galactose or tetracycline-inducible promoter, which is less recommended as the basal expression of the factor of interest could be increased, and the phenotype may take longer to manifest itself.

7. For pure RNA samples, the 260/280 ratio should oscillate around 2 and not exceed 2.4. Lower values, equal to 1.8, will indicate contamination with DNA and suggest that additional DNase treatment is required. The 260/230 ratio should be between 2.0 and 2.2. Lower values indicate contamination of the sample with phenol or other organic compounds originating from the RNA extraction procedure. For accurate measurement the sample concentration should not exceed 1 μg/μl. Therefore, for samples that have higher concentrations, we recommend preparing dilutions. If RNA enrichment is performed (poly(A)+ selection or 4tU-labeling), remember to measure both the input and the poly(A)+ fraction, note down all volumes, and calculate the recovered RNA fraction (weight of poly(A)+/total). This should be used to verify the end efficiency of purification and reproducibility.
8. Preparing the library without poly(A) enrichment on oligo-$dT_{(25)}$ usually leads to lower output obtained from the single library. This is due to the presence of non-polyadenylated RNA species, which lowers the efficiency of adapters' ligation and leads to nanopore blocking due to secondary structures.
9. The human eIF4E-K119A mutant [31], which is commonly used to enrich human samples for coding transcripts [12], does not bind yeast RNAs. To this end, another enrichment strategy based on the human cap 0-binding properties of the IFIT1 protein was developed and validated [32].
10. Metabolically labeled RNA (e.g., with 4tU/4sU) is typically a suitable substrate for the DRS library synthesis without additional purification.
11. For typical analysis, 0.3–0.5 million reads for a single sample are enough to describe RNA abundance/poly(A) lengths. There is no official multiplexing approach for direct RNA sequencing, so multiple samples cannot be sequenced together. If such multiplexing is desired one can try the unofficial (not supported by ONT) approach [33], but processing of such generated data is not covered in this protocol. Instead of multiplexing, the same flow cell can be used multiple times (assuming the presence of active pores) to sequence additional libraries.
12. In such cases, special care should be taken to ensure that organisms are evolutionarily distant, so sequencing reads can be easily assigned to given organisms by simply mapping to organism-specific reference sequences. We recommend cross-comparing the two organisms′ transcriptomes before experimenting.

13. Those reads are usually very short and rather not suitable for subsequent analyses. However, when needed they still can be used, with caution about their quality.
14. Unless it is announced by ONT that fast5 files will not be written by production base-callers (like Guppy) anymore. In such case, it will be required to use separate software provided by ONT (not described in this protocol), or older versions of Guppy.
15. Read mapping to the transcriptome, rather than the genome, ensures their better identification. The yeast genome is well-annotated and not expected to contain unknown protein-coding genes. If one needs to include noncoding RNAs in the analysis, they can be incorporated into existing annotations [4]. Please note that most yeast transcriptome references do not include UTR sequences (*see* **Note 17** and point 6 in Subheading 3.7 for additional comments).
16. Tailfindr is implemented in R, whereas Nanopolish is written in C++, which is usually much quicker. Also, the HMM algorithm used by Nanopolish is more computationally efficient than signal thresholding applied in tailfindr. In our hands, nanopolish polya performed significantly better, but the choice is up to the user.
17. Qc_tag column of nanopolish polya output summarizes the reliability of poly(A) lengths calculation for each read. Possible categories include PASS—read with correct segmentation, reliable calculation of poly(A) length, ADAPTER—the read was stuck for too long at the adapter part, so the read rate estimation, and thus poly(A) length calculation may be unreliable, SUFFCLIP—3′ terminal part of a read (before poly(A) tail) was not mapped to a reference, so the assignment to a reference may be inaccurate, NOREGION—no poly(A) region was found in the raw signal, READ_FAILED_LOAD—there was an issue with reading raw data from the fast5 file.
18. Metadata for multiple samples can be stored in data.frame object, with the required columns: polya_path (containing the path to the nanopolish output file) and sample_name. Additional columns may contain supplementary information describing experimental conditions and will be included in the analysis table loaded in the next step of the protocol.
19. There is also a read_polya_single() function, which loads data from the single experiment to the environment.
20. If there is a large proportion (>5%) of READ_FAILED_LOAD, this means there were issues with raw data accessibility. After checking that all raw fast5 files are accessible, nanopolish analysis should be run again. There should not be many NOREGION reads as the library preparation protocol selects poly

(A)-containing RNAs. Such a category usually means that the sequencing read was incorrectly processed by the sequencing software (MinKNOW) and should be discarded from the analysis. A large proportion of ADAPTER reads means that there are problems with the sequencing itself and, even if such reads could be considered for further analysis, it is advisable to run the sequencing of such samples again.

21. If spike-ins were added to the library, their poly (A) distributions should also be examined to identify any technical issues that can bias the DRS output. Samples should be discarded for which *S. pombe* (or other spike-in) poly(A)-tail lengths strongly diverge.

22. Most yeast cellular poly(A) tail profiles are close to normal with the right arm of the distribution (designating mRNAs with poly(A) tail lengths longer than the median) being extended. For yeast the mean and median poly(A) tail lengths are between 20 and 40 adenosines. Since from other studies, we know that the newly made poly(A)-tail length is around 50–60 adenosines with the upper adenylation limit of 200 [4, 15], the mean and median values most likely represent RNAs deadenylated in the cytoplasm. Distribution of yeast poly(A) tail quantiles values will range between 50 and 60 for the upper quantiles (e.g., 75–99th), and as follows those values will represent newly made mRNAs, which might be nuclear or freshly exported to the cytoplasm. The 50th and neighboring quantiles will be similar to the mean and median. In contrast, the lower quantiles $5–20^{th}$ will represent old and strongly deadenylated mRNAs, for which the poly(A) tail lengths will be close to the DRS detection limit (around 5–10 adenosines). We routinely examine quantiles: 0.05, 0.10, 0.15, 0.50, 0.75, 0.80, 0.85, 0.90, and 0.95.

23. Another option is to use genome-mapped reads, count with other tools (like featureCounts), import counts to R, and normalize using software packages for expression analysis (like DESeq2 or EdgeR).

24. If large changes are consistently noted in the poly(A)+ RNA levels recovered from total in a certain condition (e.g., mutant) compared to wild-type, special care should be taken when the DRS library is normalized to sequencing depth. In this case, the use of external spike-ins is paramount to compare changes in RNA abundance. In the absence of spike-ins, a coefficient derived from the % of poly(A)+ RNA recovered from total (derived from concentration measurement) can also be used to normalize DRS libraries.

Acknowledgments

This work was supported by National Science Center [OPUS 17 UMO-2019/33/B/NZ2/01773 to A.D]. A.D. is also a recipient of the ERA Chair's position, funded by the EU [agreement no. 810425]. Work by AT was supported by National Science Centre Poland, SONATA grant [2020/39/D/NZ2/02174 to A.T.]. SM was supported by National Science Centre Poland, SONATA BIS 10 [2020/38/E/NZ2/00372 to S.M.].

Conflict of Interest PSK received the reimbursement of registration fees from ONT to present data at an event. The remaining authors declare no conflict of interest.

References

1. Dower K, Kuperwasser N, Merrikh H, Rosbash M (2004) A synthetic A tail rescues yeast nuclear accumulation of a ribozyme-terminated transcript. RNA 10:1888–1899. https://doi.org/10.1261/rna.7166704
2. Chlebowski A, Lubas M, Jensen TH, Dziembowski A (2013) RNA decay machines: the exosome. Biochim Biophys Acta 1829:552–560. https://doi.org/10.1016/j.bbagrm.2013.01.006
3. De Magistris P (2021) The great escape: mRNA export through the nuclear pore complex. Int J Mol Sci 22:11767. https://doi.org/10.3390/ijms222111767
4. Tudek A, Krawczyk PS, Mroczek S et al (2021) Global view on the metabolism of RNA poly (A) tails in yeast Saccharomyces cerevisiae. Nat Commun 12:4951. https://doi.org/10.1038/s41467-021-25251-w
5. Webster MW, Chen Y-H, Stowell JAW et al (2018) mRNA deadenylation is coupled to translation rates by the differential activities of Ccr4-not nucleases. Mol Cell 70:1089–1100.e8. https://doi.org/10.1016/j.molcel.2018.05.033
6. Schäfer IB, Yamashita M, Schuller JM et al (2019) Molecular basis for poly(A) RNP architecture and recognition by the Pan2-Pan3 deadenylase. Cell 177:1619–1631.e21. https://doi.org/10.1016/j.cell.2019.04.013
7. Brouze A, Krawczyk PS, Dziembowski A, Mroczek S (2022) Measuring the tail: methods for poly(A) tail profiling. Wiley Interdiscip Rev RNA:e1737. https://doi.org/10.1002/wrna.1737
8. Chang H, Lim J, Ha M, Kim VN (2014) TAIL-seq: genome-wide determination of poly (A) tail length and 3′ end modifications. Mol Cell 53:1044–1052. https://doi.org/10.1016/j.molcel.2014.02.007
9. Subtelny AO, Eichhorn SW, Chen GR et al (2014) Poly(A)-tail profiling reveals an embryonic switch in translational control. Nature 508:66–71. https://doi.org/10.1038/nature13007
10. Woo YM, Kwak Y, Namkoong S et al (2018) TED-Seq identifies the dynamics of Poly (A) length during ER stress. Cell Rep 24:3630–3641.e7. https://doi.org/10.1016/j.celrep.2018.08.084
11. Gewartowska O, Aranaz-Novaliches G, Krawczyk PS et al (2021) Cytoplasmic polyadenylation by TENT5A is required for proper bone formation. Cell Rep 35:109015. https://doi.org/10.1016/j.celrep.2021.109015
12. Bilska A, Kusio-Kobiałka M, Krawczyk PS et al (2020) Immunoglobulin expression and the humoral immune response is regulated by the non-canonical poly(A) polymerase TENT5C. Nat Commun 11:2032. https://doi.org/10.1038/s41467-020-15835-3
13. Parker MT, Knop K, Sherwood AV et al (2020) Nanopore direct RNA sequencing maps the complexity of Arabidopsis mRNA processing and m6A modification. elife 9:e49658. https://doi.org/10.7554/eLife.49658
14. Workman RE, Tang AD, Tang PS et al (2019) Nanopore native RNA sequencing of a human poly(A) transcriptome. Nat Methods:1–9. https://doi.org/10.1038/s41592-019-0617-2
15. Turtola M, Manav MC, Kumar A et al (2021) Three-layered control of mRNA poly(A) tail synthesis in Saccharomyces cerevisiae. Genes

Dev 35:1290–1303. https://doi.org/10.1101/gad.348634.121
16. Schmid M, Tudek A, Jensen TH (2018) Simultaneous measurement of transcriptional and post-transcriptional parameters by 3′ end RNA-Seq. Cell Rep 24:2468–2478.e4. https://doi.org/10.1016/j.celrep.2018.07.104
17. Schmid M, Tudek A, Jensen TH (2019) Preparation of RNA 3′ end sequencing libraries of total and 4-thiouracil labeled RNA for simultaneous measurement of transcription, RNA synthesis and decay in S. cerevisiae. Bio Protoc 9:e3189. https://doi.org/10.21769/BioProtoc.3189
18. Barrass JD, Reid JEA, Huang Y et al (2015) Transcriptome-wide RNA processing kinetics revealed using extremely short 4tU labeling. Genome Biol 16:282. https://doi.org/10.1186/s13059-015-0848-1
19. Miller C, Schwalb B, Maier K et al (2011) Dynamic transcriptome analysis measures rates of mRNA synthesis and decay in yeast. Mol Syst Biol 7:458. https://doi.org/10.1038/msb.2010.112
20. Alexander RD, Barrass JD, Dichtl B et al (2010) RiboSys, a high-resolution, quantitative approach to measure the in vivo kinetics of pre-mRNA splicing and 3′-end processing in Saccharomyces cerevisiae. RNA 16:2570–2580. https://doi.org/10.1261/rna.2162610
21. Li H (2018) Minimap2: pairwise alignment for nucleotide sequences. Bioinformatics 34:3094–3100. https://doi.org/10.1093/bioinformatics/bty191
22. Krause M, Niazi AM, Labun K et al (2019) tailfindr: alignment-free poly(A) length measurement for Oxford Nanopore RNA and DNA sequencing. RNA 25:1229–1241. https://doi.org/10.1261/rna.071332.119
23. Niazi AM, Krause M, Valen E (2021) Transcript isoform-specific estimation of poly(A) tail length by nanopore sequencing of native RNA. In: Picardi E (ed) RNA bioinformatics. Springer US, New York, pp 543–567
24. Neymotin B, Athanasiadou R, Gresham D (2014) Determination of in vivo RNA kinetics using RATE-seq. RNA 20:1645–1652. https://doi.org/10.1261/rna.045104.114
25. Sun M, Schwalb B, Schulz D et al (2012) Comparative dynamic transcriptome analysis (cDTA) reveals mutual feedback between mRNA synthesis and degradation. Genome Res 22:1350–1359. https://doi.org/10.1101/gr.130161.111
26. Hardwick SA, Chen WY, Wong T et al (2016) Spliced synthetic genes as internal controls in RNA sequencing experiments. Nat Methods 13:792–798. https://doi.org/10.1038/nmeth.3958
27. Begik O, Diensthuber G, Liu H et al (2023) Nano3P-seq: transcriptome-wide analysis of gene expression and tail dynamics using end-capture nanopore cDNA sequencing. Nat Methods 20:75–85. https://doi.org/10.1038/s41592-022-01714-w
28. Jia J, Long Y, Zhang H et al (2020) Post-transcriptional splicing of nascent RNA contributes to widespread intron retention in plants. Nat Plants 6:780–788. https://doi.org/10.1038/s41477-020-0688-1
29. Haruki H, Nishikawa J, Laemmli UK (2008) The anchor-away technique: rapid, conditional establishment of yeast mutant phenotypes. Mol Cell 31:925–932. https://doi.org/10.1016/j.molcel.2008.07.020
30. Nishimura K, Fukagawa T, Takisawa H et al (2009) An auxin-based degron system for the rapid depletion of proteins in nonplant cells. Nat Methods 6:917–922. https://doi.org/10.1038/nmeth.1401
31. Bajak EZ, Hagedorn CH (2008) Efficient 5′ cap-dependent RNA purification: use in identifying and studying subsets of RNA. Methods Mol Biol 419:147–160. https://doi.org/10.1007/978-1-59745-033-1_10
32. Nowacka M, Latoch P, Izert MA et al (2022) A cap 0-dependent mRNA capture method to analyze the yeast transcriptome. Nucleic Acids Res 50:e132. https://doi.org/10.1093/nar/gkac903
33. Smith MA, Ersavas T, Ferguson JM et al (2020) Molecular barcoding of native RNAs using nanopore sequencing and deep learning. Genome Res 30:1345–1353. https://doi.org/10.1101/gr.260836.120

Chapter 13

Sequencing of Transcriptome-Wide Poly(A) Tails by PAIso-seq

Yusheng Liu, Falong Lu, and Jiaqiang Wang

Abstract

Poly(A) tails are added to most eukaryotic mRNA and have essential regulatory functions. However, due to its homopolymeric nature, the sequence information in poly(A) tails is challenging to obtain in transcriptome measurement studies. In this chapter, we describe the detailed procedures of poly(A) inclusive full-length RNA isoform-sequencing (PAIso-seq), a method that can measure transcriptome-wide poly(A) tails from as low as nanogram level of total RNA based on the PacBio HiFi sequencing platform. The accurate length and base composition of poly(A) tails can be obtained along with the full-length cDNA.

Key words Poly(A) tail, PAIso-seq, Full-length RNA isoform, Transcriptome-wide, Mixed tailing

1 Introduction

New methods to accurately sequence and measure the lengths of RNA poly(A) tails have advanced our understanding of RNA biology. The preservation of poly(A) tails during sequencing library preparation can be achieved in three ways: (a) templated end extension (PAIso-seq [10, 12, 13], PAT-seq [4]), (b) oligo ligation (PAL-seq [18], PAL-seq-v2 [2, 3], TAIL-seq [1], mTAIL-seq [9], TED-seq [19], FLEP-seq [5, 14], and PAIso-seq2 [11, 13]), and (c) GI tailing (FLAM-seq [8]). Among these methods, end extension is relatively more sensitive, suitable for nanogram-level total RNA, such as mammalian oocytes and early embryos. In contrast, the oligo ligation and GI tailing methods need a microgram level of total RNA. Methodologies that measure the RNA poly(A) tails are mainly based on three kinds of platforms: (a) Illumina platform (PAL-seq [18], PAL-seq-v2 [2, 3], TAIL-seq [1], mTAIL-seq [9], PAT-seq [4], TED-seq, [19] and poly(A)-seq [21]), (b) PacBio platform (PAIso-seq [10, 12], PAIso-seq2 [11], and FLAM-seq [8]), and (c) Nanopore platform [7, 15, 17, 20]. PacBio platform is most suitable for poly(A) tail measurement

Eugene Valkov and Aaron C. Goldstrohm (eds.), *Deadenylation: Methods and Protocols*, Methods in Molecular Biology, vol. 2723,
https://doi.org/10.1007/978-1-0716-3481-3_13,

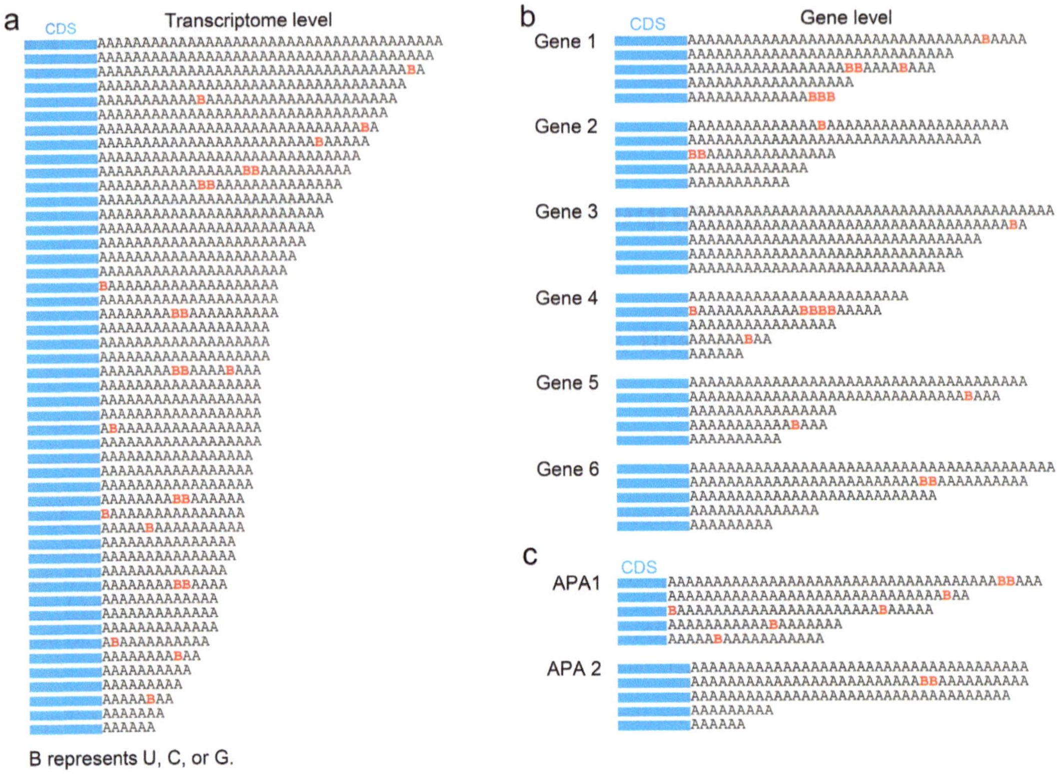

Fig. 1 Application of PAIso-seq to sequence the RNA poly(A) tails. Illustration of the application of PAIso-seq that can measure the global distribution of poly(A) tail lengths of all transcripts (CCSs) in a given sample (**a**), the distribution of poly(A) tail lengths of each individual gene (**b**), and the distribution of poly(A) tail lengths of each alternative polyadenylated isoforms of genes (**c**). "B" in the poly(A) tails represents U, C, or G residues that PAIso-seq can sequence

because of its better accuracy for homopolymeric sequences than the other two platforms. Illumina platform cannot sequence homopolymeric sequences longer than ~30 nt [1]. The Nanopore platform can quantify the length of poly(A) tails, but cannot measure non-A residues within the poly(A) tails [7, 15, 17, 20]. In this chapter, we present the detailed method for measurement of the poly(A) tail length and non-A residues within the poly(A) tails from single mouse oocytes through PAIso-seq and the associated bioinformatic pipeline. We were able to accurately measure the RNA poly(A) tail length for each of the sequenced transcripts transcriptome-wide (Fig. 1a) and at the gene level (Fig. 1b), compare the length of poly(A) tails for alternative polyadenylation isoforms of the same gene (Fig. 1c), and analyze non-A residues within RNA poly(A) tails (Fig. 1a–c), from single mouse germinal vesicle (GV) oocyte as well as the extracted total RNA.

2 Materials

Prepare all solutions using nuclease-free water and store all reagents at room temperature (unless indicated otherwise).

2.1 Preparing Mouse GV Oocytes

1. Seven- to eight-week-old CD1 (ICR) mice: Mice were housed in Individually Ventilated Cage (IVC) systems in specific pathogen-free (SPF) rooms in compliance with the institutional guidelines (*see* **Note 1**).
2. 1× Phosphate-buffered saline (PBS). Store at 4 °C.
3. Pregnant mare serum gonadotropin (PMSG): 0.1 IU/μL PMSG in 1× PBS. Store at −20 °C (*see* **Note 2**).
4. Disposable plastic syringes.
5. Surgical scissors.
6. Surgical forceps.
7. Needle, Lavender, 30-gauge 0.5 inch.
8. M2 medium. Store at 4 °C.
9. Embryo washing buffer: 0.1% (w/v) bovine serum albumin (BSA) in 1× PBS. Store at −20 °C (*see* **Note 3**).
10. Cell lysis buffer: 0.2% (v/v) Triton X-100 in 1× PBS. Store at 4 °C (*see* **Note 4**).

2.2 Clean Working Bench

1. DNA-OFF reagent.
2. RNaseZap reagent.

2.3 Templated End Extension

1. Direct-zol RNA MicroPrep kit.
2. TRIzol reagent: Store at 4 °C (*see* **Note 5**).
3. Ethanol, 200 proof, molecular biology grade (*see* **Note 6**).
4. Bioanalyzer.
5. Spectrophotometer.
6. Thin-walled PCR tube.
7. Templated end extension oligos (PAT_end-extend_2dU-BC* in Table 1): 50 μM in nuclease-free water. Store at −20 °C (*see* **Note 7**).
8. SuperScript II reverse transcriptase (Invitrogen, Thermo Fisher Scientific, Waltham, Massachusetts, USA). Store at −20 °C (*see* **Note 8**).
9. DTT (100 mM). Store at 4 °C (*see* **Note 9**).
10. Templated end extension mix: 1 × SuperScript II first-strand buffer, 5 mM DTT, 0.5 mM dNTP mix, 2 U/μL Recombinant RNase inhibitor, and 0.5 U/μL Klenow fragment, 3′ → 5′ exo$^-$. Prepare fresh every time.

Table 1
Oligos used for PAIso-seq

Used for	Name	Sequence (5′ → 3′)
Templated end extension	PAT_end-extend_2dU-BC1	AAGCAGTGGTATCAACGCAGAGTACTCAGACGATGCGTCATdUTTTTTTTTdUTTTTTTTT
	PAT_end-extend_2dU-BC2	AAGCAGTGGTATCAACGCAGAGTACCTATACATGACTCTGCdUTTTTTTTTdUTTTTTTTT
	PAT_end-extend_2dU-BC3	AAGCAGTGGTATCAACGCAGAGTACTACTAGAGTAGCACTCdUTTTTTTTTdUTTTTTTTT
	PAT_end-extend_2dU-BC4	AAGCAGTGGTATCAACGCAGAGTACTGTGTATCAGTACATGdUTTTTTTTTdUTTTTTTTT
	PAT_end-extend_2dU-BC5	AAGCAGTGGTATCAACGCAGAGTACGATCTCTACTATATGCdUTTTTTTTTdUTTTTTTTT
	PAT_end-extend_2dU-BC6	AAGCAGTGGTATCAACGCAGAGTACACAGTCTATACTGCTGdUTTTTTTTTdUTTTTTTTT
	PAT_end-extend_2dU-BC7	AAGCAGTGGTATCAACGCAGAGTACATGATGTGCTACATCTdUTTTTTTTTdUTTTTTTTT
	PAT_end-extend_2dU-BC8	AAGCAGTGGTATCAACGCAGAGTACCTGCGTGCTCTACGACdUTTTTTTTTdUTTTTTTTT
	PAT_end-extend_2dU-BC9	AAGCAGTGGTATCAACGCAGAGTACGCGCGATACGATGACTdUTTTTTTTTdUTTTTTTTT
	PAT_end-extend_2dU-BC10	AAGCAGTGGTATCAACGCAGAGTACCGCGCTCAGCTGATCGdUTTTTTTTTdUTTTTTTTT
	PAT_end-extend_2dU-BC11	AAGCAGTGGTATCAACGCAGAGTACGCGCACGCACTACAGAdUTTTTTTTTdUTTTTTTTT
	PAT_end-extend_2dU-BC12	AAGCAGTGGTATCAACGCAGAGTACACACTGACGTCGCGACdUTTTTTTTTdUTTTTTTTT
	PAT_end-extend_2dU-BC13	AAGCAGTGGTATCAACGCAGAGTACCGTCTATATACGTATAdUTTTTTTTTdUTTTTTTTT
	PAT_end-extend_2dU-BC14	AAGCAGTGGTATCAACGCAGAGTACATAGAGACTCAGAGCTdUTTTTTTTTdUTTTTTTTT
	PAT_end-extend_2dU-BC15	AAGCAGTGGTATCAACGCAGAGTACTAGATGCGAGAGTAGAdUTTTTTTTTdUTTTTTTTT
Reverse transcription	RT primer	AAGCAGTGGTATCAACGCAGAGTAC
	TSO	iCiGiCAAGCAGTGGTATCAACGCAGAGTACATrGrG+G/NH_2-C3
PCR amplification	IS PCR primer	AAGCAGTGGTATCAACGCAGAG

11. USER enzyme.
12. Thermal cycler.
13. RNA Clean and Concentrator-5 kit.

2.4 Reverse Transcription (RT)

1. RT primer (Table 1): 100 μM in nuclease-free water. Store at −20 °C (*see* **Note 10**).
2. Template Switching Oligo (TSO, Table 1): 100 μM in nuclease-free water. Store at −80 °C (*see* **Note 11**).

3. RT reaction mix: 10 U/μL SuperScript II reverse transcriptase, 1 × SuperScript II first-strand buffer, 1 U/μL Recombinant RNase inhibitor, 5 mM DTT, 1 M Betaine, 6 mM $MgCl_2$, and 0.98 μM TSO. Prepare freshly every time.

2.5 PCR Amplification

1. KAPA HiFi HotStart ReadyMix.
2. IS PCR primer (Table 1): 100 μM in nuclease-free water. Store at −20 °C (*see* **Note 10**).

2.6 Clean-Up and Quantification of Amplification Product

1. SPRIselect beads.
2. Magnetic stand.
3. DNA washing buffer: 80% (v/v) ethanol in nuclease-free water. Prepare freshly every time.
4. DeNovix DS-11 FX+ or Qubit 3.0.
5. Agilent TapeStation or other similar DNA fragment size analyzing tools.

2.7 SMRTbell Library Preparation and PacBio Sequencing

1. SMRTbell Template Prep Kit 1.0-SPv3.
2. PacBio system at HiFi mode, such as Sequel or Sequel II System.

2.8 Data Analysis

1. Computational Hardware (Computing server or workstation running Linux)
2. Software and code functions (Table 2).

3 Methods

Carry out all procedures at room temperature unless otherwise specified.

3.1 Preparing Mouse GV Oocytes

1. Prepare the Embryo washing buffer. Weigh 50 mg BSA and transfer to a 50 mL polystyrene conical tube containing 50 mL of 1× PBS. Filter through a 0.22 μm syringe filter and aliquot in 1 mL aliquots.
2. Add 1× PBS to dissolve the lyophilized PMSG at a concentration of 0.1 IU/μL. Filter through a 0.22 μm syringe filter and aliquot in 0.5 mL aliquots.
3. Intraperitoneal injection of 10 U of PMSG to seven- to eight-week-old CD1 (ICR) mice with disposable plastic syringes.
4. Humanely kill mice 48 h post-PMSG injection and collect ovaries with surgical scissors and forceps.
5. Release the GV oocytes with a 30-gauge needle from the ovaries.
6. Wash the GV oocytes with M2 medium three times and the Embryo washing buffer three times (*see* **Note 12**).

Table 2
Software and code functions

Software	Package	Link	Function
ccs, version 5.0.0		https://github.com/PacificBiosciences/ccs	Convert subreads to CCS reads.
subread	bam2fastq	http://subread.sourceforge.net/	Convert *bam* file to *fastq* file.
minimap2, version v2.15		https://github.com/lh3/minimap2	Align the clean-reads to the genome.
SAMtools, version 1.9		http://samtools.sourceforge.net,	Manipulate the *sam/bam* files produced by minimap2.
Parafly		https://github.com/ParaFly/ParaFly	Run a predefined list of *unix* commands in parallel
perl	GetCCSpass.pl	https://github.com/Lulab-IGDB/PAIso-seq_scripts/blob/main/GetCCSpass.pl	Get the number of passes for each CCS read.
	fastq-splitter.pl	http://kirill-kryukov.com/study/tools/fastq-splitter/fastq-splitter-0.1.2.zip	Split one *fastq* file into multiple *fastq* files.
python, version 3.7 or above	pysam	https://github.com/pysam-developers/pysam	Read, manipulate, and write genomic data sets.
	regex	https://bitbucket.org/mrabarnett/mrab-regex	Match text.
	parasail	https://github.com/jeffdaily/parasail-python	A library containing implementations of the Smith-Waterman (local), Needleman-Wunsch (global), and semi-global pairwise sequence alignment algorithms.
	itertools	https://github.com/more-itertools/more-itertools	Create iterators for efficient looping
	numpy	https://numpy.org/	Scientific computing.
	statistics	https://pypi.python.org/pypi/statistics	Calculate mathematical statistics of numeric data.

(continued)

Table 2
(continued)

Software	Package	Link	Function
	CCS_split_clean_end_extension_v1.py	https://github.com/Lulab-IGDB/PAIso-seq_scripts/blob/main/CCS_split_clean_end_extension_v1.py	Extract clean reads from CCS reads
	PolyA_trim_V5.4.1.py	https://github.com/Lulab-IGDB/PAIso-seq_scripts/blob/main/PolyA_trim_V5.4.1.py	Extract poly(A) tails from clean-reads.
	PolyA_note_V2.1.py	https://github.com/Lulab-IGDB/PAIso-seq_scripts/blob/main/PolyA_note_V2.1.py	Annotate poly(A) tails from clean-reads

3.2 Preparing a Clean Working Bench

1. Clean the bench with DNA-OFF reagent first, then with RNaseZap reagent.
2. Spray the surface of pipettes with RNaseZap (*see* **Note 13**).

3.3 Total RNA Isolation from Bulk GV Oocytes with Direct-zol RNA MicroPrep Kit

1. Add additional TRIzol reagent to bulk GV oocytes to achieve the volume of 500 μL (*see* **Note 14**).
2. Add 500 μL of absolute ethanol and mix thoroughly.
3. Transfer the mixture into a Zymo-Spin IC Column in a Collection Tube, centrifuge at 10,000 g for 30 s at 4 °C, and discard the flow-through.
4. Add 400 μL of RNA Wash PreBuffer to the column and centrifuge at 10,000 g for 30 s at 4 °C. Discard the flow-through.
5. Add 700 μL of RNA Wash Buffer to the column and centrifuge at 10,000 g for 30 s at 4 °C. Discard the flow-through. Repeat this step once.
6. Add 12 μL nuclease-free water directly to the column matrix. Place the column into a 1.5 mL tube and centrifuge at 10,000 g for 60 s at 4 °C to elute RNA.
7. Check the quality of the extracted total RNA with Bioanalyzer to measure the RNA integrity number (RIN). And check the purity of the total RNA with a spectrophotometer to measure the A260/A280 ratio (*see* **Note 15**).
8. Take 50 ng total RNA to use in **step 1** in Subheading 3.5.

3.4 Single Oocyte Lysis

1. Prepare the cell lysis buffer. Add 20 μL of Triton X-100 to 15 mL polystyrene conical tube containing 10 mL of 1× PBS. Filter through a 0.22 μm syringe filter and aliquot in 1 mL aliquots.
2. Transfer the single GV oocytes from **step 6** in Subheading 3.1 using a microcapillary pipette in the lowest possible volume (around 0.5 μL) into a 0.2 mL thin-walled PCR tube containing 2.5 μL of cell lysis buffer to a final volume of around 3 μL (*see* **Note 16**).
3. Vortex the tubes briefly to mix, then spin at 2000 g for 5 s at room temperature to collect the solution at the bottom of the tubes.
4. Incubate the tubes at 85 °C for 5 min, followed by holding at 4 °C. Place the tubes on ice immediately after the temperature reaches 4 °C (*see* **Note 17**).

3.5 Templated End Extension

1. Add 1 μL templated end extension oligo (50 μM) to the 50 ng of total RNA from **step 8** in Subheading 3.3 or single GV oocyte lysate from **step 4** in Subheading 3.4, then add nuclease-free water to a total volume of 5.5 μL (*see* **Note 18**).
2. Vortex the tubes briefly to mix, then spin at 2000 g for 5 s at room temperature to collect the solution at the bottom of the tubes.
3. Incubate the reaction at 80 °C for 5 min, then at 37 °C for 10 min in a thermal cycler with a heated lid (105 °C). Put the tubes immediately back on ice.
4. Add 4.5 μL of the templated end extension mix to each tube to obtain a final reaction volume of 10 μL.
5. Vortex the tubes briefly to mix, then spin at 2000 g for 5 s at room temperature to collect the solution at the bottom of the tubes.
6. Incubate the reactions at 37 °C for 1 h, then at 80 °C for 10 min in a thermal cycler with a heated lid (105 °C). Put the sample tubes back on ice immediately after the completion of the reaction.
7. Add 1 μL USER enzyme and 15 μL nuclease-free water to each tube.
8. Vortex the tubes briefly to mix, then spin at 2000 g for 5 s at room temperature to collect the solution at the bottom of the tubes.
9. Incubate the reaction at 37 °C for 30 min in a thermal cycler with a heated lid (105 °C). Put the sample tubes back on ice immediately after the completion of the reaction (*see* **Note 19**).
10. Mix the end-extended RNA from different single GV oocyte samples (*see* **Note 20**).

3.6 Purification of End-Extended RNA with RNA Clean and Concentrator-5 Kit

1. Add two volumes of RNA Binding Buffer to each sample. Vortex the tubes briefly to mix, then spin at 2000 g for 5 s at room temperature to collect the solution at the bottom of the tubes.
2. Add an equal volume of absolute ethanol. Vortex the tubes briefly to mix, then spin at 2000 g for 5 s at room temperature to collect the solution at the bottom of the tubes.
3. Transfer the mixture to the Zymo-Spin IC Column in a Collection Tube and centrifuge at 10,000 g for 30 s at 4 °C. Discard the flow-through.
4. Add 400 μL RNA Prep Buffer to the column and centrifuge at 10,000 g for 30 s at 4 °C. Discard the flow-through.
5. Add 700 μL RNA Wash Buffer to the column and centrifuge at 10,000 g for 30 s at 4 °C. Discard the flow-through. Repeat this step.
6. Add 7 μL nuclease-free water directly to the column matrix and centrifuge at 10,000 g for 60 s at 4 °C (*see* **Note 19**).

3.7 Reverse Transcription

1. Add 0.4 μL of RT primer (100 μM) and 1 μL of dNTP mix (10 mM each) to the samples from **step 6** in Subheading 3.6.
2. Vortex the tubes briefly to mix, then spin at 2000 g for 5 s at room temperature to collect the solution at the bottom of the tubes.
3. Incubate the tubes at 72 °C for 3 min in a thermal cycler with a heated lid (105 °C). Put the tubes immediately back on ice.
4. Dilute the TSO to 20 μM by adding 10 μL of 100 μM TSO and 40 μL of nuclease-free water to a tube and mix well.
5. Add 11.6 μL of the RT reaction mix to the tubes to obtain a final reaction volume of 20 μL.
6. Vortex the tubes briefly to mix, then spin at 2000 g for 5 s at room temperature to collect the solution at the bottom of the tubes.
7. Incubate the tubes in a thermal cycler with a heated lid (105 °C) following the program below: 42 °C 90 min; 50 °C 2 min, 42 °C 2 min, for 10 cycles; 70 °C 15 min; 4 °C hold (*see* **Notes 19, 21**).

3.8 PCR Pre-amplification

1. Add 25 μL of 2× KAPA HiFi HotStart ReadyMix and 5 μL of IS PCR primer (10 μM) to each tube containing 20 μL first-strand reaction from **step 7** in Subheading 3.7.
2. Vortex the tubes briefly to mix, then spin at 2000 g for 5 s at room temperature to collect the solution at the bottom of the tubes.

3. Perform the PCR in a thermal cycler with a heated lid (105 °C) following the program below: 98 °C 3 min; 98 °C 20 s, 67 °C 15 s, 72 °C 6 min, for 15 cycles; 72 °C 10 min; 4 °C hold (*see* **Note 22**).

3.9 Clean-Up and Quantification of Pre-amplification Product

1. Vortex SPRIselect beads thoroughly to achieve even suspension.
2. Add 0.8× volume (40 μL) SPRIselect beads, and mix by pipetting up and down until the mixture looks homogeneous (*see* **Note 23**).
3. Incubate the mixture undisturbed for 8 min at room temperature to bind the DNA to the beads.
4. Place tubes on a magnetic stand for 5 min. The solution shall become clear with the beads collected at one corner of the tube. Keep tubes on the magnetic stand and carefully remove most of the liquid without taking any beads.
5. Wash the beads with 150 μL of DNA washing buffer with the tubes on the magnetic stand. Incubate for 30 s without disturbing the beads, then remove the DNA washing buffer. Repeat this step twice.
6. Remove any trace amount of liquid and let the beads air dry at room temperature on the magnetic stand for 5 min or until small cracks are visible on the surface of the beads (*see* **Note 24**).
7. Add 20 μL of nuclease-free water to each tube on the magnetic stand. Take the tubes off the magnetic stand. Mix ten times to resuspend the beads completely. Then, incubate the tubes on a regular stand for 2 min to elute the DNA.
8. Place the tubes on the magnetic stand and leave them until the solution becomes clear, with beads collected in the corner of each tube.
9. Collect the liquid which contains the purified pre-amplification product without taking any beads and transfer it to a fresh 1.5 mL tube (*see* **Note 25**).
10. Measure the DNA concentration of the purified pre-amplification product using fluorometer-based methods, such as DeNovix DS-11 FX+ or Qubit 3.0 (*see* **Notes 22, 26**).
11. Take 20 ng purified PCR pre-amplification product for large-scale PCR.

3.10 Large-Scale PCR

1. Add 400 μL of 2× KAPA HiFi HotStart ReadyMix and 80 μL of IS PCR primer to 20 ng purified PCR pre-amplification product (**step 11** in Subheading 3.9), add nuclease-free water to a total volume of 800 μL, and prepare aliquots of 50 μL PCR mix into sixteen 0.2 mL thin wall PCR tubes (*see* **Note 27**).

2. Vortex the tubes briefly to mix, then spin at 2000 g for 5 s at room temperature to collect the solution at the bottom of the tubes.
3. Perform the PCR with the following program: 98 °C 3 min; 98 °C 20 s, 67 °C 15 s, 72 °C 6 min, for 10 cycles; 72 °C 10 min; 4 °C hold (*see* **Notes 22**, **28**).

3.11 Clean-Up of Large-Scale PAIso-seq Double-Stranded cDNA

1. Similar to Subheading 3.9 with 0.8× volume (640 μL) SPRI-select beads, wash with 900 μL DNA washing buffer and elute in 100 μL nuclease-free water. The expected yield shall be around 1–10 μg. Agilent TapeStation or other similar DNA fragment size analyzing tools can measure the size distribution of the PAIso-seq double-stranded cDNA. Typical PAIso-seq double-stranded cDNA shows a peak around ~1–2 kb [12]. However, for some precious in vivo samples in which RNA fragmentation cannot be avoided, PAIso-seq can still be performed with the ability to analyze the poly(A) tails while sacrificing its ability to measure full-length RNA isoforms. The size of this type of PAIso-seq double-stranded cDNA is generally shorter, with a peak of less than 500 bp [12] (*see* **Notes 22**, **29**).

3.12 SMRTbell Library Preparation and Long-Read Sequencing on the PacBio Platform

1. Perform SMRTbell library construction using the PAIso-seq double-stranded cDNA according to the instructions of SMRTbell Template Prep Kit 1.0-SPv3 with 160–500 ng input DNA.
2. Sequence the SMRTbell library on a PacBio Sequel or Sequel II System according to the standard PacBio Iso-Seq procedures (*see* **Note 30**).

3.13 Convert Subreads to CCS Reads

1. Generate accurate single-molecule consensus reads using ccs software (*see* **Note 31**).

```
$ ccs subreads.bam ccs.bam -j 30 &> ccs.log
```

2. Check the successful CCS reads conversion (*see* **Note 32**).

```
$ cat ccs.log
```

3. Convert the ccs.bam file to the fastq.gz file (*see* **Note 33**).

```
$ bam2fastq -o ccs ccs.bam > ccs.bam2fastq.log
```

4. Get the number of passes for each CCS read.

```
$ perl GetCCSpass.pl ccs.bam > ccs.pass.txt
```

3.14 Extract the Clean Reads Containing the Poly(A) Inclusive Full-Length cDNA from the CCS Reads Based on the Barcodes for Each Sample

1. Prepare files for splitting ccs and cleaning the reads (*see* **Note 34**).

```
$ gunzip ccs.fastq.gz
$ fastq-splitter.pl --n-parts 100 ccs.fastq
$ vim barcode.fa
>BC1
ATGACGCATCGTCTGAGTACTCTGCGTTGATACCACTGCTT
>BC2
GCAGAGTCATGTATAGGTACTCTGCGTTGATACCACTGCTT
>BC3
GAGTGCTACTCTAGTAGTACTCTGCGTTGATACCACTGCTT
>BC4
CATGTACTGATACACAGTACTCTGCGTTGATACCACTGCTT
>BC5
GCATATAGTAGAGATCGTACTCTGCGTTGATACCACTGCTT
>BC6
CAGCAGTATAGACTGTGTACTCTGCGTTGATACCACTGCTT
>BC7
AGATGTAGCACATCATGTACTCTGCGTTGATACCACTGCTT
>BC8
GTCGTAGAGCACGCAGGTACTCTGCGTTGATACCACTGCTT
>BC9
AGTCATCGTATCGCGCGTACTCTGCGTTGATACCACTGCTT
>BC10
CGATCAGCTGAGCGCGGTACTCTGCGTTGATACCACTGCTT
>BC11
TCTGTAGTGCGTGCGCGTACTCTGCGTTGATACCACTGCTT
>BC12
GTCGCGACGTCAGTGTGTACTCTGCGTTGATACCACTGCTT
>BC13
TATACGTATATAGACGGTACTCTGCGTTGATACCACTGCTT
>BC14
AGCTCTGAGTCTCTATGTACTCTGCGTTGATACCACTGCTT
>BC15
TCTACTCTCGCATCTAGTACTCTGCGTTGATACCACTGCTT
```

2. Extract clean reads from the CCS reads (*see* **Note 35**).

```
$ for i in {001..100}; do echo 'python CCS_split_-
clean_end_extension_v1.py ccs.part-'${i}'.fastq
ccs.pass.txt barcode.fa 2 1> '${i}'.out.txt 2>
'${i}'.err.txt'; done > split_clean.sh
$ ParaFly -c split_clean.sh -CPU 25
```

3. Convert the clean reads into fastq format (*see* **Note 36**).

```
$ cat *.out.txt > all.out.txt
$ cat *.err.txt > all.err.txt
$ awk '{print "@"$2"\n"$6"\n+\n"$7}' all.out.txt |
gzip -nc > clean.fastq.gz
$ rm *.err.txt *.out.txt *fastq.log.txt ccs.part-*.
fastq
```

3.15 Align the Clean-Reads to the Genome

1. Generate reference files for genome alignment (*see* **Note 37**).

```
$ paftools.js gff2bed genome/gencode.vM25.pri-
mary_assembly.annotation.gtf > gencode.vM25.pri-
mary_assembly.annotation.bed
$ minimap2 -x splice -t 20 -d gencode.vM25.mmi
genome/GRCm38.primary_assembly.genome.fa.gz
```

2. Mapping the clean-reads to the genome.

```
$ minimap2 -ax splice -uf --secondary=no -t 40 -L --
MD --cs --junc-bed gencode.vM25.primary_assembly.
annotation.bed gencode.vM25.mmi clean.fastq.gz 2>
align.log | samtools view -F 3844 -bS > clean.
filter.bam
```

3.16 Extract and Annotate Poly(A) Tails from Mapped Clean-Reads

1. Extract and annotate poly(A) tails from clean reads (*see* **Note 38**).

```
$ python PolyA_trim_V5.4.1.py clean.filter.bam >
polyA_trim.out.txt
$ featureCounts -L -g gene_id -t exon -s 1 -R CORE -a
genome/gencode.vM25.primary_assembly.annotation.
gtf -o featureCounts clean.filter.bam &> feature-
Counts.log
$ python PolyA_note_V2.1.py polyA_trim.out.txt
clean.filter.bam.featureCounts 1> polyA_note.txt
2> polyA_note.err.txt
```

4 Notes

1. All mouse procedures comply with the Animal Care and Use Committee of the Institute of Genetics and Development Biology, Chinese Academy of Sciences guidelines.
2. The PMSG solution can be stored at −20 °C for 12 months. Then, remove the required amount and thaw on ice.

3. The PBS with 0.1% (w/v) BSA solution can be stored at −20 °C for 6 months. Thaw on ice before use.
4. The cell lysis buffer can be stored at 4 °C for 6 months. Add 5% (v/v) of 40 U/μL TaKaRa Recombinant RNase inhibitor before use—wear gloves, eye protection, face protection, and a lab coat when handling Triton X-100.
5. Wear gloves, eye protection, face protection, and a lab coat when handling the TRIzol reagent. In addition, avoid breathing the fumes of the TRIzol reagent.
6. Avoid heat/sparks/open flames/hot surfaces when handling ethanol.
7. The templated end extension oligos contain an oligo(dT) sequence, which can anneal to the 3′-end of RNA containing a poly(A) tail. Each oligo contains two dU that can be cleaved by the USER enzyme. Stored at −20 °C in aliquots for at least 12 months and keep the freeze-thaw cycles no more than five.
8. Superscript II works well for template switching. Replacing with another reverse transcriptase is only recommended if tested for the performance in template switching.
9. Wear gloves, eye protection, face protection, and a lab coat when handling the DTT solution.
10. Stored at −20 °C in aliquots for at least 12 months and keep the freeze-thaw cycles no more than five.
11. The TSO carries two iso-deoxycytosine (iC) and one iso-deoxyguanosine (iG) at the 5′-end to inhibit the reverse transcriptase from extending the cDNA beyond the TSO and avoid the formation of concatamers of TSOs and increase cDNA yield [6]. The TSO carries two riboguanosines (rG) and one locked nucleic acid (LNA)-modified guanosine (+G) at the 3′-end to increase the affinity of TSO toward the non-templated CCC residues added to the ends of reverse transcription product and increase the success rate of template switching [16]. The TSO is 3′-aminated (/NH_2-C3–3′) to block it from working as an RT primer in reverse transcription. Dissolve the TSO in nuclease-free water at 100 μM. Stored at −80 °C in aliquots for at least 6 months and keep the freeze-thaw cycles no more than five.
12. The GV oocytes can either be used immediately after preparation or stored in bulk in Trizol reagent at −80 °C for at least 4 weeks or stored in single cells individually in PCR tubes at −80 °C for no more than 2 weeks. In addition, oocytes can be collected in batches.
13. A clean bench in a separate room is preferred to prevent potential unwanted degradation of RNA in the samples.
14. Oocytes collected in different batches can be combined.

15. About 50–100 ng total RNA could be extracted from 200 oocytes. Therefore, total RNA with RIN >8 and A260/A280 ratio in the 1.8–2.0 indicate that it is reasonable to proceed. The RNA can be used immediately or stored at −80 °C for at least 2 weeks.
16. Single mouse GV oocytes are used as an example in this protocol. Other samples of a limited number of cells can also be processed with this protocol.
17. After each collection, single oocyte lysate can be collected in batches and stored in the −80 °C refrigerator. Single oocyte lysate can be used immediately or stored at −80 °C for 2 weeks.
18. Use different barcodes for each single GV oocyte lysate because different single GV oocyte samples should be mixed after templated end extension.
19. The samples can be stored at −80 °C for at least 2 weeks.
20. The users can change the number of samples mixed according to their experimental design. Make sure that the samples have different barcodes.
21. Additional ten cycles of 50 °C and 42 °C after the initial incubation at 42 °C can slightly increase the final cDNA yield [16].
22. The samples can be stored at −80 °C for at least 12 months.
23. Keep the volume of beads at a 0.8× ratio, which is suitable for the purification of poly(A) inclusive full-length cDNA to avoid very short fragments.
24. The overdried beads are challenging to resuspend and lead to reduced yield, thus, do not overdry the beads.
25. Do not take all but leave about 1 μL of the solution in the tube to minimize bead carryover.
26. Combining 15 single mouse GV oocytes can yield around 25 ng pre-amplification product, and 50 ng total RNA can yield 200 ng pre-amplification product. However, suppose the yield of pre-amplification is much lower than expected. In that case, the user should increase the amount of starting material, increase the PCR cycles for pre-amplification, use cells in good condition, or change new aliquots of reagents for templated end-extension and reverse transcription.
27. Use all of the purified pre-amplification product as input for large-scale PCR if the total amount is lower than 20 ng.
28. Ten cycles are typically used for large-scale PCR for about 20 ng purified pre-amplification product. If the input amount is smaller than 20 ng, do more cycles depending on the amount of input.

29. If the overall double-stranded cDNA size is short, it may be due to RNA fragmentation, and the user should use RNase-free aliquots of reagents and consumables and use fresh samples or samples stored at −80 °C for no more than 2 weeks. Unfortunately, due to the nature of sample collection, RNA fragmentation cannot be avoided for some rare samples. However, the cDNA can still be sequenced to get the poly(A) tail information, while sacrificing the full-length RNA isoform information.
30. If the sequencing output is too low, the user should run the remaining SMRTbell library under HiFi mode on a new SMRT cell.
31. It is computationally intensive and time-consuming that converts subreads to CCS. Therefore, doing this on a multicore server in parallel is better. The number of threads (−j option here) depends on the available CPU cores on the computing server.
32. It is important to check that the conversion is successful. If the CCS reads conversion is successful, the *ccs.log* file shall be empty.
33. It means no errors occurred if the *ccs.bam2fastq.log* file is empty. Instead, a *ccs.fastq.gz* file is expected.
34. It is time-consuming. We separate the *.fastq* file into multiple (100 here) files for execution in parallel. This will create *ccs.part-001.fastq*; *ccs.part-002.fastq* ... and *ccs.part-100.fastq*. According to the computing resource, the number of parts can be increased or lowered. The user should create the *barcode.fa* file according to the actual barcode sequences used for the samples.
35. This step removes the 5′-adapter and 3′-adapter and outputs the poly(A) inclusive full-length cDNA sequence in the 5′-end to 3′-end orientations. Parameter 2 means the number of mismatches allowed for matching barcodes in the *barcode.fa* file with reads, and we recommend it no more than 4. The *out.txt* files contain clean reads, while the *err.txt* files contain reads with incomplete structure or containing no barcodes. The *err.txt* files can be used for potential troubleshooting.
36. If some barcodes for single-cell samples produce none or a very low number of reads, it might mean that cells were dead or damaged, and the user should use cells in good condition and analyze more single cells if possible.
37. The *paftools.js* is a script provided by *minimap2* software that converts the *.gtf* format annotation file into *.bed* format. This step maps the clean reads to the reference genome and gets the mapped clean reads in bam format.

38. It means no error happens during the step if the *polyA_note.err.txt* file is empty; otherwise, an error message will be reported. The *polyA_note.txt* contains all the poly(A) tail information for the reads assigned a poly(A) tail, including reads without tails that give tail size as 0 nt and the sequence as "No". Each line in the *polyA_note.txt* file contains the following 13 columns of information for one read: barcode, read_id, ensembl_id, pass_-number, "1", number_of_residue_A, number_of_residue_T, number_of_residue_C, number_of_residue_G, number_of_re-sidue_T+C+G, "0", poly(A)_tail_sequence, and average_qua-lity_value_of_poly(A)_tail_bases.

Acknowledgments

This work was supported by the National Key Research and Development Program of China (2018YFA0107001), the Strategic Priority Research Program of the Chinese Academy of Sciences (XDA24020203), National Natural Science Foundation of China (32201060, 31970588, 32170606, 81891001), Natural Science Foundation of Heilongjiang province (YQ2020C003), and the State Key Laboratory of Molecular Developmental Biology.

References

1. Chang H, Lim J, Ha M et al (2014) TAIL-seq: genome-wide determination of poly(A) tail length and 3′ end modifications. Mol Cell 53: 1044–1052
2. Eisen TJ, Eichhorn SW, Subtelny AO et al (2020) MicroRNAs cause accelerated decay of short-tailed target mRNAs. Mol Cell 77(775–785):e778
3. Eisen TJ, Eichhorn SW, Subtelny AO et al (2020) The dynamics of cytoplasmic mRNA metabolism. Mol Cell 77:786–799 e710
4. Harrison PF, Powell DR, Clancy JL et al (2015) PAT-seq: a method to study the integration of 3'-UTR dynamics with gene expression in the eukaryotic transcriptome. RNA 21: 1502–1510
5. Jia J, Lu W, Liu B et al (2021) An atlas of plant full-length RNA reveals tissue-specific and monocots-dicots conserved regulation of poly (A) tail length. Nat Plants 8:1118–1126
6. Kapteyn J, He R, Mcdowell ET et al (2010) Incorporation of non-natural nucleotides into template-switching oligonucleotides reduces background and improves cDNA synthesis from very small RNA samples. BMC Genomics 11:413
7. Kim D, Lee JY, Yang JS et al (2020) The architecture of SARS-CoV-2 transcriptome. Cell 181(914–921):e910
8. Legnini I, Alles J, Karaiskos N et al (2019) FLAM-seq: full-length mRNA sequencing reveals principles of poly(A) tail length control. Nat Methods 16:879–886
9. Lim J, Lee M, Son A et al (2016) mTAIL-seq reveals dynamic poly(A) tail regulation in oocyte-to-embryo development. Genes Dev 30:1671–1682
10. Liu Y, Nie H, Liu H et al (2019) Poly (A) inclusive RNA isoform sequencing (PAIso-seq) reveals wide-spread non-adenosine residues within RNA poly(A) tails. Nat Commun 10:5292
11. Liu Y, Nie H, Zhang Y et al (2023) Comprehensive analysis of mRNA poly(A) tails by PAIso-seq2. Sci China Life Sci 66:187–190
12. Liu Y, Zhang Y, Wang J et al (2022) Transcriptome-wide measurement of poly (A) tail length and composition at subnanogram total RNA sensitivity by PAIso-seq. Nat Protoc 17:1980–2007
13. Liu Y, Zhao H, Shao F et al (2023) Remodeling of maternal mRNA through poly(A) tail

orchestrates human oocyte-to-embryo transition. Nat Struct Mol Biol 30:200–215

14. Long Y, Jia J, Mo W et al (2021) FLEP-seq: simultaneous detection of RNA polymerase II position, splicing status, polyadenylation site and poly(A) tail length at genome-wide scale by single-molecule nascent RNA sequencing. Nat Protoc 16(9):4355–4381
15. Parker MT, Knop K, Sherwood AV et al (2020) Nanopore direct RNA sequencing maps the complexity of Arabidopsis mRNA processing and m(6)A modification. elife 9:e49658
16. Picelli S, Faridani OR, Bjorklund AK et al (2014) Full-length RNA-seq from single cells using Smart-seq2. Nat Protoc 9:171–181
17. Roach NP, Sadowski N, Alessi AF et al (2020) The full-length transcriptome of C. elegans using direct RNA sequencing. Genome Res 30:299–312
18. Subtelny AO, Eichhorn SW, Chen GR et al (2014) Poly(A)-tail profiling reveals an embryonic switch in translational control. Nature 508:66–71
19. Woo YM, Kwak Y, Namkoong S et al (2018) TED-Seq identifies the dynamics of poly (A) length during ER stress. Cell Rep 24(3630–3641):e3637
20. Workman RE, Tang AD, Tang PS et al (2019) Nanopore native RNA sequencing of a human poly(A) transcriptome. Nat Methods 16:1297–1305
21. Yu F, Zhang Y, Cheng C et al (2020) Poly(A)-seq: a method for direct sequencing and analysis of the transcriptomic poly(A)-tails. PLoS One 15:e0234696

Chapter 14

Nano3′RACE: A Method to Analyze Poly(A) Tail Length and Nucleotide Additions at the 3′ Extremity of Selected mRNAs Using Nanopore Sequencing

Pietro Giraudo, Quentin Simonnot, David Pflieger, Jackson Peter, Dominique Gagliardi, and Hélène Zuber

Abstract

Deadenylation is a major process that regulates gene expression by shaping the length of mRNA poly(A) tails. Deadenylation is controlled by factors *in trans* that recruit or impede deadenylases, by the incorporation of non-adenosines during poly(A) tail synthesis, and by the posttranscriptional addition of 3′ nucleotides to poly(A) tails. Deciphering the regulation of poly(A) tail shortening requires both transcriptome-wide approaches and more targeted methodologies, allowing deep analyses of specific mRNAs. In this chapter, we present Nano3′RACE, a nanopore-based cDNA sequencing method that allows in-depth analysis to precisely measure poly(A) tail length and detect 3′ terminal nucleotide addition, such as uridylation, for mRNAs of interest.

Key words Poly(A) tail, Deadenylation, Uridylation, Nanopore sequencing, mRNA isoforms

1 Introduction

The poly(A) tail serves major functions in the metabolism of mRNAs, including export from the nucleus, translation, and protection against degradation [1]. The shortening of the poly(A) tail is called deadenylation and is carried out by specific enzymes known as deadenylases. Deadenylation can occur both in the nucleus and in the cytoplasm and is mainly catalyzed by the multi-subunit protein complex CCR4-NOT [2, 3]. In most species, the CCR4-NOT complex comprises a conserved core with two catalytic subunits, CCR4 and CAF1, each having distinct and overlapping roles [2–6]. Deadenylation has long been exclusively associated with mRNA decay. However, the most stable and well-translated eukaryotic mRNAs are actively deadenylated and bear short poly(A) tails, which can bind only one or two poly(A) binding proteins (PABP)

Eugene Valkov and Aaron C. Goldstrohm (eds.), *Deadenylation: Methods and Protocols*, Methods in Molecular Biology, vol. 2723, https://doi.org/10.1007/978-1-0716-3481-3_14,

[7]. This accumulation of short poly(A) tail mRNAs may reflect a change in the deadenylation rate [8]. Additionally, other 3′ terminal posttranscriptional modifications, such as guanylation or uridylation, can influence the deadenylation rate and the mRNA steady-state by slowing the removal of the poly(A) tail [9, 10]. The function of deadenylation and tailing, along with their dynamic, is dependent on the organism, cell type, and cellular context. This dynamic can affect poly(A) tail metabolism, thus deeply influencing the transcript's fate. A precise profiling of the poly(A) tails of target transcripts, at the steady-state and over time, is necessary to understand how deadenylation rates impact the metabolism of these mRNAs. Hence, several sequencing-based methods to estimate the poly(A) tail length and composition were designed and extensively reviewed by Brouze et al. [11].

TAIL-seq and PAL-seq were the first methods using Illumina sequencing that were specifically implemented to measure the poly (A) tail length at the transcriptome level, shedding light on the complexity of poly(A) tail length distribution [12–15]. However, technical complexities related to library preparation and data analysis have limited their usage. As an alternative to these global methods to study low expressed transcripts, 3′RACE-seq was used to provide in-depth profiling of the 3′ end of selected target transcripts with very little starting material [16–18]. However, as with all Illumina-based sequencing methods, fluorescence carryover during amplification cycles leads to overestimating the poly(A) tail length [13]. Additionally, the sequencing of DNA spike-in standards with different poly(A) tail sizes revealed a systematic decrease of read recovery for spike-ins with long poly(A) tails compared to those with short tails, resulting in a strong bias toward short tails [14]. Those biases complicate the poly(A) tail length estimation, making the Illumina platform a less appropriate technology for poly (A) profiling compared to third-generation sequencing methods.

The advent of nanopore sequencing developed by Oxford Nanopore Technologies (ONT) has greatly impacted high-throughput RNA sequencing by allowing direct RNA sequencing (DRS), the first efficient PCR-free method used for poly(A) tail profiling [19]. DRS involves real-time sequencing of single RNA molecules and generates full-length, strand-specific reads, enabling the detection of base modifications and the study of transcript isoforms and alternative polyadenylation sites at the single-molecule level [20, 21]. Providing long reads with DRS offers versatility and allows for the precise measurement of poly(A) tails [10, 11, 21, 22]. Indeed, the poly(A) tail length measurement does not rely on the nucleotide basecalling. Rather, the length is calculated based on the poly(A) region signal and is corrected using the time taken by the poly(A) tail to translocate through the pore, overcoming the issues with the basecalling of homopolymeric sequences. However, the use of the technique is limited by the relatively large amount of RNA needed for library preparation (typically more than 100 μg are needed for current protocols), making it unsuitable for many applications. An interesting alternative, also based on sequencing with nanopores, is FLEP-seq2. This PCR-based cDNA-sequencing method provides full-length transcript information and poly(A) profiling at the transcriptome level, starting from 500 ng to 3 μg of total RNA [23]. Additionally, this method has the advantage of allowing the detection of 3′ terminal non-adenosine nucleotides, such as uridines, which is not possible with the currently used DRS protocols because of the RNA ligation of an oligo(dT) splint adapter. On the downside, as for DRS and other transcriptome-wide poly(A) sequencing methods, the depth obtained for transcript expressed at low level is poor. It may not be sufficient to generate accurate poly(A) tail size distributions or to study transcript isoforms.

Here, we describe Nano3′RACE, a nanopore cDNA sequencing method designed to perform in-depth analysis for specific transcripts of interest and to accurately assess the poly(A) tail length, polyadenylation sites, mRNA isoform variety, and the presence of 3′ terminal non-adenosine nucleotides (Fig. 1). This method enables the selection of target transcripts through PCR amplification after 3′ adapter ligation and cDNA synthesis steps, using 5 μg of total RNA as starting material. Barcodes used for multiplexing are added during a second PCR amplification, allowing the simultaneous study of several mRNA targets for different replicates and conditions. More than 50 distinct samples can be analyzed in a single MinION run. The reliability of the Nano3′RACE method is confirmed by a strong correlation of the poly(A) tail size estimation with both DRS and FLEP-seq2 for target transcripts (Fig. 2). Overall, Nano3′RACE represents a cost-effective tool to offer a simple and accurate method for poly(A) tail measurement and the detection of eventual 3′ terminal non-adenosine nucleotides for mRNAs of interest.

2 Materials

2.1 RNA Extraction

1. TRI Reagent (Molecular Research Center).
2. Chloroform.
3. Water-saturated biophenol (pH 4):chloroform:isoamyl alcohol solution (25:24:1, v/v/v).
4. Propan-2-ol.
5. Absolute ethanol.
6. 3 M sodium acetate (pH 5.2).
7. 20 mg/mL glycogen.
8. Refrigerated microcentrifuge reaching 16,000 g.

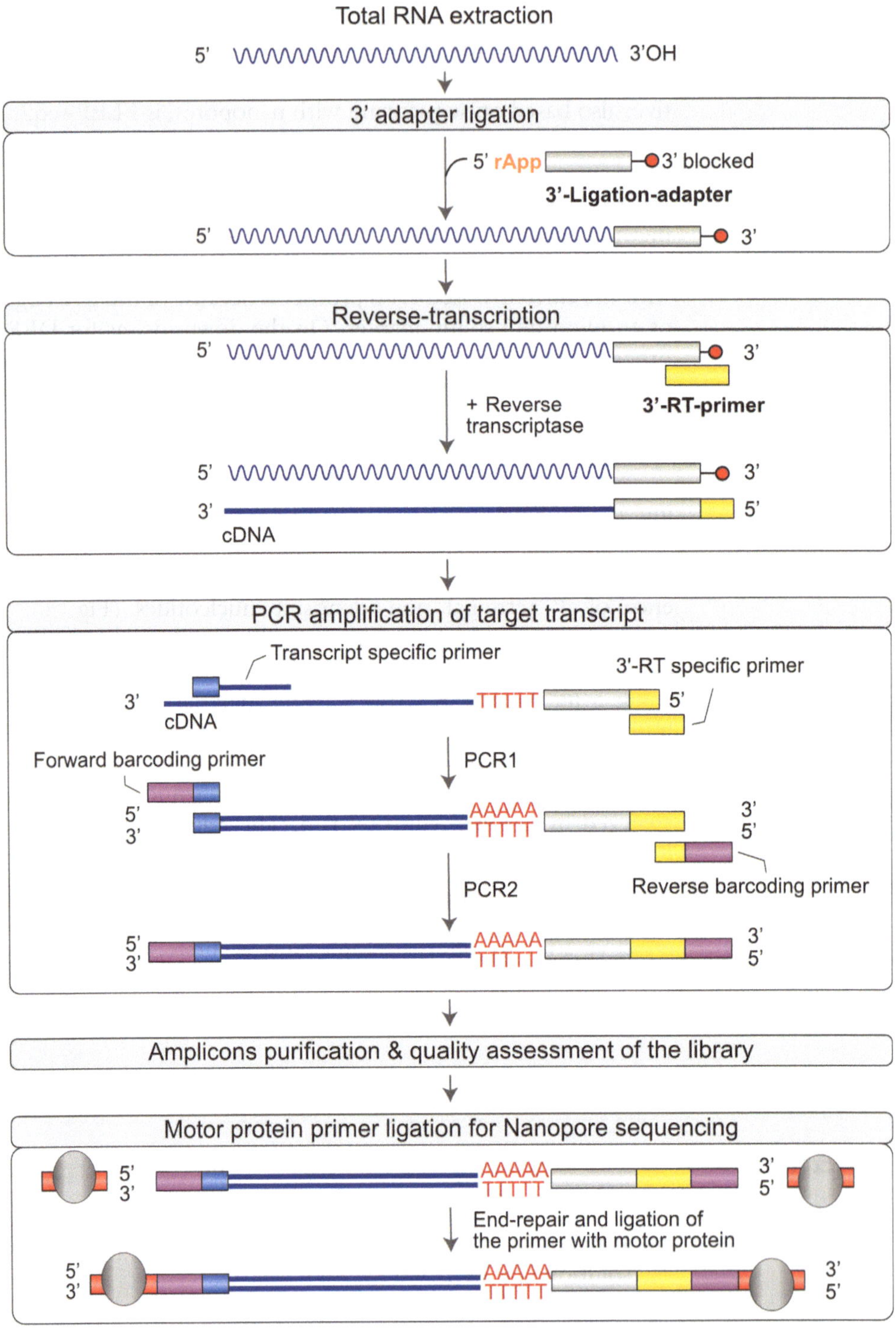

Fig. 1 A flowchart of the Nano3′RACE procedure. After total RNA extraction, an adapter is ligated to the 3′ OH end of RNAs (Subheading 3.2 in the text). This adapter contains a random sequence unique to each RNA molecule, and that may be used to deduplicate amplicon reads (*see* **Note 7**). This deduplication script will be implemented in the next update of the pipeline. A primer complementary to the end of the adapter is then used to reverse-transcribe RNAs (Subheading 3.3 in the text). A first PCR step (PCR1) is performed to specifically amplify the selected targets (Subheading 3.4 in the text). A second round of PCR (PCR2) uses primers that

9. 70% (v/v) ethanol.
10. Nuclease-free water.
11. 2x RNA denaturing loading dye: 0.1% (v/v) formamide, 0.025% (w/v) bromophenol blue, 0.025% (w/v) xylene cyanol FF, 5 mM EDTA, 0.025% (w/v) SDS (pH 8.5).
12. 0.5x TBE buffer: 44.5 mM Tris base, 44.5 mM boric acid, 1 mM EDTA.
13. Gel system for agarose electrophoresis (well combs, casting tray, gel box) and electrophoresis power supply.
14. Agarose.
15. 10 mg/mL ethidium bromide (EtBr).
16. (*Optional*) Agilent 2100 Bioanalyzer.

2.2 3′ Adapter Ligation

1. 3′-Ligation-adapter oligonucleotide (50 μM) (Table 1).
2. Nuclease-free water.
3. Heating block for 16 °C and 65 °C incubation.
4. 10x T4 RNA Ligase Reaction buffer (NEB): 10 mM $MgCl_2$, 50 mM Tris-HCl, 1 mM DTT (pH 7.5).
5. T4 RNA Ligase 2 truncated KQ (20 U/μL).
6. RNA Clean & Concentrator-5 kit (ZYMO).
7. 40 U/μL RNaseOUT.
8. 2x RNA denaturing loading dye: 0.1% (v/v) formamide, 0.025% (w/v) bromophenol blue, 0.025% (w/v) xylene cyanol FF, 5 mM EDTA, 0.025% (w/v) SDS (pH 8.5).
9. 0.5x TBE buffer: 44.5 mM Tris base, 44.5 mM boric acid, 1 mM EDTA.
10. Gel system for agarose electrophoresis (well combs, casting tray, gel box) and electrophoresis power supply.
11. Agarose.
12. 10 mg/mL EtBr.

2.3 cDNA Synthesis

1. 3′-RT-primer (100 μM) (Table 1).
2. 10 mM dNTP mix (dATP, dGTP, dCTP and dTTP, each at 10 mM).
3. Nuclease-free water.

Fig. 1 (continued) include ONT barcodes complementary to the common end sequences added during PCR1 (Subheading 3.6 in the text). Amplicons are then purified and quantified, and library sizes are checked (Subheadings 3.7 and 3.8 in the text). Finally, the amplicons are pooled in a final library, and the primers carrying the motor proteins are ligated. After a final purification step, the library is ready for sequencing on a MinION flow cell (Subheading 3.9 in the text)

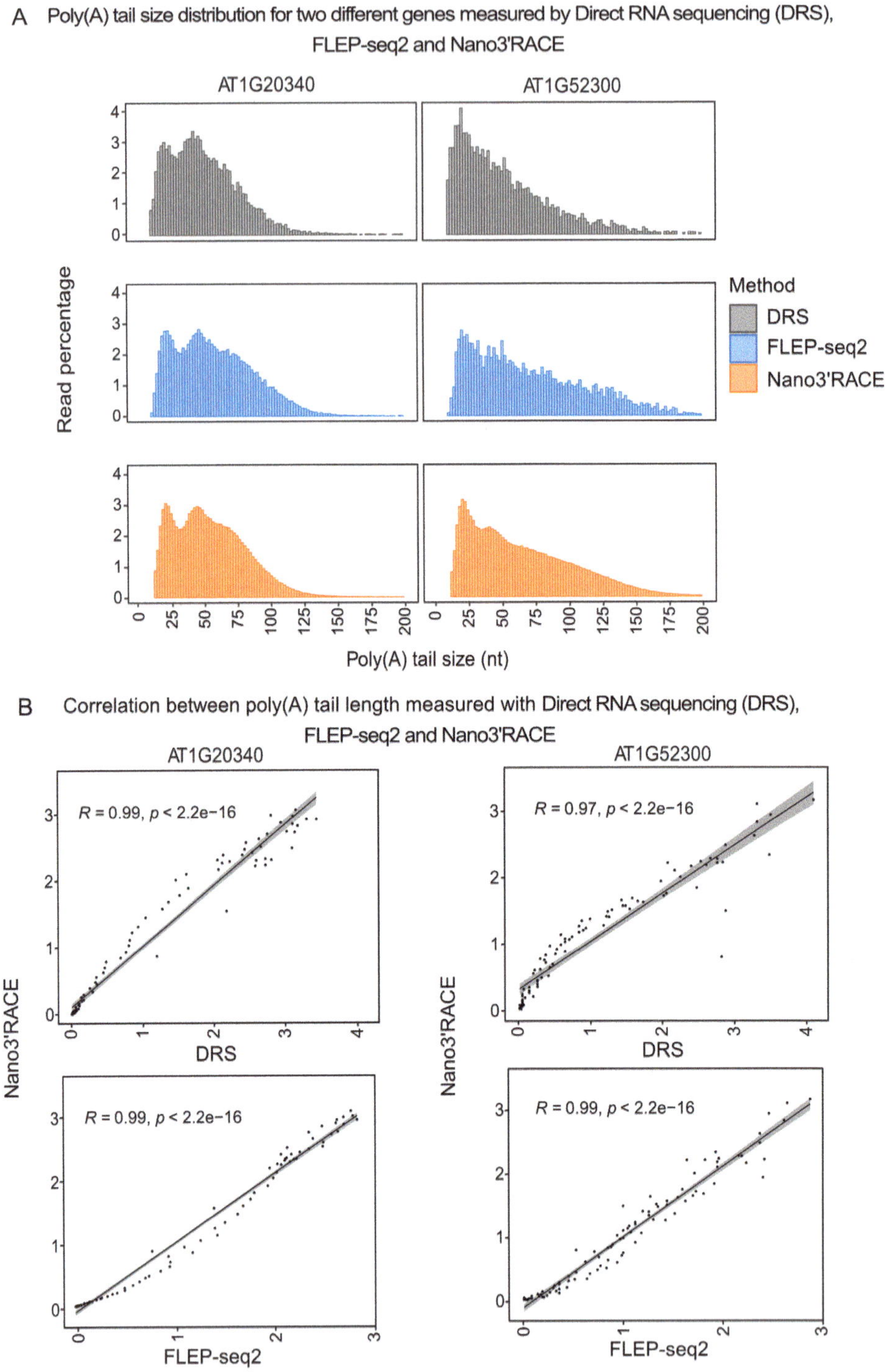

Fig. 2 Poly(A) tail sizes of two representative mRNAs measured using Nano3′RACE, FLEP-seq2, or Direct RNA Sequencing (DRS) datasets are highly similar. (**a**) The percentage of reads according to poly(A) tail size was calculated for each dataset and represented as a distribution. (**b**) Correlation of poly(A) tail size measurements between Nano3′RACE and DRS or FLEP-seq2. Correlations were calculated between the percentages of reads according to poly(A) tail size ($r > 0.95$, Spearman's test p-value $<2.2e-16$)

Table 1
Oligonucleotides used in Nano3′RACE. All oligonucleotides are listed in the 5′ to 3′ orientation

Oligonucleotide name	5′-3′ sequence
3′-Ligation-adapter	5′rApp/CTGAC**NNNNNNNNNN**CTGTAGGCACCATCAAT/3′AmMO
3′-RT-primer	5′Phos/**ACTTGCCTGTCGCTCTATCTTC**ATTGATGGTGCCTACAG/3′
PCR1_Rev	5′ – **ACTTGCCTGTCGCTCTATCTTC** – 3′
Transcript specific_Fw	5′– TTTCTGTTGGTGCTGATATTGC – transcript specific (20–22 nt) – 3′

3′-Ligation-adapter contains two modifications: *5rApp* 5′, 5′-adenyl pyrophosphoryl moiety, *3′AmMO* 3′-Amino Modification (IDT), *5′Phos* 5′ phosphate. The 3′- Ligation-adapter should be purified by HPLC

4. PCR thermal cycler.
5. 0.2 mL strip PCR tubes.
6. 200 U/μL Maxima H Minus Reverse Transcriptase.
7. 5x RT Maxima Buffer: Tris-HCI 250 mM, KCI 375 mM, MgCl2 15 mM, DTT 50 mM (pH 8.3).
8. 40 U/μL RNaseOUT.

2.4 PCR Amplification and Quality Assessment by Electrophoresis

1. Primers: forward PCR primer (Target-specific, *see* **Note 1** and Table 1) and reverse PCR primer (Adapter specific, Table 1) (100 μM each).
2. Forward and reverse barcoding primers (BP, *see* **Note 2**) (100 μM each).
3. 1.25 U/μL PrimeSTAR GXL DNA polymerase (Takara) supplied with 10 x PrimeSTAR GXL DNA polymerase buffer and 2.5 mM dNTP mix (dATP, dGTP, dCTP, and dTTP, each at 2.5 mM) (*see* **Note 3**).
4. Nuclease-free water.
5. PCR thermal cycler.
6. 0.2 mL strip PCR tubes.
7. 6x DNA loading buffer: 10 mM Tris-HCl 60% (v/v) glycerol, 0.03% (w/v) bromophenol blue, 0.03% (w/v) xylene cyanol FF, 60 mM EDTA (pH 7.6).
8. 0.5x TAE buffer: 4 mM Tris-acetate, 0.1 mM EDTA (pH 8.3).
9. Agarose.
10. Gel system for agarose electrophoresis (well combs, casting tray, gel box) and electrophoresis power supply.
11. 10 mg/mL EtBr.

2.5 PCR Product Purification

1. AMPure XP beads (Agencourt) (*see* **Note 4**).
2. Benchtop mini-centrifuge.
3. Magnetic stand compatible with 1.5 mL microtubes (e.g., DynaMag-2).
4. 70% (v/v) ethanol.
5. Nuclease-free water.
6. Microcentrifuge.
7. UV spectrophotometer (e.g., NanoDrop 2000).

2.6 Qubit and Bioanalyzer Analysis of Purified Amplicons

1. Qubit fluorometric quantitation system.
2. Agilent 2100 Bioanalyzer.
3. Agilent High Sensitivity DNA kit for Agilent 2100 Bioanalyzer.
4. Microcentrifuge.

2.7 ONT Library Preparation and Flowcell Loading

1. NEBNext Companion Module for ONT Ligation Sequencing (NEB).
2. AMPure XP beads (Agencourt).
3. 1.5 mL Eppendorf DNA LoBind tubes.
4. 0.2 mL thin-walled PCR tubes.
5. Nuclease-free water.
6. 70% (v/v) ethanol.
7. Ligation Sequencing Kit (SQK-LSK109, ONT).
8. Flow Cell Priming Kit (EXP-FLP002, ONT).
9. Thermal cycler.
10. Magnetic stand compatible with 1.5 mL microtubes (e.g., DynaMag-2).
11. Benchtop mini-centrifuge.
12. ONT R9.4 MinION flowcell.

3 Methods

3.1 RNA Extraction

1. Extract total RNA using TRI Reagent:
   (a) Add 1 mL of TRI Reagent to 100 mg of tissue powder (for cultured cell samples, see the manufacturer's recommendations) and mix by vortexing for 30 s.
   (b) Incubate 5 min at room temperature and centrifuge for 15 min at 16,000 g (4 °C).
   (c) Transfer the aqueous phase into a new tube and add 0.5 mL of propan-2-ol to precipitate the RNA.

   (d) Mix by inversion and incubate for 10 min at room temperature.
   (e) Centrifuge 10 min at 16,000 g (4 °C).
   (f) Discard the supernatant and wash the pellet with 70% (v/v) ethanol to remove residual salt.
   (g) Discard, spin down, and aspirate residual ethanol droplets and dissolve RNA pellet in 100 μL of nuclease-free water.

2. (*Optional*). To further remove contaminants, a second purification step using acid phenol:chloroform:isoamyl alcohol followed by ethanol precipitation can be performed:
   (a) Add 1 volume of biophenol:chloroform:isoamyl alcohol solution (25:24:1, v/v/v) to your sample (*see* **Note 5**).
   (b) Mix by vortexing and centrifuge for 15 min at 16,000 g (4 °C).
   (c) Transfer the aqueous phase into a new tube and precipitate RNA by adding 0.5 μL of glycogen (20 mg/mL), 0.1 volume of 3 M sodium acetate (pH 5.2), and 2.5 volumes of absolute ethanol.
   (d) Mix by inversion and incubate for 1 h at −80 °C.
   (e) Centrifuge for 30 min at 16,000 g (4 °C).
   (f) Discard the supernatant and wash the pellet with 70% (v/v) ethanol to remove residual salt.
   (g) Discard, spin down, and aspirate residual ethanol droplets and dissolve RNA pellet in 20 μL of nuclease-free water.

3. Measure the quantity and purity of your RNA sample with a UV spectrophotometer.

4. To quickly and inexpensively evaluate the integrity of your total RNA sample, examine its profile on a 1% (w/v) agarose 0.5x TBE buffer gel containing EtBr (0.5 μg/ml):
   (a) Aliquot between 500 ng and 1 μg of RNA for each sample in separate tubes and add 1 volume of 2x RNA denaturing loading dye.
   (b) Denature for 5 min at 65 °C and chill on ice before loading into the agarose gel.
   (c) Run agarose gel at 135 V for 25–30 min and visualize using a UV transilluminator.

   Alternatively, RNA quality can be assessed on the Agilent Bioanalyzer system, RNA 6000 Nano kit.

3.2 3′ Adapter Ligation

1. To allow for the analysis of RNA 3′ extremities, a 5′--riboadenylated DNA oligonucleotide (3′-Ligation-adapter, Table 1) is ligated at the RNA 3′ end (*see* **Notes 6**, 7).
   - (a) Take 5 μg of total RNA for each sample.
   - (b) Add nuclease-free water to a final volume of 5 μL.
   - (c) Add 1 μL of the 3′-Ligation-adapter (50 μM).
   - (d) Denature sample for 5 min at 65 °C.
   - (e) Put on ice for at least 2 min.
   - (f) Add 2 μL of 10x T4 RNA Ligase Reaction buffer.
   - (g) Add 1 μL RNAseOUT (40 U/μL).
   - (h) Add 1 μL of T4 RNA Ligase 2 truncated KQ (20 U/μL).
   - (i) Add 10 μL 50% (w/v) PEG8000.
   - (j) Mix well by pipetting and incubate for 10 h at 16 °C.
   - (k) Add 30 μL of nuclease-free water.
2. Purify RNA using the RNA Clean and Concentrator-5 kit:
   - (a) Per sample, mix 50 μL of RNA Binding Buffer with 50 μL of 100% (v/v) ethanol.
   - (b) Add 100 μL of the mixed buffer to the sample and mix by pipetting.
   - (c) Transfer the mixture to a Zymo-Spin Column, centrifuge for 30 s at 16,000 g and discard the flow-through.
   - (d) Add 400 μL RNA Prep Buffer to the column, centrifuge for 30 s at 16,000 g, and discard the flow-through.
   - (e) Add 700 μL RNA Wash Buffer to the column, centrifuge for 30 s at 16,000 g, and discard the flow-through.
   - (f) Add 400 μL RNA Wash Buffer to the column and centrifuge for 1 min at 16,000 g.
   - (g) Transfer the column to a new tube and add 12 μL of nuclease-free water.
   - (h) Centrifuge for 1 min at 16,000 g.
3. Measure the quantity and purity of your RNA sample with a UV spectrophotometer.
4. To quickly and inexpensively evaluate the integrity of your ligated RNA sample, examine its profile on a 1% (w/v) agarose 0.5x TBE buffer gel containing EtBr (0.5 μg/ml):
   - (a) Aliquot between 100 ng and 300 ng of RNA for each sample in separate tubes and add 1 volume of 2x RNA denaturing loading dye.
   - (b) Denature for 5 min at 65 °C and chill on ice before loading onto the agarose gel.

(c) Run agarose gel at 135 V for 25–30 min and visualize using a UV transilluminator.

Alternatively, RNA quality can be assessed on the Agilent Bioanalyzer system using an RNA 6000 Nano kit.

3.3 cDNA Synthesis

cDNA synthesis is initiated using a primer (3′-RT-primer; Table 1) complementary to the last 3′ 20 nucleotides of the 3′-Ligation-adapter. This primer also includes a portion complementary to the primers used for the first round of PCR amplification (Fig. 1). All cDNA synthesis steps are carried out in a PCR thermal cycler using 0.2 mL strip PCR tubes.

1. To 2 μg of adapter-ligated RNA, add 0.5 μL of 100 μM 3′-RT primer and 1 μL of 10 mM dNTP mix.
2. Add nuclease-free water to a final volume of 14 μL.
3. Denature 5 min at 65 °C.
4. Chill samples on ice for at least 2 min.
5. Add 6 μL of RT Mix composed of 4 μL of Maxima RT 5x Buffer, 1 μL of RNaseOUT (40 U/μL) and 1 μL of Maxima H minus Reverse Transcriptase (200 U/μL).
6. Incubate for 30 min at 50 °C.
7. Inactivate the reaction by incubating for 5 min at 85 °C.

3.4 PCR Amplification for Target Selection (PCR1)

Two consecutive PCR rounds are performed: the first PCR allows the amplification of the transcripts of interest. The second PCR allows for adding the barcodes necessary for multiplexing (*see* **Notes 1**, **2**).

A first PCR step of 15 cycles is performed using the cDNA molecules as templates and a forward primer specific to the target sequence combined with a reverse primer complementary to the 3′-Ligation-adapter (*see* **Note 1**, Table 1, Fig. 1). Part of these oligos is complementary to the primers used for the second round of PCR. PCR reactions are performed in a PCR thermal cycler in 0.2 mL strip PCR tubes.

1. Set up the PCR reaction by mixing the following components for each reaction:
   (a) 5 μL of 5x PrimeSTAR GXL DNA polymerase buffer.
   (b) 2 μL of dNTP (2.5 mM each, provided with the GXL).
   (c) 1 μL of 10 μM Transcript_specific_Fw primer (*see* **Note 1**, Table 1).
   (d) 1 μL of 10 μM PCR1_Rev primer (Table 1).
   (e) 0.5 μL of PrimeSTAR GXL DNA polymerase (1.25 U/μ L).
   (f) Add nuclease-free water to a final volume of 24 μL.

2. Distribute PCR mix in strip PCR tubes and add 1 μL of template cDNA.
3. Spin down and place the reaction in a thermal cycler.
4. Run PCR reaction with the following settings:
   (a) Initial denaturation step: 30 s at 98 °C.
   (b) 15 cycles composed of:
      (i) Denaturation step: 10 s at 98 °C.
      (ii) Hybridization step: 15 s at 65 °C.
      (iii) Elongation step: 15 s/kilobases (kb) at 68 °C.
   (c) Final extension: 5 min at 68 °C.

3.5 PCR2 for Cycle Optimization

Optimizing the number of PCR2 cycles is best to minimize over-cycling bias (*see* **Note 8**). We suggest starting the optimization with 5, 10, 15, and 20 cycles and further optimizing the number of cycles if necessary (Fig. 3).

1. For each target and condition, prepare five PCR reactions (four different cycle numbers plus a negative control) by mixing the following components:
   (a) 5 μL of PrimeSTAR GXL DNA polymerase buffer.
   (b) 2 μL of dNTP (2.5 mM each, provided with the PrimeSTAR GXL DNA polymerase).
   (c) 1 μL of 10 μM reverse barcoding primer [24].

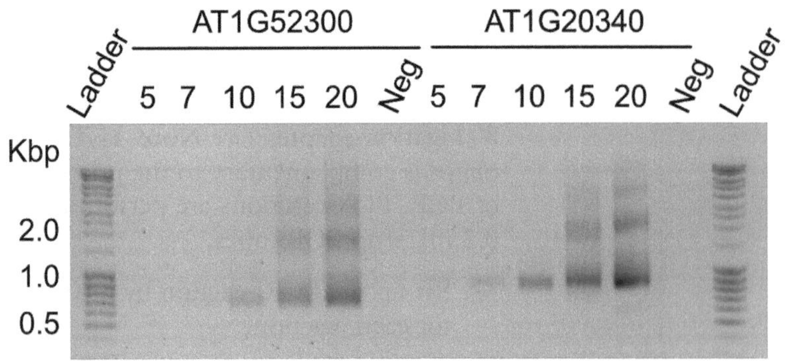

Fig. 3 Cycle optimization for the PCR2 amplification of chosen targets. After the first PCR step of 15 cycles (PCR1), the amplicons are amplified with the barcoding primers (Subheading 3.6 in the text). To choose the optimal cycle condition, amplifications from different cycles are run on a 1% (w/v) agarose 0.5x TAE buffer gel with MassRuler 1 kb DNA Ladder Mix (ThermoFisher). The numbers above the lanes represent the number of cycles. In this example, the recommended number of cycles for the selected targets is ten. Profiles for 15 and 20 cycles show smears indicating over-cycling. Kbp kilobase pairs, Neg negative control

(d) 1 μL of 10 μM forward barcoding primer [24].
(e) 0.5 μL of PrimeSTAR GXL DNA polymerase (1.25 U/μL).
(f) Add nuclease-free water to a final volume of 24 μL.

2. Distribute PCR mix in strip PCR tubes and add 1 μL of PCR1.
3. Spin down and place the reaction in a thermal cycler.
4. Run PCR reaction with the following settings:
   (a) Initial denaturation step: 30 s at 98 °C.
   (b) 5 cycles composed of:
      (i) Denaturation step: 10 s at 98 °C.
      (ii) Hybridization step: 15 s at 65 °C.
      (iii) Elongation step: 15 s/kb at 68 °C.
   (c) Intermediate elongation step: 1 min at 68 °C. At this point, remove the first tube of the thermocycler as the first cycle condition.
   (d) Repeat **step b** and **c** 4 times in total (take out of the thermal cycler the corresponding tube every 5 cycles).
   (e) Final extension: 5 min at 68 °C.

Visualize PCR products by loading 3 μL of the PCR amplicon with 1 μL DNA loading dye 6x and 2 μL of nuclease-free water on a 1% (w/v) agarose 0.5x TAE buffer gel containing EtBr (0.5 μg/ml) (*see* **Note 9**).

3.6 PCR2 for Barcoding Samples

Repeat the PCR amplification using the optimized number of cycles and the different barcode primers complementary to the adapters used for PCR1.

1. Set up the PCR reaction by mixing the following components for each reaction:
   (a) 5 μL of 5x PrimeSTAR GXL buffer.
   (b) 2 μL of dNTP (2.5 mM each, provided with the GXL).
   (c) 1 μL of 10 μM reverse barcoding primer [24].
   (d) 1 μL of 10 μM forward barcoding primer [24].
   (e) 0.5 μL of PrimeSTAR GXL DNA polymerase (1.25 U/μL).
   (f) Add nuclease-free water to a final volume of 24 μL.
2. Distribute PCR mix in strip PCR tubes and add 1 μL of template cDNA.
3. Spin down and place the reaction in a thermal cycler.
4. Run PCR reaction with the following settings:

(a) Initial denaturation step: 30 s at 98 °C.
(b) X (optimized) cycles composed of:
    (i) Denaturation step: 10 s at 98 °C.
    (ii) Hybridization step: 15 s at 65 °C.
    (iii) Elongation step: 15 s /kb at 68 °C.
(c) Final extension: 5 min at 68 °C.

Visualize PCR products by loading 3 μL of the PCR amplicon with 1 μL DNA loading dye 6x and 2 μL of nuclease-free water on a 1% (w/v) agarose 0.5x TAE buffer gel (*see* **Note 9**).

3.7 PCR Product Purification

Purification of PCR products is performed with AMPure XP beads in 1.5 mL tubes compatible with the magnetic stand.

1. Allow the AMPure XP beads to equilibrate to room temperature for at least 10 min and shake the Agencourt AMPure XP bottle before pipetting to resuspend magnetic particles.
2. Transfer each PCR reaction to individual 1.5 mL tubes.
3. Add 1 volume of beads to 1 volume of PCR reaction and mix well by pipetting (*see* **Note 10**).
4. Incubate 5 min at room temperature.

   Tubes are kept on the magnetic stand for the next **steps (5–9)**.
5. Transfer tubes to a magnetic stand and wait 5 min to pellet the beads or until the solution appears clear.
6. Discard the supernatant without disturbing the beads.
7. Wash with 500 μL of 70% (v/v) ethanol without disturbing the beads.
8. Incubate for 1 min and carefully remove the ethanol.
9. Repeat the washing step.
10. Short-spin the tubes to pellet droplets, put them back on the magnetic stand, and remove any remaining ethanol.
11. Air-dry beads for a maximum of 1 min.
12. Add 12 μL of nuclease-free water, remove tubes from the magnetic stand, and resuspend beads by pipetting.
13. Incubate for 5 min at room temperature.
14. Put tubes back on the magnetic stand and wait 5 min to pellet the beads or until the solution appears clear.
15. Transfer 11 μL of supernatant in a new, clean tube and avoid pipetting beads.

3.8 Qubit and Bioanalyzer Analysis of Purified Amplicons

To quantify the libraries, we use the Qubit fluorometric quantitation system. We also verified the size of our purified amplicons using Agilent 2100 Bioanalyzer, to check for any contamination by primer dimers.

1. Determine the Qubit concentration of each library (Qubit fluorometric quantitation system [25]) (*see* **Note 11**):
   (a) Prepare the Qubit Working Solution by diluting Qubit creagent 1:200 in Qubit buffer. Prepare 200 μL of Working Solution for each standard and sample.
   (b) Add 190 μL of Working Solution to 10 μL of each standard.
   (c) Add 199 μL of Working Solution to 1 μL of each sample.
   (d) Mix by vortexing for 5 s and incubate for 2 min at room temperature.
   (e) Insert the tubes in the Qubit Fluorometer and take readings.
2. Check library sizes using Agilent 2100 Bioanalyzer according to the manufacturer's instructions [26]. Choose the reagent kit according to the range of concentration of your samples (*see* **Note 12**).

Use Qubit concentration and Bioanalyzer size estimation to calculate the molarity of each sample library (*see* **Note 13**).

3.9 ONT Library Preparation and Flowcell Loading

Mix each amplicon in an equimolar pool of 200 fmol and follow the Ligation sequencing amplicons (SQK-LSK109) protocol by ONT according to the manufacturer's instructions [27] (*see* **Notes 14, 15**).

3.10 Data Analysis

Process the obtained fast5 files using a pipeline based on [23] and adapted for Nano3′RACE available on GitHub (https://github.com/jackson-peter/NanoRACE). Figure 4 illustrates the bioinformatic workflow (*see* **Note 16**).

4 Notes

1. Nano3′RACE allows for the analysis of full-length transcripts. Design a PCR1 forward primer that is complementary to the 5′ region of your target mRNA. This primer should be between 19 and 22 nucleotide long, have a GC content of around 45–50% and a melting temperature of around 60 °C. To this gene-specific primer, add the sequence 5′-TTTCTGTTGGTGCTGATATTGC-3′ that is necessary for PCR2 primer anchoring (Fig. 1).

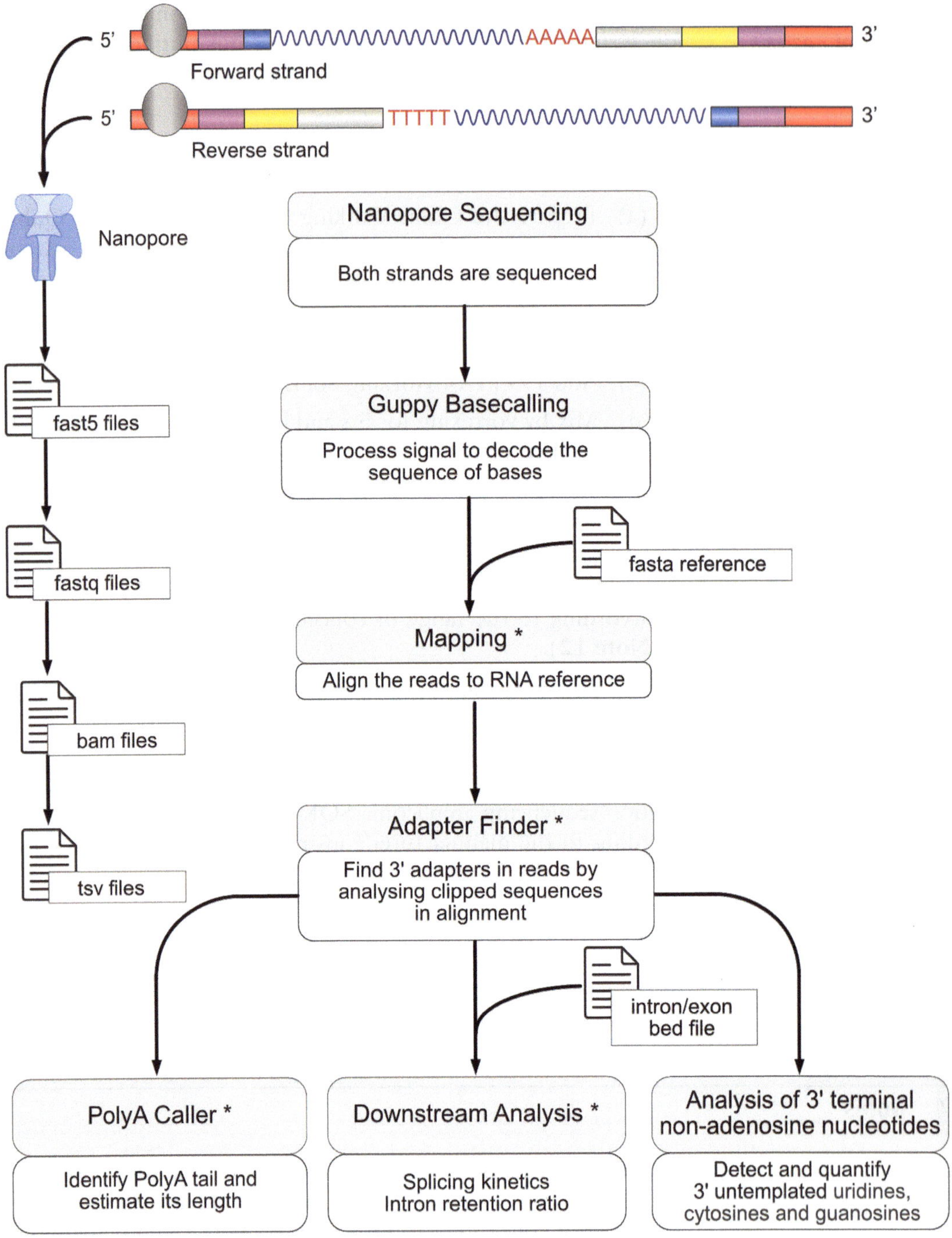

Fig. 4 Schematic overview of the analysis pipeline for Nano3′RACE data. Nanopore sequencing produces fast5 files containing raw signal data. Those files are basecalled using ONT Guppy basecalling software (https:/nanoporetech.com/) to obtain .fastq files. Reads are then mapped to a reference sequence provided as a .fasta file. Adapter sequences are searched in the clipped sequences of the alignment, and only reads corresponding to full-length amplicons are retained for the following analysis. The pipeline allows the measurement of poly (A) tail size, the detection of 3′-terminal non-adenosines, and the identification of different RNA isoforms. The poly(A) tail length is measured based on the basecalling sequence. The length is then corrected using the raw signal information of the poly(A) region in the .fast5 file. At the same time, the 3′-terminal non-adenosines are

2. The classical SQK-LSK109 kit from ONT does not include barcodes. However, an expansion kit with 96 barcodes (EXP-PBC096) can be purchased directly from ONT. As a cheaper alternative, a possibility is to order custom primers for the corresponding barcodes using the sequences available in the ONT documentation [24]. These primers are used during PCR2 and are partially complementary to the sequences added during PCR1 (Fig. 1).
3. We use PrimeSTAR GXL DNA polymerase by Takara as it provides an efficient amplification even with homopolymeric sequences such as poly(A) tails.
4. To purify PCR products, ONT recommend using the AMPure XP beads (Agencourt). These magnetic beads can bind PCR products of at least 100 bp, allowing the purification of amplicons from primer dimers, nucleotides, salts, and other reagents.
5. In alkaline pH conditions, DNA and RNA molecules are highly soluble and retained in the aqueous phase. In acidic pH conditions, DNA molecules are more soluble in the organic phase and interphase, while RNA remains in the aqueous phase. Hence, acid phenol is used for RNA isolation, whereas DNA isolation is best performed with buffer-saturated phenol (pH ~ 8).
6. The 3′-Ligation-adapter contains a 5′,5′-adenyl pyrophosphoryl moiety that the T4 RNA ligase 2 KQ can ligate to a 3′-hydroxyl terminated single-strand RNA without the need for free ATP (Fig. 1). This allows for the specific ligation of the adapter and prevents the ligation of endogenous RNA having a 5′ phosphate.
7. The 3′-Ligation-adapter contains a random sequence of 10N, which can discriminate reads originating from a single RNA molecule from PCR duplicates. This 10N sequence offers around a million of unique molecular indexes (UMI) for each target. Recently, UMIs were specifically designed for ONT and PacBio platforms to improve UMI recognition accuracy and reliability [28]. In the future, we will optimize the UMI sequence included in the 3′-Ligation-adapter sequence. We will implement an additional step to identify and tag PCR duplicates in the next update of the analysis pipeline (Fig. 4).
8. To determine the best PCR2 cycle condition for each target, repeat the PCR optimization with a narrower range (e.g., every cycle).

Fig. 4 (continued) detected based on the fastq sequence. The steps marked with an asterisk (*) are adapted from the original FLEP-seq2 pipeline [22]. All the scripts are available on GitHub (https:/github.com/jackson-peter/NanoRACE)

9. To analyze amplicons made with PrimeSTAR GXL DNA polymerase by agarose gel electrophoresis, avoid using Tris-borate EDTA (TBE) buffer and use Tris-acetate EDTA buffer instead (0.5× TAE buffer: 4 mM Tris-acetate, 0.1 mM EDTA (pH 8.3)), as recommended by the manufacturer.
10. Adjusting the beads/sample ratio allows recovery of other size ranges of DNA fragments. It can help better to remove primers or contaminating bands during PCR amplicon purification. According to the manufacturer's instructions, a beads/sample ratio of 0.8 should effectively remove bands under 300 base pairs (bp).
11. The Invitrogen Qubit is a fluorescence-based quantification method that accurately measures RNA, DNA, or protein concentration using target-selective dyes. It allows for a more specific and sensitive quantification compared to spectrophotometry-based devices. Various assay kits are available, but for the needs of this protocol, we suggest using the dsDNA HS Assay kit as the quantification range starts from 10 pg/μL up to 100 ng/μL.
12. While analyzing many samples, it is possible to pool PCR amplicons of each target before Bioanalyzer analysis. Adjust the samples to the same concentration after Qubit measurement, pool them in an equimolar ratio, and perform the analysis.
13. According to the SQK-PCB109 ligation sequencing kit protocol [29], 100 ng of a 1.5 kb amplicon correspond to 103 fmol of DNA.
14. The more recent ligation sequencing kits, SQK-LSK110, SQK-LSK112, and SQK-LSK114, are also compatible with this protocol. Of note, SQK-LSK114 is compatible with the R10 flowcell chemistry.
15. In the SQK-LSK109 ligation sequencing kit, the primers carrying the motor protein are ligated by an enzymatic reaction to the ends of the amplicons after a step of end-repair [27]. The SQK-PCB109/111 is a faster but more expensive alternative to the SQK-LSK109. This kit provides barcoding primers for the PCR2 with the same sequence as the one used for the LSK109 but with the addition of a 5′ modification [29]. This modification allows for rapid and enzyme-free ligation of the motor protein with the Rapid Attachment Primer (RAP) included in the kit.
16. The normal output for a MinION flow cell is around 10–20 M reads. In our hands, analyzing an average of 50 different samples per run gives a depth of 200,000 reads per sample.

Acknowledgments

This work of the Interdisciplinary Thematic Institute IMCBio, as part of the ITI 2021–2028 program of the University of Strasbourg, CNRS and Inserm, was supported by IdEx Unistra (ANR-10-IDEX-0002), and by SFRI-STRAT'US project (ANR 20-SFRI-0012) and EUR IMCBio (ANR-17-EURE-0023) under the framework of the French Investments for the Future program.

References

1. Nicholson AL, Pasquinelli AE (2019) Tales of detailed poly(A) tails. Trends Cell Biol 29:191–200. https://doi.org/10.1016/j.tcb.2018.11.002
2. Collart MA (2016) The Ccr4-Not complex is a key regulator of eukaryotic gene expression. WIREs RNA 7:438–454. https://doi.org/10.1002/wrna.1332
3. Doidge R, Mittal S, Aslam A et al (2012) Deadenylation of cytoplasmic mRNA by the mammalian Ccr4–Not complex. Biochem Soc Trans 40:896–901. https://doi.org/10.1042/BST20120074
4. Raisch T, Chang C-T, Levdansky Y et al (2019) Reconstitution of recombinant human CCR4-NOT reveals molecular insights into regulated deadenylation. Nat Commun 10:3173. https://doi.org/10.1038/s41467-019-11094-z
5. Tucker M, Valencia-Sanchez MA, Staples RR et al (2001) The transcription factor associated Ccr4 and Caf1 proteins are components of the major cytoplasmic mRNA deadenylase in Saccharomyces cerevisiae. Cell 104:377–386. https://doi.org/10.1016/s0092-8674(01)00225-2
6. Webster MW, Chen Y-H, Stowell JAW et al (2018) mRNA deadenylation is coupled to translation rates by the differential activities of Ccr4-Not nucleases. Mol Cell 70:1089–1100.e8. https://doi.org/10.1016/j.molcel.2018.05.033
7. Lima SA, Chipman LB, Nicholson AL et al (2017) Short poly(A) tails are a conserved feature of highly expressed genes. Nat Struct Mol Biol 24:1057–1063. https://doi.org/10.1038/nsmb.3499
8. Park J, Kim M, Yi H et al (2023) Short poly(A) tails are protected from deadenylation by the LARP1–PABP complex. Nat Struct Mol Biol 30:330–338. https://doi.org/10.1038/s41594-023-00930-y
9. Lim J, Kim D, Lee Y-S et al (2018) Mixed tailing by TENT4A and TENT4B shields mRNA from rapid deadenylation. Science 361:701–704. https://doi.org/10.1126/science.aam5794
10. Scheer H, de Almeida C, Ferrier E et al (2021) The TUTase URT1 connects decapping activators and prevents the accumulation of excessively deadenylated mRNAs to avoid siRNA biogenesis. Nat Commun 12:1298. https://doi.org/10.1038/s41467-021-21382-2
11. Brouze A, Krawczyk PS, Dziembowski A et al (2023) Measuring the tail: methods for poly(A) tail profiling. WIREs RNA 14:e1737. https://doi.org/10.1002/wrna.1737
12. Subtelny AO, Eichhorn SW, Chen GR et al (2014) Poly(A)-tail profiling reveals an embryonic switch in translational control. Nature 508:66–71. https://doi.org/10.1038/nature13007
13. Chang H, Lim J, Ha M et al (2014) TAIL-seq: genome-wide determination of poly(A) tail length and 3′ end modifications. Mol Cell 53:1044–1052. https://doi.org/10.1016/j.molcel.2014.02.007
14. Zuber H, Scheer H, Ferrier E et al (2016) Uridylation and PABP cooperate to repair mRNA deadenylated ends in Arabidopsis. Cell Rep 14:2707–2717. https://doi.org/10.1016/j.celrep.2016.02.060
15. Morgan M, Much C, DiGiacomo M et al (2017) mRNA 3′ uridylation and poly(A) tail length sculpt the mammalian maternal transcriptome. Nature 548:347–351. https://doi.org/10.1038/nature23318
16. Wilusz JE, Whipple JM, Phizicky EM et al (2011) tRNAs marked with CCACCA are targeted for degradation. Science 334:817–821. https://doi.org/10.1126/science.1213671
17. Warkocki Z, Krawczyk PS, Adamska D et al (2018) Uridylation by TUT4/7 restricts retrotransposition of human LINE-1s. Cell 174:1537–1548.e29. https://doi.org/10.1016/j.cell.2018.07.022
18. Zuber H, Scheer H, Joly A-C et al (2018) Respective contributions of URT1 and

HESO1 to the uridylation of 5′ fragments produced from RISC-cleaved mRNAs. Front Plant Sci 9. https://doi.org/10.3389/fpls.2018.01438

19. Garalde DR, Snell EA, Jachimowicz D et al (2018) Highly parallel direct RNA sequencing on an array of nanopores. Nat Methods 15: 201–206. https://doi.org/10.1038/nmeth.4577
20. Lorenz DA, Sathe S, Einstein JM et al (2020) Direct RNA sequencing enables m6A detection in endogenous transcript isoforms at base-specific resolution. RNA 26:19–28. https://doi.org/10.1261/rna.072785.119
21. Parker MT, Knop K, Sherwood AV et al (2020) Nanopore direct RNA sequencing maps the complexity of Arabidopsis mRNA processing and m6A modification. eLife 9:e49658. https://doi.org/10.7554/eLife.49658
22. Bilska A, Kusio-Kobiałka M, Krawczyk PS et al (2020) Immunoglobulin expression and the humoral immune response is regulated by the non-canonical poly(A) polymerase TENT5C. Nat Commun 11:2032. https://doi.org/10.1038/s41467-020-15835-3
23. Jia J, Lu W, Liu B et al (2022) An atlas of plant full-length RNA reveals tissue-specific and monocots–dicots conserved regulation of poly (A) tail length. Nat Plants 8:1118–1126. https://doi.org/10.1038/s41477-022-01224-9
24. Oxford Nanopore Technologies (2023) Chemistry technical document. Accessed 14 Apr 2023
25. Thermo Fisher Scientific (2023) Qubit 4 Fluorometer guide. Accessed 14 Apr 2023
26. Agilent (2023) Agilent 2100 Bioanalyzer user guide. Accessed 14 Apr 2023
27. Oxford Nanopore Technologies (2023) SQK-LSK109 protocol for MinION. Accessed 14 Apr 2023
28. Karst SM, Ziels RM, Kirkegaard RH et al (2021) High-accuracy long-read amplicon sequences using unique molecular identifiers with Nanopore or PacBio sequencing. Nat Methods 18:165–169. https://doi.org/10.1038/s41592-020-01041-y
29. Oxford Nanopore Technologies (2023) SQK-PCB109 protocol for MinION. Accessed 14 Apr 2023

Chapter 15

Quantification of Poly(A) Tail Length and Terminal Modifications Using Direct RNA Sequencing

Ankit Gupta, Brian N. Papas, Marine Baptissart, and Marcos Morgan

Abstract

Poly(A) tail metabolism is critical for various biological processes, including early embryogenesis and cell differentiation. While traditional biochemical methods to measure poly(A) tail length allow for the study of selected transcripts, the advent of long-read sequencing technologies enabled the development of simple and robust protocols to measure poly(A) tail length at the transcriptome level. Here, we describe a direct RNA sequencing protocol to capture poly(A) tail terminal additions based on the splint ligation of barcoded oligos compatible with terminal guanylation and uridylation. We cover how to prepare the libraries and perform the bioinformatics analysis to simultaneously determine the length of the transcripts' poly(A) tails and detect the presence of terminal guanylation and uridylation.

Key words Direct RNA-seq, Poly(A) tail, Terminal guanylation, Terminal uridylation, Adenylation

1 Introduction

The metabolic state of a transcript is reflected by the length of its poly(A) tail [1, 2]. Studies from early embryo development and germ cells have shown dynamic changes in the poly(A) tail length to be relevant for cell differentiation [3, 4]. However, measuring poly (A) tail lengths has been challenging. Traditional biochemical approaches, such as northern blot analysis coupled with targeted poly(A) tail degradation with RNase H, have been used to determine the poly(A) tail length of specific transcripts [5]. Although these techniques can reliably measure the length of the poly (A) tails, they are limited to a few selected transcripts. With the expansion of short RNA sequencing capacity, researchers developed methods to capture poly(A) tails at the transcriptome level. However, the implementation of these methods often required interventions on the hardware or access to raw image data from the sequencing instrumentation [3, 6–8]. Only a handful of laboratories use these methods, and some are no longer being developed.

Eugene Valkov and Aaron C. Goldstrohm (eds.), *Deadenylation: Methods and Protocols*, Methods in Molecular Biology, vol. 2723,
https://doi.org/10.1007/978-1-0716-3481-3_15,

The systematic analysis of poly(A) tails from the whole transcriptome became more accessible with the advent of long-read sequencing technologies.

During the last decade, two long-read platforms, Oxford Nanopore Technologies (ONT) and Pacific Biosciences (PacBio), facilitated the precise determination of poly(A) tails at the transcriptome level [5]. PacBio relies on detecting the activity of a DNA polymerase as it incorporates nucleotides complementary to a template sequence [9]. ONT sequencing instead determines the sequence of DNA or RNA as the molecule is threaded through a pore inducing changes to the flow of ions [10]. Since multiple nucleotides occupy the channel at any given time, the current depends on the identity of all nucleotides transiting through the pore. Thus, base-calling algorithms integrate the change in the ionic current signal through time. Heterogenous sequences can be determined unambiguously since every nucleotide going through the pore changes the current. However, the length of homopolymer sequences, such as the poly(A) tail, cannot be inferred by the number of current fluctuations because the transition from one base to the next does not alter the sequence of nucleotides in the pore. Thus, the length of the poly(A) tail is estimated by its retention time within the pore, which typically corresponds to the longest homopolymeric stretch close to the read's 3′ end [11].

The nanopore technology is the only one capable of directly sequencing RNA without cDNA conversion. The relatively short and straightforward RNA sequencing protocol allowed for its widespread adoption. Moreover, direct RNA sequencing avoids PCR steps, which can also introduce amplification biases affecting the poly(A) tail length estimation [5]. For the bioinformatics analysis, robust programs have been developed to accurately determine the length of the poly(A) from the ionic current data, which is easily accessible after the run is completed [11].

The poly(A) tails can be extended by terminal nucleotidyl transferases impacting the metabolism of the transcripts [12, 13]. For example, the terminal nucleotidyltransferase 4A (TENT4A) and the terminal nucleotidyltransferase 4B (TENT4B) introduce Gs into the poly(A) tail promoting transcript stability [14], while the terminal uridylyl transferases 4 (TUT4) and the terminal uridylyl transferase 7 (TUT7) add terminal uracils to the 3′ end of the poly(A) tail promoting transcript decay [15]. However, not all protocols that measure tail lengths capture the terminally modified poly(A) tails [5].

Here, we present a direct RNA sequencing protocol to simultaneously determine poly(A) tail length and terminal additions at a transcriptome level. The approach uses a combination of splint ligation adaptors to capture terminal uridylation and guanylation (Fig. 1). The adaptors are barcoded by modification type, and the

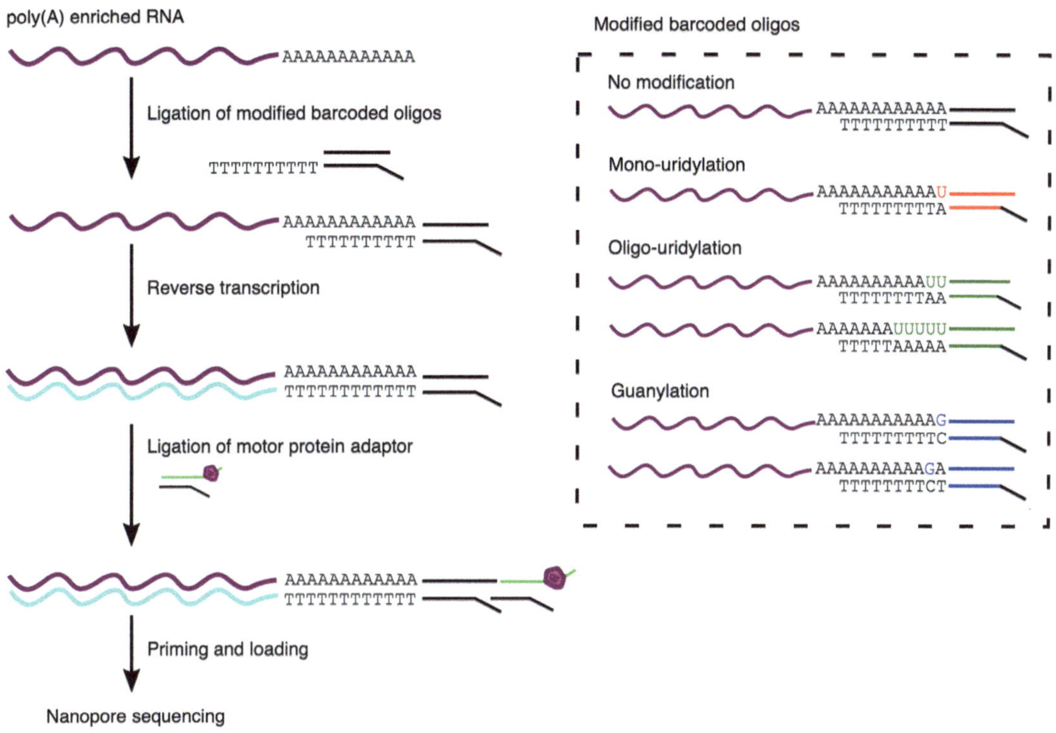

Fig. 1 Schematic representation of the splint ligation-based approach to detect poly(A) tail terminal additions. mRNAs are ligated to barcoded oligos that capture different terminal additions, reverse transcribed, and ligated to the motor protein adapter. The library is then run on an ONT device

identity of the terminal addition is determined, not by sequencing of the 3′ end, but by the read's barcode. The protocol covers all relevant steps, from poly(A)+ RNA selection to data generation. The wet lab part of the protocol includes the synthesis of barcoded poly(A) tail standards and library preparation using the nanopore direct RNA sequencing kit with custom oligos designed to capture the terminal additions. The bioinformatics part covers the determination of the poly(A) tail length and the identification of terminally modified 3′ ends. Finally, we show some typical results obtained using this approach (Figs. 2 and 3)

2 Materials

Prepare solutions using nuclease-free water.

2.1 Standard Preparation

1. Oligos for standard preparation (*see* Table 1).
2. Template plasmid for standards: pSpCas9(BB)-2A-GFP (PX458) [16].
3. Phusion High-Fidelity DNA polymerase.

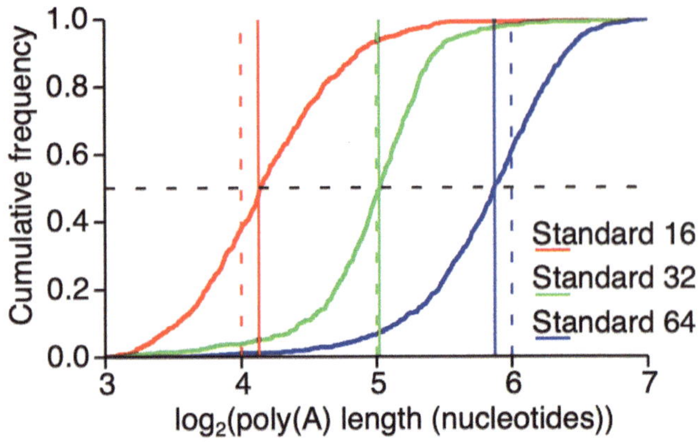

Fig. 2 Cumulative distribution plot of in vitro synthesized RNA standards. A solid vertical line indicates the median poly(A) tail length for each standard

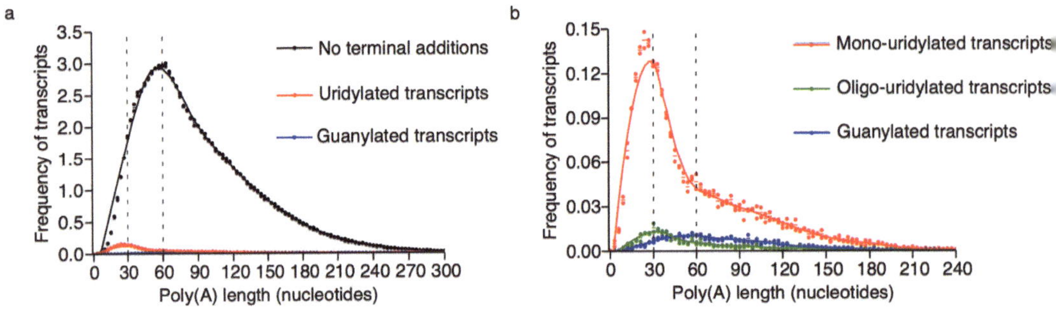

Fig. 3 Typical results obtained using this protocol. (**a**) Poly(A) tail length distribution of not terminally modified (black), uridylated (red), and guanylated (blue) transcripts from C2C12-derived myotubules. (**b**) Poly(A) tail length distribution as in (**a**) with uridylated tails split in mono-uridylated (red) and oligo-uridylated (green). The distribution of guanylated transcripts (blue) is the same as in (**a**), and the non-modified tails are omitted

Table 1
Oligos used to generate standards with different poly(A) tail lengths

Oligo name	Oligo sequence
RNA standard Forward	CGTCGAGGAGTAATACGACTCACTATAGAATCCTGGCCCAG TGAGCAA
RNA standard 16 Reverse	T(16)GGCACAGTCGGCTCGTCGCGCACAAGGCTGATCAGC GAGCTCTA
RNA standard 32 Reverse	T(32)GGCACAGTCGACAGTGCGCTGTCTATAGGCTGATCAGC GAGCTCTA
RNA standard 64 Reverse	T(64)GGCACAGTCGTCACACTCTAGAGCGAAGGCTGATCAGC GAGCTCTA

4. 5× Phusion HF Buffer.
5. QIAquick PCR Purification Kit.
6. T7 RNA Polymerase Mix.
7. 10× T7 RNA polymerase reaction buffer: 0.4 M Tris-HCl, 60 mM $MgCl_2$, 10 mM DTT, 20 mM spermidine (pH 7.9).
8. RNA Clean and Concentrator kit (Zymo Research).

2.2 Selection of Poly (A)+ RNA

1. Oligo$(dT)_{25}$ magnetic beads.
2. Binding buffer: 20 mM Tris-HCl, 1.0 M LiCl, 2 mM EDTA (pH 7.5). Store at room temperature.
3. Wash buffer: 10 mM Tris-HCl, 0.15 M LiCl, 1 mM EDTA (pH 7.5). Store at room temperature.
4. Elution buffer: 10 mM Tris-HCl (pH 7.5). Store at room temperature.

2.3 Library Preparation

1. Annealing buffer: 10 mM Tris-HCl, 50 mM NaCl (pH 7.5). Store at room temperature.
2. Oligos for barcoded adaptor preparation (*see* Table 2).
3. Direct RNA Sequencing Kit (ONT).
4. MinION Flow Cell (*see* **Note 1**).
5. Flow Cell Priming Kit (ONT).
6. Flow Cell Wash Kit (ONT).
7. T4 DNA Ligase.
8. 10× T4 DNA ligase buffer: 0.5 M Tris-HCl, 0.1 M $MgCl_2$, 0.1 M DTT, 10 mM ATP (pH 7.5).
9. SuperScript II Reverse Transcriptase.
10. 5× First-strand buffer: 0.25 M Tris-HCl, 375 mM KCl, 15 mM $MgCl_2$ (pH 8.3).
11. Ligation buffer: 6% PEG8000 in 10× T4 DNA ligase buffer.
12. AMPure beads (Beckman Coulter).

2.4 Software

1. MinKNOW software version 5.2.15.
2. DeePlexiCon version 1.2.0.
3. SAMtools version 1.7 using htslib 1.7–2.
4. Nanopolish version 0.13.3.
5. Minimap2 version 2.20-r1061.
6. Guppy version 5.0.10+4b5802d0b.

Table 2
Oligos used to capture terminal RNA additions

Oligo name	Oligo sequence	Barcode name	Percent
BC1 anneal 1F BC1 anneal 1R	/5Phos/GGCTTCTTCTTGCTCTTAGGTAGTAGGTTC GAGGCGAGCGGTCAATTTTCCTAAGAGCAAGAAGAAGCCTTTTTTTTTT	Annealed oligo 1	90%
BC2 anneal 2F BC2 anneal 2R	/5Phos/GTGATTCTCGTCTTTCTGCGTAGTAGGTTC GAGGCGAGCGGTCAATTTTCGCAGAAAGACGAGAATCACATTTTTTTTT	Annealed oligo 2	3%
BC3 anneal 3F BC3 anneal 3R	/5Phos/GTACTTTTCTCTTTGCGCGGTAGTAGGTTC GAGGCGAGCGGTCAATTTTCCGCGCAAAGAGAAAAGTACAATTTTTTTT	Annealed oligo 3	2.40%
BC3 anneal 4F BC3 anneal 4R	/5Phos/GTACTTTTCTCTTTGCGCGGTAGTAGGTTC GAGGCGAGCGGTCAATTTTCCGCGCAAAGAGAAAAGTACAAATTTTTTT	Annealed oligo 4	0.45%
BC3 anneal 5F BC3 anneal 5R	/5Phos/GTACTTTTCTCTTTGCGCGGTAGTAGGTTC GAGGCGAGCGGTCAATTTTCCGCGCAAAGAGAAAAGTACAAAATTTTTT	Annealed oligo 5	0.08%
BC3 anneal 6F BC3 anneal 6R	/5Phos/GTACTTTTCTCTTTGCGCGGTAGTAGGTTC GAGGCGAGCGGTCAATTTTCCGCGCAAAGAGAAAAGTACAAAAATTTTT	Annealed oligo 6	0.08%
BC4 anneal 7F BC4 anneal 7R	/5Phos/GGTCTTCGCTCGGTCTTATTTAGTAGGTTC GAGGCGAGCGGTCAATTTTAATAAGACCGAGCGAAGACCCTTTTTTTTT	Annealed oligo 7	3%
BC4 anneal 8F BC4 anneal 8R	/5Phos/GGTCTTCGCTCGGTCTTATTTAGTAGGTTC GAGGCGAGCGGTCAATTTTAATAAGACCGAGCGAAGACCTCTTTTTTTT	Annealed oligo 8	0.75%
BC4 anneal 9F BC4 anneal 9R	/5Phos/GGTCTTCGCTCGGTCTTATTTAGTAGGTTC GAGGCGAGCGGTCAATTTTAATAAGACCGAGCGAAGACCTTCTTTTTTT	Annealed oligo 9	0.20%
BC4 anneal 10F BC4 anneal 10R	/5Phos/GGTCTTCGCTCGGTCTTATTTAGTAGGTTC GAGGCGAGCGGTCAATTTTAATAAGACCGAGCGAAGACCTTTCTTTTTT	Annealed oligoOligo(dT) 10	0.05%

3 Methods

Carry all steps at room temperature, but keep the reagents for the enzymatic reactions, the enzymatic reaction mixtures, and the RNA on ice in between steps. All centrifugation steps are carried out on a tabletop centrifuge at 17,900 g. To avoid sample loss, use low-binding tubes throughout the protocol.

3.1 Preparation of Standards

1. To generate in vitro transcription templates, prepare a PCR amplification mix for each standard containing 20 μL of 5× Phusion HF Buffer, 1 μL of 2000 U/mL Phusion High Fidelity DNA polymerase, 10 μL of 10 μM forward and standard-specific reverse primers each (Table 1), 10 μL of 10 pg/μL template plasmid, and 10 μL of 10 mM dNTP mix in a final volume of 100 μL. Then, run an amplification reaction starting with 3 min of denaturation at 95 °C, followed by 34 amplification cycles of 30 s of denaturation at 95 °C, 30 s of annealing at 58 °C and 2 min of elongation at 72 °C. After the amplification, do a final extension at 72 °C for 30 min before cooling the samples to 12 °C (*see* **Note 2**).
2. Purify the amplicons using the QIAquick PCR Purification Kit. Add 500 μL of buffer PB to the 100 μL of PCR reaction, mix by pipetting, and transfer to the provided column. Spin down for 1 min and discard the flow through. Next, wash the column by adding 750 μL of buffer PE. After removing the flow through, centrifuge the tube for two additional minutes (*see* **Note 3**). Finally, add 30 μL of buffer EB to the center of the column, centrifuge for 1 min, and recover the flow through with the purified amplicon.
3. Perform in vitro transcription on each amplicon to obtain the RNA standards. For each template, prepare a reaction mix containing 2 μL of 10× T7 RNA polymerase reaction buffer, 2 μL of the T7 RNA Polymerase Mix, 2 μL of 100 mM ATP, GTP, CTP, and UTP each, and 1 μg of the amplicon in a final volume of 20 μL. Mix the reaction by pipetting and incubate for 2 h at room temperature.
4. To remove the template from the reaction using DNase treatment, prepare a mix with 5 μL of DNase I (1 U/μL) and 5 μL of 10× DNA Digestion Buffer from the RNA Clean and Concentrator kit in a final volume of 30 μL and add it to the in vitro transcribed RNA. Incubate at room temperature for 15 min.
5. Purify the DNase-treated RNA with the RNA Clean and Concentrator kit by first adding 50 μL of RNA binding buffer followed by 50 μL of 100% ethanol to the DNase reaction. Then, mix by pipetting, transfer to the provided column, and spin for 30 s to allow binding of the RNA to the column. Wash

the column with 400 μL of RNA Prep Buffer, followed by a wash with 700 μL of RNA Wash Buffer and a final wash with 400 μL of RNA Wash Buffer. Remove the flow through after each step and centrifuge for 2 min after the last wash (*see* **Note 3**). Finally, add 30 μL of nuclease-free water to the center of the column, centrifuge, and recover the eluate with the purified RNA.

6. Quantify the RNA and prepare an equimolar mix of all standards with a final concentration of 50 ng/μL (100×).
7. Dilute some of the standard mix to 0.5 ng/μL (1×) and aliquot in batches of 10 μL ready to use. Store the 100× mix and the 1× aliquots at −80 °C.

3.2 Selection of Poly (A)+ RNA

1. To prepare the oligo(dT) beads, transfer 40 μL of the beads' suspension per sample to a low-binding tube and remove the storage buffer using a magnetic rack. Wash the beads with 40 μL of binding buffer and resuspend them in 40 μL of binding buffer (*see* **Note 4**).
2. To purify the poly(A)+ RNA, dilute 1–5 μg of total RNA in nuclease-free water to a final volume of 40 μL and equilibrate by adding 40 μL of binding buffer. Denature the diluted RNA by incubating at 65 °C for 2 min and then place it immediately on ice. Combine 40 μL of pre-equilibrated oligo(dT) beads with the denatured RNA and incubate the mixture at room temperature for 5 min. After the incubation, place the mix on a magnetic rack and remove the supernatant. Wash the beads with wash buffer twice. Remove the supernatant after the last wash and resuspend the beads in 10 μL of elution buffer. Finally, elute the poly(A) selected RNA by incubating for 2 min at 80 °C and removing the beads using a magnetic rack (*see* **Note 5**).

3.3 Preparation of Libraries

1. Prepare ten sets of 1 μM annealed oligos by mixing corresponding forward and reverse strand oligos (Table 2). For each adaptor, incubate the oligos at 94 °C for 5 min and allow them to cool to room temperature in the annealing buffer. Prepare a final 100 nM mixture of barcoded oligos according to the ratios of A, U, and G capturing oligos in Table 2. This mixture is used to detect non-modified poly (A) tails (Annealed oligo 1) as well as terminally uridylated (Annealed oligos 2–6), and guanylated (Annealed oligos 7–10) poly(A) tails (Fig. 1) (*see* **Note 6**).
2. Prepare a ligation reaction in a 15 μL volume by mixing 3 μL of ligation buffer, 1.5 μL of 400,000 U/mL T4 DNA Ligase, 1 μL of 0.5 ng/μL RNA standard mix, 1.5 μL of barcoded

oligo mix (100 nM), and the ~8 μL of poly(A)+ RNA (*see* **Note 7**). Incubate the ligation reaction at room temperature for 60 min.

3. To reverse transcribe the adaptor-ligated RNA, mix 15 μL of splint-ligated RNA with 9 μL of nuclease-free water, 2 μL of 10 mM dNTPs, 8 μL of 5× First-strand buffer, 4 μL of 100 mM DTT, and 2 μL of 200 U/μL SuperScript II Reverse Transcriptase. Incubate the reaction at 50 °C for 50 min, followed by 10 min at 70 °C and cooling at 4 °C. Add 72 μL of AMPure beads to each reaction and incubate for 5 min at room temperature. Wash the beads with 150 μL of 70% ethanol by rotating the tubes twice 180 degrees on the magnetic rack and then remove the supernatant. Resuspend the beads in 20 μL of nuclease-free water and incubate them at room temperature for 5 min. Elute the DNA/RNA hybrid from the beads using a magnetic rack (*see* **Note 8**).
4. To add the motor protein adaptors to the libraries, first mix 20 μL of the purified cDNA/RNA hybrid with 8 μL of ligation buffer, 6 μL of RNA Adapter Mix (from the Direct RNA Sequencing Kit), 3 μL of nuclease-free water, and 3 μL of 400,000 U/mL T4 DNA Ligase to a final volume of 40 μL. Incubate the mixture for 15 min at room temperature to allow the motor protein adaptors to be ligated to the RNA. Then add 40 μL of AMPure beads to the motor protein adaptor-ligated RNA and incubate for 5 min at room temperature. Wash the beads twice with 150 μL of Wash Buffer (from the Direct RNA Sequencing Kit) removing the supernatant using a magnetic rack. After washing, resuspend the beads in 21 μL of Elution Buffer (from the Direct RNA Sequencing Kit) and incubate for 20 min at room temperature. Finally, elute the samples from the beads using a magnetic rack and keep them at 4 °C.
5. To ensure the flow cell is in good condition before loading the library (*see* **Note 9**), place the flow cell into the instrument and launch the quality check using the MinKNOW software. The flow cell should have more than 800 active pores. Otherwise, return the defective flow cell and request a replacement from ONT.
6. Prime the flow cells using the Flow Cell Priming Kit. Remove air in the flow cell by reverse pipetting with the pipette dial 10–30 μL of storage buffer from the priming port using a P1000 pipette to ensure no bubbles come into contact with the pores. Then mix 30 μL of Flush Tether (FLT) solution with 1 mL of Flush Buffer (FB) (from the Flow Cell Priming Kit) and load 800 μL of the FLT-FB mix onto the flow cell through the priming port. Leave the flow cell at room temperature for 5 min. Then, open the SpotON port and load an additional 200 μL of the FLT-FB mix through the priming port.

7. To prepare the RNA for loading, add to the samples 17.5 μL of nuclease-free water and 37.5 μL of RNA Running Buffer (from the Direct RNA Sequencing Kit). Then load the 75 μL of the RNA library through the SpotON port drop by drop (*see* **Note 10**). Finally, close both ports after loading the library.
8. Insert the flow cell into the MinION and select *run sequencing* in the MinKNOW software. Select the direct RNA kit option and request the fast5 file (*see* **Note 11**). Do not select base calling (*see* **Note 12**). Run the library for 16 h (*see* **Note 13**).
9. After the run is completed, wash partially used flow cells for reuse. First, add 2 μL of Wash Mix (from the Flow Cell Wash Kit) solution to 398 μL of Diluent (from the Flow Cell Wash Kit) and mix well by pipetting. Then drain the waste channel using a P1000 pipette with the SpotON and priming ports closed. Make sure there is no air between the array and the priming port. Pull back up to 30 μL with a P1000 to remove any air bubbles, as in the sample loading step. Next, load 400 μL of the wash mix into the flow cell through the priming port. Close the priming port and wait for 60 min. Then, remove all liquid from the waste channel through waste port 1 using a P1000 pipette. Add 500 μL of Storage Buffer (from the Flow Cell Wash Kit) through the priming port, close the priming port, and store the flow cell at 4 °C (*see* **Note 14**).

3.4 Bioinformatics Analysis

1. Base call the fast5 files using Guppy (*see* **Note 15**).

```
$ guppy_basecaller \
--device auto \
--fast5_out \
--flowcell FLO-MIN106 \
--kit SQK-RNA002 \
--compress_fastq \
--trim_strategy none \
-i flowcell1/batch0/ \
-r -s flowcell1/batch0
```

2. Map the reads to the reference transcriptome using Minimap2 [17] and sort the bam files with SAMtools [18] (*see* **Note 16**).

```
$ minimap2 -a -x map-ont\
 --secondary=no \
data/reference.fa \
flowcell1/batch0/batch0.fastq.gz | \
samtools sort \
-O BAM \
-o flowcell1/batch0/batch0.bam
```

3. Index the mapped reads with SAMtools.

```
$ samtools index flowcell1/batch0/batch0.bam \
flowcell1/batch0/batch0.bam.bai
```

4. Index the fastq files and determine the length of the poly (A) tails using Nanopolish [19].

```
$ nanopolish index --directory flowcell1/batch0 \
flowcell1/batch0/batch0.fastq.gz
```

```
$ nanopolish polya \
--threads=5 \
--reads=flowcell1/batch0/batch0.fastq.gz \
--bam=flowcell1/batch0/batch0.bam \
--genome=data/reference.fa > \
flowcell1/batch0/batch0.polya_results.tsv
```

5. Split the reads by barcodes using DeePlexiCon [20] to determine the type of terminal addition (*see* **Note 17**).

```
$ deeplexicon_sub.py dmux -p flowcell1/batch0 \
-m /deeplexicon/models/resnet20-final.h5 > \
flowcell1/batch0/batch0.barcodes.tsv
```

4 Notes

1. While the MinION and GridION use the same flow cells, the GridION can run up to five flow cells in parallel and each one can be loaded and unloaded independently.
2. Synthetic DNA oligos are usually reliable to generate poly (A) standards up to 64 nucleotides long. They can also generate poly(A) tails longer than 100 nucleotides, although with low yields.
3. Adding an extra centrifugation step after the wash ensures that no ethanol traces are carried on, which can interfere with subsequent enzymatic reactions.
4. Prepare the beads for all samples in batch and keep them at room temperature while preparing the RNA samples.
5. The poly(A) tail selection is a cost-effective way of obtaining polyadenylated RNA from total RNA. However, it has been shown to introduce biases against short-tailed transcripts [21]. Some alternatives have been discussed in the literature, including mRNA enrichment by ribosomal RNA depletion or

no selection [7, 19]. Ribosomal RNA depletion can be done using different kits, such as riboPOOLs (siTOOLs Biotech), which requires 10 ng to 3 μg of the starting material.

6. The splint ligation oligos have been designed to capture terminal uridylation as deposited by the mammalian enzymes TUT4 and TUT7 [15]. The splint ligation oligos to detect terminal guanylation have been designed to capture guanylation as deposited by TENT4A and TENT4B, two poly (A) polymerases that occasionally incorporate G into the poly (A) tail [14]. Note that Gs present in the poly(A) tail that do not anneal to the barcoded oligos will not be detected by this method.
7. With the amount of standard mix added, the proportion of reads coming from the standards typically represents less than 1% of the library.
8. The cDNA conversion avoids RNA secondary structures, which facilitates RNA transit through the pore. Although the step could be skipped, its inclusion is recommended since it takes little time and increases the number of reads obtained.
9. The number of pores in each chip can vary substantially due to manufacturing or handling problems. Therefore, before loading any library, the chip must be verified for the number of pores available. If the number of pores falls below the 800 active pores threshold, the chip should not be used, and a replacement needs to be requested.
10. We recommend loading the sample in three steps since washing the flow cells helps the recovery of the pores. To this end, at the last step of the library preparation, we typically split the sample into three aliquots of 25 μL; we then load one of the samples and store the other two at −80 °C. The first run will start with ~1600 pores, so we run it for only 12 h, and we typically obtain around 600,000–900,000 reads. The next day, we take the flow cell, wash it, and keep it in the Storage Buffer (from the Flow Cell Wash Kit) until the afternoon when we load the second third of the library. Approximately 1000 pores are left after the first run. The second part of the library is run for 14 h to adjust for the number of pores lost. We launch the library's third part on the third day and run it overnight. If there are less than 50 active pores the following morning, we stop the run. Alternatively, we leave the sample running and monitor every 2–3 h. If a flow cell's starting pores are below 1200 and above 800, only two loadings are recommended, and an additional flow cell is needed to complete the run.
11. The fast5 files occupy several Gigabytes for each run. Ensure that there is 100 Gb of space for the run to complete. Although the run will start, in any case, the software will show a warning

message every few minutes. If the sample does not finish running and terminates prematurely, the files can still be used. The fast5 files with the ionic current data are required to determine the poly (A) tail length and the presence of terminal modifications. Therefore, always select the option to include the fast5 files.

12. Base calling can be done with Guppy while the run is in progress. The approach is used to save time or to do adaptive sampling and benefits significantly if done with a GPU-powered computer. We recommend avoiding this step if running Guppy on a CPU.

13. Twenty minutes after starting the sequencing, monitor the run to verify that the library is of good quality. Pore occupancy should be around 90% of the available pores. Also, monitor the length of the reads. We typically get read lengths of 1.1 kb (N50) or more. Although the length of the reads could be further optimized, in our case, we prefer more reads than longer reads to capture more poly(A) tails; thus, the 1.1 kb read length is a good compromise. If the library preparation is defective, the number of pores occupied can be less than 50% of the available pores. These occupancy levels are typically associated with an increase in short reads dominated by adaptor-only reads. The pore monitoring will also inform how long the flow cell needs to be run to detect a certain number of reads. ONT claims that 3–6 million reads can be obtained from a direct RNA library with one flow cell, but we rarely get over 3 million reads. The typical yield in our hands is between 1.5 and 2 million reads.

14. Alternatively, the washed flow cell can be primed again and loaded with another library.

15. The Guppy and DeePlexiCon steps are the most time-consuming steps of the pipeline. However, both can be run using GPUs rather than CPUs. Running the steps on GPUs speeds up data processing 5× times or more. In addition, MinKNOW generates multiple fast5 files, each one containing 4000 reads if generated with default settings. Therefore, the pipeline can also be parallelized to increase the data processing speed further.

16. For the identification of standards, select reads with a Minimap2 mapping quality score (MAPQ) greater than 1 to avoid cross-contamination.

17. DeePlexiCon's neural network training has been done using adaptors for unmodified poly(A) tails [20]. Incorporating different bases to the 3′ end of the poly(A) tail may interfere with identifying the barcode location, which could result in an underestimation of the terminally modified reads, particularly oligo-uridylation.

Acknowledgments

This study was supported by the research project ZIA ES103355 funded by the Division of Intramural Research of the National Institute of Environmental Health Sciences, National Institutes of Health, which was awarded to MM. The authors would like to thank Yolanda L. Jones for her help in editing the manuscript.

References

1. Raisch T, Valkov E (2022) Regulation of the multisubunit CCR4-NOT deadenylase in the initiation of mRNA degradation. Curr Opin Struct Biol 77:102460
2. Passmore LA, Coller J (2022) Roles of mRNA poly(A) tails in regulation of eukaryotic gene expression. Nat Rev Mol Cell Biol 23(2): 93–106
3. Subtelny AO, Eichhorn SW, Chen GR et al (2014) Poly(A)-tail profiling reveals an embryonic switch in translational control. Nature 508(7494):66–71
4. Morgan M, Kumar L, Li Y et al (2021) Post-transcriptional regulation in spermatogenesis: all RNA pathways lead to healthy sperm. Cell Mol Life Sci 78(24):8049–8071
5. Brouze A, Krawczyk PS, Dziembowski A et al (2023) Measuring the tail: methods for poly (A) tail profiling. Wiley Interdiscip Rev RNA 14(1):e1737
6. Chang H, Lim J, Ha M et al (2014) TAIL-seq: genome-wide determination of poly(A) tail length and 3′ end modifications. Mol Cell 53(6):1044–1052
7. Lim J, Lee M, Son A et al (2016) mTAIL-seq reveals dynamic poly(A) tail regulation in oocyte-to-embryo development. Genes Dev 30(14):1671–1682
8. Nicholson AL, Pasquinelli AE (2019) Tales of detailed poly(A) tails. Trends Cell Biol 29(3): 191–200
9. Rhoads A, Au KF (2015) PacBio sequencing and its applications. Genomics Proteomics Bioinformatics 13(5):278–289
10. Wang Y, Zhao Y, Bollas A et al (2021) Nanopore sequencing technology, bioinformatics and applications. Nat Biotechnol 39(11): 1348–1365
11. Krause M, Niazi AM, Labun K et al (2019) Tailfindr: alignment-free poly(A) length measurement for Oxford Nanopore RNA and DNA sequencing. RNA 25(10):1229–1241
12. Yu S, Kim VN (2020) A tale of non-canonical tails: gene regulation by post-transcriptional RNA tailing. Nat Rev Mol Cell Biol 21(9): 542–556
13. Warkocki Z, Liudkovska V, Gewartowska O et al (2018) Terminal nucleotidyl transferases (TENTs) in mammalian RNA metabolism. Philos Trans R Soc Lond Ser B Biol Sci 373(1762):20180162
14. Lim J, Kim D, Lee YS et al (2018) Mixed tailing by TENT4A and TENT4B shields mRNA from rapid deadenylation. Science 361(6403):701–704
15. Lim J, Ha M, Chang H et al (2014) Uridylation by TUT4 and TUT7 marks mRNA for degradation. Cell 159(6):1365–1376
16. Ran FA, Hsu PD, Wright J et al (2013) Genome engineering using the CRISPR-Cas9 system. Nat Protoc 8(11):2281–2308
17. Li H (2018) Minimap2: pairwise alignment for nucleotide sequences. Bioinformatics 34(18): 3094–3100
18. Danecek P, Bonfield JK, Liddle J et al (2021) Twelve years of SAMtools and BCFtools. Gigascience 10(2):giab008
19. Simpson JT, Workman RE, Zuzarte PC et al (2017) Detecting DNA cytosine methylation using nanopore sequencing. Nat Methods 14(4):407–410
20. Smith MA, Ersavas T, Ferguson JM et al (2020) Molecular barcoding of native RNAs using nanopore sequencing and deep learning. Genome Res 30(9):1345–1353
21. Viscardi MJ, Arribere JA (2022) Poly (a) selection introduces bias and undue noise in direct RNA-sequencing. BMC Genomics 23(1):530

Chapter 16

Differential Poly(A) Tail Length Analysis Using Nanopore Sequencing

Showkat A. Dar, Sulochan Malla, Cedric Belair, and Manolis Maragkakis

Abstract

The poly(A) tail is a sequence of several adenosine nucleotides added to the 3′ end of RNA molecules transcribed by polymerase II. The dynamics of poly(A) tail length play a significant role in regulating post-transcriptional gene expression by regulating the stability, translation, and decay of messenger RNAs. As a result, an accurate measurement of poly(A) tail length changes is important for understanding its regulatory function in different cellular contexts. Here, we outline a method for using nanopore sequencing and linear mixed models to analyze differences in poly(A) tail length across conditions.

Key words Nanopore direct RNA sequencing, Poly(A) tail, NanopLen, Deadenylation, Differential analysis

1 Introduction

The poly(A) tail is a regulatory element in almost all eukaryotic mRNAs. It corresponds to a stretch of adenines added to an mRNA during transcription in the nucleus. mRNAs are then exported to the cytoplasm, where the poly(A) tail plays a key role in post-transcriptional gene regulation, particularly translation and mRNA decay [1, 2]. Interestingly, the length of the poly(A) tail can vary significantly, both within mRNAs of the same gene and among different genes [3]. Deadenylation is considered the rate-limiting step for mRNA decay as it promotes decapping and decay [4, 5]. Although mRNAs with fast deadenylation rates degrade faster [3], short poly(A) tails do not always correlate with reduced mRNA stability [6]. Therefore, quantifying and comparing poly (A) tail length dynamics across conditions is an important step toward discovering new biology.

Advances in high-throughput sequencing approaches have allowed the quantitative measurement of poly(A) tail lengths and have provided unprecedented information on this regulatory

Eugene Valkov and Aaron C. Goldstrohm (eds.), *Deadenylation: Methods and Protocols*, Methods in Molecular Biology, vol. 2723,
https://doi.org/10.1007/978-1-0716-3481-3_16,

element's biological role and dynamics. The field has developed several methods [7–11], but most typically involve laborious experimental processes. However, new technologies have allowed the direct sequencing of native RNA molecules, including their poly (A) tails, using nanopores [12]. Since homopolymers still pose a major limitation for all sequencing platforms, the poly(A) tail length is indirectly quantified by modeling the dwell time of the poly(A) tail in the nanopore [13]. The principal advantage of nanopore sequencing is the simplicity of the library preparation, the accessibility to the technology, and the low cost.

In this work, we describe the use of the nanopore platform to perform True End-to-End RNA Sequencing (TERA-Seq) coupled with statistical analysis to compare the poly(A) tail length changes across conditions. We describe the bioinformatics steps for the computational analysis of these datasets to discover robust poly (A) tail length changes (Fig. 1). TERA-Seq allows the tagging of the exact 5′ ends of mRNA molecules by ligating uniquely designed adapters to it [14]. This allows the direct comparison of the association between poly(A) tail length dynamics and 5′ end mRNA decay. However, this is not strictly required, and the computational steps described here also apply to the ONT direct RNA-Seq protocol. The presented workflow uses NanopLen, a statistical package we developed for differential length analysis, to directly compare poly(A) tail length dynamics across different conditions. We will use HeLa cells as an example, but the approach is generalizable to other cell types and tissues.

2 Materials

2.1 Equipment

1. 50 and 15 mL falcon tubes.
2. Incubator (37 °C).
3. Pipettes.
4. Pipettor.
5. Gel-loading pipette tips.
6. Benchtop minicentrifuge.
7. Digital Dry Baths/Block Heater.
8. ThermoMixer.
9. Centrifuge.
10. NanoDrop.
11. Agilent BioAnalyzer.
12. Nuclease-free/sterile pipette filter tips.
13. Standard DNase/RNase-free microcentrifuge tube (1.5 mL).
14. Standard thin-wall PCR tubes (0.2 mL).
15. Thermocycler.

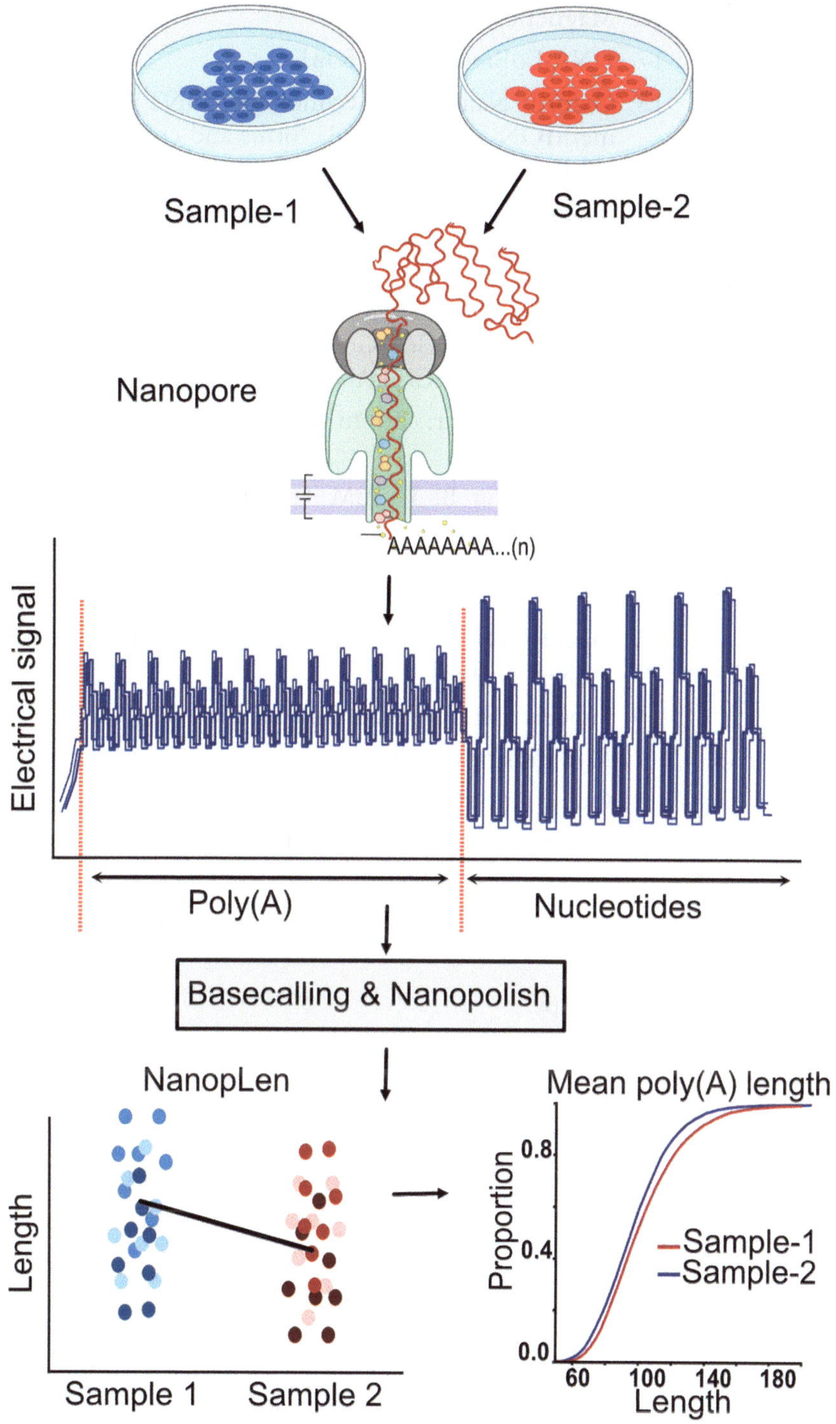

Fig. 1 Workflow schematic. Starting with two conditions, polyadenylated RNA is extracted and sequenced with nanopores. The electrical signal for the poly(A) tail is distinct from other nucleotides allowing the quantification of the poly(A) tail length. Following data processing, the poly(A) tail lengths for all replicates of two conditions are fed into NanopLen for differential length analysis

16. Magnetic rack.
17. DNA LoBind Tube 1.5 mL.
18. Oxford Nanopore Technology (ONT) MinION or PromethION sequencing platforms and corresponding flow cells (FLO-MIN106 or FLO-PRO002).
19. A sequencing and analysis computer. A GPU-equipped computer will significantly improve basecalling speed.

2.2 Buffers and Reagents

1. 5P-ONT-linker (/5PCBio/rArArUrGrArUrArCrGrGrCrGrArCrCrArCrCrGrArGrArUrCrUrArCrArCrUrCrUrUrUrCrCrCrUrArCrArCrGrArCrGrCrUrCrUrUrCrCrGrArUrCrU).
2. Binding buffer: 20 mM Tris-HCl (pH 7.5), 1 M LiCl, 2 mM EDTA.
3. Wash buffer B: 10 mM Tris-HCl (pH 7.5), 150 mM LiCl, 1 mM EDTA.

2.3 Commercially Available Kits

4. NEBNext Quick ligation module.
5. Direct RNA sequencing kit (Oxford Nanopore Technology, SQK-RNA002).
6. SuperScript III First-Strand Synthesis System.
7. Mycoplasma detection kit.
8. Qubit Fluorometer.
   (a) Qubit RNA IQ assay kit.
   (b) Qubit RNA high sensitivity (HS) assay kit.
   (c) Qubit 1× dsDNA high sensitivity (HS) assay kit.

2.4 Required Software Installation

1. Guppy.
2. cutadapt (https://github.com/marcelm/cutadapt).
3. minimap2 (https://github.com/lh3/minimap2).
4. samtools (http://www.htslib.org/).
5. nanopolish (https://github.com/jts/nanopolish).
6. R.
7. Python.
8. NanopLen (https://github.com/maragkakislab/nanoplen).

2.5 Reference Data Files

1. Transcript sequences in FASTA format.

   Example command to download the ENSEMBL human transcriptome from ENSEMBL:

```
curl https://ftp.ensembl.org/pub/release-108/fas-
ta/homo_sapiens/cdna/Homo_sapiens.GRCh38.cdna.
all.fa.gz \
  | zcat > transcripts.fa
```

3 Methods

Perform all the procedures at room temperature unless otherwise indicated. Maintain an RNase-free workstation.

3.1 Cell Culture and RNA Isolation

1. Culture HeLa cells at 37 °C, 5% CO_2, 90% humidity in Dulbecco's Modified Eagle Medium supplemented with 10% heat-inactivated fetal bovine serum, 1% MEM-nonessential amino acids, and 2 mM L-Glutamine. Test cells regularly for potential mycoplasma contamination using the mycoplasma detection kit.
2. Extract total RNA with the TRIzol reagent following the manufacturer's recommendation. Remove potential genomic DNA contamination by treating total RNA with 20 U DNase I for 20 min at 37 °C followed by purification with 1 volume of the acid phenol:chloroform:iso-amyl alcohol (125:24:1, pH 4.5) mixture and precipitation with 0.3 M sodium acetate (pH 5.5) and 2.5 volumes ethanol at −20 °C. Quantify the RNA using High Sensitivity RNA Qubit or Nanodrop. Measure the RNA integrity on BioAnalyzer or Qubit RNA IQ assay (Fig. 2a–c). RIN values greater than 8.0 are suggested for direct RNA sequencing (*see* **Note 1**).

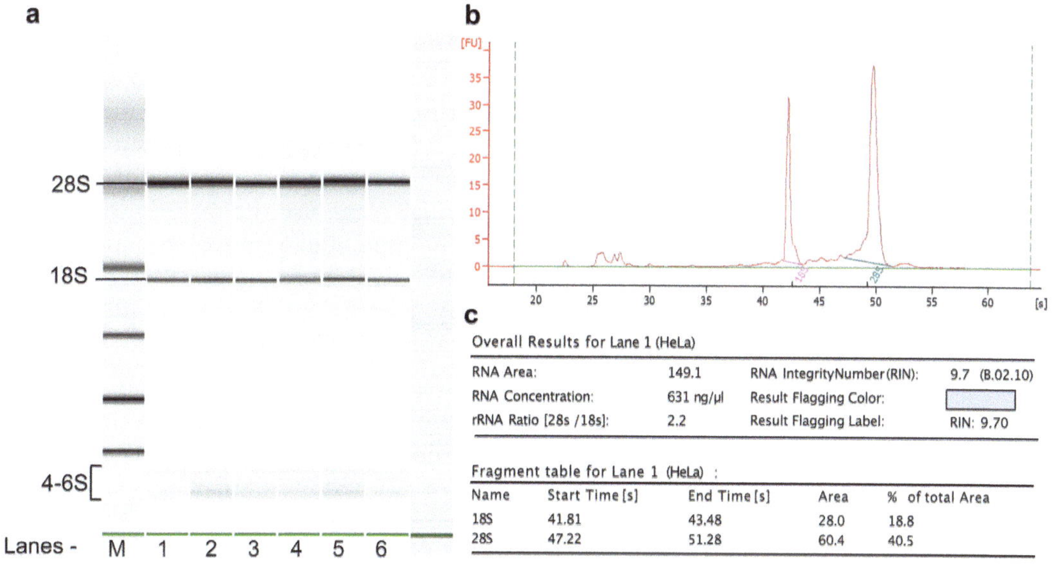

Fig. 2 Example of Agilent 2100 Bioanalyzer output for RNA quality evaluation. (**a**) Gel-like image showing well-defined bands for 28S and 18S rRNA. (**b**) Representative electropherogram peaks for 28S and 18S rRNA as shown in the first lane from (**a**). (**c**) RNA quality metrics

Table 1
Indicative measurements of RNA quality and amount during protocol steps

Total RNA	Absorbance (A260/280)	Qubit value	RIN Value	Amount of poly(A) RNA	Amount of final library
75 μg	2.0	8.3	9.7	1000 ng	250 ng

3.2 TERA-Seq Library Preparation and Sequencing

1. Total RNA preparation: Adjust the volume of the RNA sample to 100 μL with nuclease-free water and add 100 μL of binding buffer. Heat to 65 °C for 2 min to disrupt any secondary structures and place immediately on ice. We typically start with 75 μg of total RNA from HeLa cells to obtain 1–2 μg of poly(A) RNA (Table 1). ONT recommends 500 ng poly (A) RNA as starting material; adjust total RNA to achieve a minimum amount of starting material (*see* **Note 2**).
2. Oligo $d(T)_{25}$ magnetic beads preparation: Resuspend the magnetic beads to obtain a uniform suspension by vortexing or flicking the bead container. Transfer 200 μL to an RNase-free 1.5 mL microcentrifuge tube. Place the tube on a magnet for 1–2 min. While the tube is on the magnet, discard the supernatant. Remove the tube from the magnet and wash the beads with 200 μL binding buffer B. Place the tube on the magnet for 1–2 min and remove the supernatant. Resuspend the beads in 100 μL of binding buffer.
3. Poly(A) RNA capture: Add 200 μL of total RNA (from **Step 1**) to the 100 μL washed beads. Mix thoroughly and anneal for 5 min at room temperature with flicking. Place the tube on the magnet for 1–2 min and carefully remove all supernatant. Remove the tube from the magnet and add 300 μL washing buffer B. Mix by pipetting. Next, use the magnet to pull the beads to the side of the tube and remove all supernatant. Repeat the washing step two more times. Resuspend the beads containing poly(A)-enriched mRNAs in 50 μL T4 ligation reaction buffer mixture described in **step 4**.
4. 5′ adapter (REL5) ligation: Prepare 50 μL of the T4 RNA ligation reaction mix containing the following components. Mix by pipetting to obtain a uniform suspension. Perform the reaction for 3 h at 37 °C with intermittent shaking (*see* **Note 3**).

5 μL	10× T4 RNA ligase buffer
5 μL	ATP (10 mM)
15 μL	Polyethylene glycol (PEG) 8000 (50% w/v)

(continued)

5 μL	10 μM 5P-ONT Linker (50 pmole)
2 μL	T4 RNA Ligase 1 (ssRNA Ligase) (10 U/μL)
2 μL	RNaseOUT Recombinant Ribonuclease Inhibitor (40 U/μL)

Up to 50 μL H_2O

5. Isolate poly(A)-enriched RNA: After 3 h of the ligation reaction, pellet the beads on the magnet for 2 min. Discard the supernatant and resuspend the beads in 300 μL of washing buffer B. Put the beads back on the magnet and remove the supernatant. Repeat the washing twice (*see* **Note 4**). Next, resuspend the beads in 12 μL of nuclease-free water, heat at 80 °C for 2 min, and place the tube immediately on the magnet. Collect the supernatant containing poly(A) mRNAs in a new RNase-free tube. Take 1 μL of the solution to measure the concentration using the Qubit RNA HS Assay (Table 1). The sample may need to be diluted 1:5.
6. ONT RTA top/bottom ligation: Assemble the following reaction in RNase-free 0.2 mL PCR tubes and incubate the reaction at room temperature for 30 min.

9.5 μL	RNA eluate (100–500 ng of poly(A) mRNA)
3 μL	5× NEBNext Quick Ligation Reaction Buffer
1 μL	RTA Adapter (SQK-RNA002)
1.5 μL	Quick T4 DNA Ligase

7. Prepare the reverse transcription master mix with the following composition.

9 μL	Nuclease-free H_2O
2 μL	10 mM dNTPs
8 μL	5× first-strand buffer
4 μL	0.1 M DTT
2 μL	Superscript III (200 U/μL)

8. Reverse transcription: Add 25 μL of the reverse transcription master mix to the RTA reaction and incubate at 50 °C for 50 min followed by 70 °C for 10 min. Keep at 4 °C until use.
9. Clean RNA/cDNA: Transfer the RT reaction to a DNA LoBind tube and mix with 1.8 volume (72 μL) of RNAclean XP beads by pipetting. Incubate for 5–15 min at room temperature with intermittent flicking. Briefly spin down the mixture using a benchtop centrifuge and place tubes on a magnet for

15 min or until the solution is clear. Remove the supernatant using a gel-loading pipette tip without disturbing the beads and wash with 150 μL of freshly prepared 70% ethanol. Keeping the magnetic rack on the bench, rotate the bead-containing tube by 180°. Wait for the beads to migrate toward the magnet and form a pellet. Rotate the tube 180° again back to the starting position and wait for the beads to pellet. Discard the 70% ethanol using a gel-loading pipette tip. Spin down and place the tube back on the magnet. Pipette off any residual 70% ethanol and air-dry. Do not let the pellet over-dry (cracked beads).

10. RNA/cDNA elution: Remove the tube from the magnetic rack and resuspend the beads in 23–24 μL of nuclease-free water. Incubate at room temperature for 5 min. Pellet the beads on the magnet until the eluate is colorless and transfer the RNA/cDNA-containing eluate to a new DNA LoBind tube.

11. Ligation of 1D Sequencing Adapter RMX (Top/Bottom): Assemble the following reaction in a DNA LoBind tube and incubate at room temperature for 30 min.

23 μL	eluate with RNA/cDNA
8 μL	5× NEBNext Quick Ligation Reaction Buffer
6 μL	RMX Adapter (SQK-RNA002)
3 μL	Quick T4 DNA Ligase

12. Clean non-ligated RMX adapter: Add 1 volume (40 μL) of RNAClean XP Beads and mix by pipetting. Incubate at room temperature for 15 min with intermittent flicking. Spin down and place the tube on a magnetic rack to separate the beads from the supernatant. When the solution is clear (~15 min), remove and discard the supernatant using a gel-loading pipette tip without disturbing the beads. Add 150 μL of WSB Wash Buffer (SQK-RNA002). Resuspend beads by flicking the tube. Place the tube on the magnet and wait until beads are pelleted (~15 min) and the solution is clear. Aspirate and discard the supernatant and repeat wash.

13. Elute RMX-ligated RNA/cDNA: Resuspend the beads in 21 μL Elution Buffer (SQK-RNA002). Incubate the mixture at room temperature for 10 min with intermittent flicking. Pellet beads using a magnetic rack until eluate is clear and colorless. Collect supernatant with a gel-loading pipette tip and use 1 μL for quantitation with the Qubit dsDNA High Sensitivity Assay (Table 1). (Note: the elution volume for PromethION is 42 μL.)

14. Preparation for library loading: Thaw the RNA Running Buffer (RRB), Flush Tether (FLT), and Flush Buffer (FB) at room temperature and place on ice.
15. Flow cell priming: Open the priming port of the flow cell and aspirate a small volume (~20 μL) slowly with a P1000 pipette; this removes potential bubbles at the port entry. Prepare the flow priming mix by adding 30 μL of FLT to 1 tube of FB. Vortex and spin down. Carefully and slowly load 800 μL of the priming mix in the priming port to avoid bubbles. Wait for 5 min. (Note: load 500 μL of priming mix for PromethION.)
16. Prepare library mix*:* Mix by pipetting. (Note: double the indicated volumes for PromethION.)

20 μL	library from **step 13**
17.5 μL	H_2O
37.5 μL	RRB

17. Finish flow cell priming: Gently lift the sample port cover and slowly load 200 μL of the priming mix via the priming port avoiding bubbles. (Note: load 500 μL of the priming mix for PromethION.)
18. Load the flow cell: Load 75 μL of the library mix in the sample port. Do this slowly and in drops to avoid bubbles on the chip. Gently snap back the sample port cover, close the priming port and close the flow cell lid.
19. Run MinION or PromethION sequencing for 72 h.

3.3 Raw Nanopore Sequencing Data Processing

The output from the sequencer consists of FAST5 files encoding the raw electrical signal. These files need to be processed to extract the nucleotide sequence and the poly(A) tail lengths. The following steps apply to each library separately and will produce a file that reports the poly(A) tail length of each transcript. These files can be used to perform differential poly(A) length analysis. The commands to be executed are shown below each step. We use dummy file paths and generic names throughout the text. The user must modify these names and paths accordingly.

1. Basecall raw FAST5 files (*see* **Note 5**): Convert the signal to basecalled sequences using guppy.

```
guppy_basecaller \
  -x "auto" \
  --flowcell FLO-MIN106 \
  --kit SQK-RNA002 \
  --hp_correct on \
```

```
--records_per_fastq 0 \
--u_substitution off \
--trim_strategy none \
--input_path raw_data_folder/ \
--save_path lib1/guppy_output_folder/ \
--recursive \
--gpu_runners_per_device 5 \
--num_callers 3 \
--chunks_per_runner 4096 \
--compress_fastq
```

2. Concatenate all FASTQ files created at previous step into a single file:

```
zcat lib1/guppy_output_folder/fastq_runid_*.
fastq.gz \
 | gzip \
 > lib1/guppy_output_folder/reads.fastq.gz
```

3. (Optional, *see* **Note 6**) Trim the REL5 adaptor from the reads: Use cutadapt to remove the REL5 sequence from the reads. This command will also divide the reads into two files. One for reads that the adaptor was present and one for reads that the adaptor was not present. Following cutadapt, merge the two files for further processing.

```
cutadapt \
  -g XAATGATACGGCGACCACCGAGATCTACACTCTTTCCCTACACG
  ACGCTCTTCCGATCT \
  --overlap 31 \
  --minimum-length 25 \
  --error-rate 0.29 \
  --cores 0 \
  --output lib1/reads.with.adaptor.fastq.gz \
  --untrimmed-output lib1/reads.without.adaptor.
  fastq.gz \
  lib1/guppy_output_folder/reads.fastq.gz
```

Merge the two output files from cutadapt into a single file.

```
zcat \
 lib1/reads.with.adaptor.fastq.gz \
 lib1/reads.without.adaptor.fastq.gz \
 | gzip \
 > lib1/reads.merged.fastq.gz
```

4. Align reads to transcriptome. Use minimap2 to align the reads to the human reference transcriptome. Use samtools to convert the output to BAM format.

```
minimap2 \
 -a \
 -x map-ont \
 -k 12 \
 -u f \
 -p 1 \
 --secondary=yes \
 reference_data/transcripts.fa \
 lib1/reads.merged.fastq.gz \
 | samtools view -b -F 256 - \
 | samtools sort \
 > lib1/aligned.toTranscriptome.bam
```

5. Index the alignment BAM file for faster information retrieval:

```
samtools index lib1/aligned.toTranscriptome.bam
```

6. Build index for nanopolish. In the command below, the --sequencing-summary is optional but highly recommended as it dramatically increases performance.

```
nanopolish index \
 --directory raw_data_folder_path_fast5/ \
 --sequencing-summary guppy_outout_folder_path/
 sequencing_summary.txt \
 guppy_outout_folder_path/reads.fastq.gz
```

7. Calculate poly(A) tail lengths. Run nanopolish to produce a tab-delimited file with the columns: "readname," "contig," "polya_length," and "qc_tag." This file will be used to perform differential poly(A) tail length analysis. The qc_tag reports the confidence in the poly(A) tail length estimate of each read. Subsequent steps will only use reads marked as "PASS."

```
nanopolish polya \
 --reads guppy_outout_folder_path/reads.fastq.gz \
 --bam lib1/aligned.toTranscriptome.uxi.bam \
 --genome hg38_data/transcripts.fa \
 --threads 5 \
 > polya_folder/polya.tab
```

3.4 Differential Analysis of Poly(A) Tail Length

To perform differential analysis of poly(A) tail length, we developed NanopLen, a statistical package that uses Linear Mixed Models (LMM) to identify robust length changes across conditions. Besides the LMM, other available options include the *t*-test and the Wilcoxon rank test. Adjustment for multiple testing is performed using the Benjamini–Hochberg method. The output of the differential length analysis consists of reported poly(A) length fold change and the corresponding statistics for each transcript. The steps below provide an example for comparing two conditions with three replicates.

1. Install NanopLen as an R package: NanopLen is available as open-source software at https://github.com/maragkakislab/nanoplen. Usage examples are available in the online repository. Download the package directly from GitHub or install it as a package in R.

   To install, open the R console and execute:

```
devtools::install_github("maragkakislab/nano-
plen")
```

2. Create NanopLen input files: Use the helper Python scripts in the NanopLen repository to process the nanopolish output and create files compatible with NanopLen. Use the following command to generate a tab-delimited file with the columns library id, transcript id, and poly(A) tail length. The file will contain one line for each read. Execute this step individually for each library.

```
python3 ./nanoplen/nanoplen_polyA_read_length.py \
  --ifile Control_1/polya.tab \
  --opt-col-name sample \
  --opt-col-val Control_1 \
  > Control_1/polya_with_lib_name.tab
```

   An example of the expected output file is shown in Table 2.

Table 2
Expected output file

library_id	Transcript id	length
Control_1	ENST00000174653	780
Control_1	ENST00000222345	955
Control_1	ENST00000222345	840
...	...	...

Table 3
Expected output

library_id	Transcript id	length
Control_1	ENST00000174653	780
...	...	...
Control_2	ENST00000174653	900
...	...	...
Control_3	ENST00000174653	880
...	...	...
Treatment_1	ENST00000174653	430
...	...	...

Concatenate all files for all the libraries into a single file: In the command below, use the files of all libraries that you want to compare.

```
awk 'FNR>1 || NR==1' \
 Control_1/polya_with_lib_name.tab \
 Control_2/polya_with_lib_name.tab \
 Control_3/polya_with_lib_name.tab \
 Treatment_1/polya_with_lib_name.tab \
 Treatment_2/polya_with_lib_name.tab \
 Treatment_3/polya_with_lib_name.tab \
 > polya_all_libs.tab
```

An example of the expected output is given in Table 3.

3. Create a metadata file with the experimental design. Use a text editor to create a tab-delimited file that describes the experiment. Specify the names of the replicates and the condition. The names in the column library_id should match the names used in **step 2** of this section (Table 4).
4. Run differential poly(A) tail length analysis. Use LMM in NanopLen to calculate poly(A) tail length differences between conditions and the corresponding statistics. Other available options for the statistical test are the *t*-test and the Wilcoxon rank test.

```
Rscript ./nanoplen/scripts/diff_length.R \
 --data_path polya_all_libs.tab \
 --metadata_path metadata.txt \
 --baseline "Control" \
 --test m \
 --logscale TRUE \
 --ofile nanoplen_output_LinearMixedModel.tab
```

Table 4
library_id

library_id	Condition
Control_1	Control
Control_2	Control
Control_3	Control
Treatment_1	Treatment
Treatment_2	Treatment
Treatment_3	Treatment

For the linear mixed model option, the output file will contain columns with the transcript identifier, log2FC of poly (A) tail length difference, p-value of the difference, and adjusted p-value. The "n.control" and "n.alt" denote the number of reads in the control and alternative condition, respectively. The two last columns, i.e., "mean.length.control" and "mean.length.alt," correspond to the mean poly(A) length in the two conditions (Table 5).

5. Visualize the results. Use the NanopLen helper scripts to visualize the results and identify possible differences between tested conditions. Advanced visualization is left to the user. Use the following command to create a scatter plot and cumulative density distribution plots.

```
. /nanoplen/scripts/nanoplen_two_col_plots.py \
 --ifile nanoplen_output_LinearMixedModel.tab \
 --treatment mean_length.alt \
 --control mean_length.control \
 --lowerlimit 50 \
 --upperlimit 200 \
 --ofigure density_scatter_plots.pdf
```

Example output plots are shown in Fig. 3. Notice that in this case, the condition has longer average poly(A) tails than the control.

4 Notes

1. TERA-Seq tags the 5′ end to provide information for full-length RNAs. It is therefore critical to use high-quality RNA to achieve maximum interpretability. A ribonuclease-free environment is essential to achieve this.

Table 5
Mean poly(A) length in the two conditions

name	log2FC	pvalue	qvalue	n. control	n. alt	mean_length. control	mean_length. alt
ENST00000174653	-1.04679122	0.02223407	0.02371634	14	17	1802.64286	1130.47059
ENST00000222345	-1.219345073	3.15E-05	0.00075559	39	33	1812.30769	1027.93939
ENST00000257468	-1.211848382	0.01234344	0.01645792	12	14	1301.25	559.142857
ENST00000258281	-1.007579667	0.01428013	0.01803806	14	15	2556.85714	1493.8

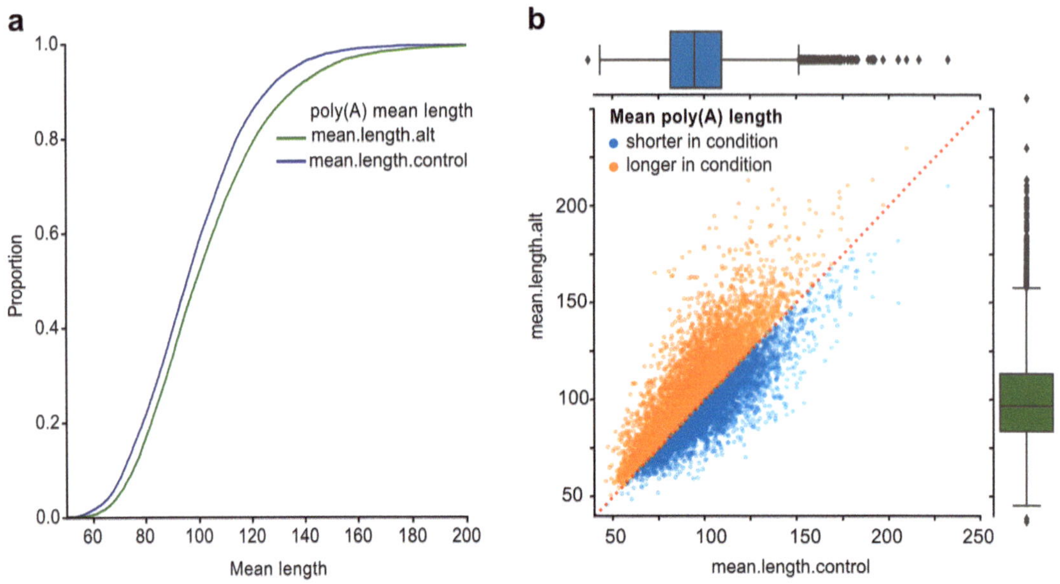

Fig. 3 Example plots for NanopLen output. (**a**, **b**) Cumulative density plot (**a**) and scatter plot (**b**) of mean transcript poly(A) tail length between two conditions. The red dashed line indicates the $y = x$ diagonal line

2. The ONT protocol suggests 500 ng of poly(A) RNA as the starting material. Successful runs with less poly(A) RNA can still be achieved but with reduced sequencing throughput.
3. During the ligation of the 5′ adapter, it is critical to periodically shake the reaction mix to avoid sedimentation of the magnetic beads. Achieve this by using a ThermoMixer with the following program: shake for 15 s at 1200 rpm every 2 min.
4. Remove the last drop of the washing buffer by spinning the tube using a benchtop centrifuge and a 10 μL pipette.
5. During sequencing, the user has the option to enable live basecalling. Typically, this option uses the low accuracy mode in guppy (quick). This step will redo basecalling with a more accurate model. The "pass" and "fail" reads from the previous quick runs will be combined.
6. To simplify and generalize the protocol, information about identified adaptors is not used in subsequent steps. The analysis of this additional data modality is left to the user. Therefore, the computational steps can be used even if the user has executed the default ONT direct RNA-Seq protocol that does not involve adaptor ligation.

Acknowledgments

This work was supported by the Intramural Research Program of the National Institute on Aging, National Institutes of Health (ZIAAG000696).

References

1. Passmore LA, Coller J (2021) Roles of mRNA poly(A) tails in regulation of eukaryotic gene expression. Nat Rev Mol Cell Biol: 23:93–106
2. Xiang K, Bartel DP (2021) The molecular basis of coupling between poly(A)-tail length and translational efficiency. elife 10. https://doi.org/10.7554/eLife.66493
3. Eisen TJ, Eichhorn SW, Subtelny AO et al (2020) The dynamics of cytoplasmic mRNA metabolism. Mol Cell 77:786–799.e10
4. Wahle E, Winkler GS (2013) RNA decay machines: deadenylation by the Ccr4-not and Pan2-Pan3 complexes. Biochim Biophys Acta 1829:561–570
5. Raisch T, Valkov E (2022) Regulation of the multisubunit CCR4-NOT deadenylase in the initiation of mRNA degradation. Curr Opin Struct Biol 77:102460
6. Legnini I, Alles J, Karaiskos N et al (2019) FLAM-seq: full-length mRNA sequencing reveals principles of poly(A) tail length control. Nat Methods 16:879–886
7. Subtelny AO, Eichhorn SW, Chen GR et al (2014) Poly(A)-tail profiling reveals an embryonic switch in translational control. Nature 508:66–71
8. Chang H, Lim J, Ha M et al (2014) TAIL-seq: genome-wide determination of poly(A) tail length and 3′ end modifications. Mol Cell 53:1044–1052
9. Lima SA, Chipman LB, Nicholson AL et al (2017) Short poly(A) tails are a conserved feature of highly expressed genes. Nat Struct Mol Biol 24:1057–1063
10. Woo YM, Kwak Y, Namkoong S et al (2018) TED-Seq identifies the dynamics of poly(A) length during ER stress. Cell Rep 24:3630–3641.e7
11. Liu Y, Zhang Y, Wang J et al (2022) Transcriptome-wide measurement of poly(A) tail length and composition at subnanogram total RNA sensitivity by PAIso-seq. Nat Protoc 17:1980–2007
12. Garalde DR, Snell EA, Jachimowicz D et al (2018) Highly parallel direct RNA sequencing on an array of nanopores. Nat Methods 15:201–206
13. Workman RE, Tang AD, Tang PS et al (2019) Nanopore native RNA sequencing of a human poly(A) transcriptome. Nat Methods: 16:1297–1305
14. Ibrahim F, Oppelt J, Maragkakis M et al (2021) TERA-Seq: true end-to-end sequencing of native RNA molecules for transcriptome characterization. Nucleic Acids Res 49:e115

Chapter 17

Single-Molecule Poly(A) Tail Sequencing (SM-PATseq) Using the PacBio Platform

James R. Iben, Tianwei Li, Sandy Mattijssen, and Richard J. Maraia

Abstract

The polyadenylation of the 3′ ends of messenger RNAs is an important regulator of stability and translation. We developed the single-molecule poly(A) tail sequencing method, SM-PATseq, to assay tail lengths of the whole transcriptome at nucleotide resolution using long-read sequencing. This method generates cDNA using an oligo-dT 3′ splint adaptor ligation to prime first-strand cDNA synthesis, followed by random hexamer priming for second-strand synthesis. By directly sequencing the cDNA on long-read platforms, we can resolve tail lengths at nucleotide resolution, identify non-A bases within the tail, and quantify transcript abundance analogous to traditional RNAseq methods. Here, we discuss the method for generating, sequencing, and primary analysis of poly(A) tail data from total RNA using the Pacific Biosciences Sequel platform.

Key words Single-molecule, Poly(A) tail, SM-PATseq, Long read sequencing, Polyadenylation

1 Introduction

Polyadenylation of the 3′ ends of nascent RNA transcripts is an important regulator of the stability, processing, and translation efficiency of messenger RNA (mRNA) [1–3]. Among eukaryotic cellular transcripts, most mRNAs, some long non-coding RNAs, and primary microRNAs are polyadenylated [4]. The deadenylation of the poly(A) tail of transcripts is linked to turnover [5], and rates of tail decay are attenuated by unique cellular conditions and developmental contexts [6]. In addition, alternative polyadenylation sites downstream of the coding regions of mRNAs can be utilized and affect the availability of trans-acting regulatory effector sites [7]. The range of factors in play that attenuate polyadenylation, tail length, and stability point toward the tails acting as major regulators of mRNA productivity and fate. Therefore, the ability to monitor poly(A) tails experimentally in an accurate fashion is of great scientific interest.

Eugene Valkov and Aaron C. Goldstrohm (eds.), *Deadenylation: Methods and Protocols*, Methods in Molecular Biology, vol. 2723, https://doi.org/10.1007/978-1-0716-3481-3_17,

Classically, poly(A) tail lengths were measured experimentally by gene-specific northern blotting +/− their poly(A) fragments [8]. Higher throughput sequencing methods have recently been leveraged to perform transcriptome-wide assays with popular Illumina short-read sequencers, such as PAL-seq [9] or TAIL-seq [10, 11]. However, Illumina short-read sequencers have difficulty reading through homopolymeric regions such as the poly(A) tail. These initial methods thus utilized strategies to infer length indirectly rather than require the sequencer to read tail length directly. These high-throughput short-read methods continue to be refined including the recent Poly(A)-seq protocol [12] that deal with inherent platform limitations related to the accuracy of homopolymer base calls.

Long-read sequencers, such as the Nanopore and Pacific Biosciences (PacBio) Sequel, have begun to address these challenges and continue to develop their respective platforms and technologies. In the case of Nanopore, a full-length native mRNA can be read in a single pass providing an estimate of poly(A) tail length in the process [13]. This long-read platform generally trades off the ability to read through much longer and potentially repetitive sequences with a lower per-base call accuracy rate. However, with circularized DNA such as that generated with the PacBio Sequel platform, the accuracy loss per base call is mitigated entirely by redundantly reading the same base during many passes of the circle. On the PacBio Sequel, these refined sequences built up from multiple passes of the same fragment (Fig. 1) are referred to as circular consensus sequences (CCS) or high fidelity reads (HiFi; implying a CCS read of minimum Phred quality 20 or greater). This redundant information allows polishing an insert sequence to a higher per-base accuracy than currently achievable by single-pass platforms, including otherwise problematic low complexity repeat expansion regions [14] and homopolymeric stretches, yielding a true nucleotide-level resolution of poly(A) tail length through direct sequencing [15].

This chapter describes using the Pacific Biosciences Sequel platform to obtain single-molecule poly(A) tail sequencing and analysis. After preparing total poly(A) enriched RNA into a "ready-to-sequence" library, the poly(A) tail lengths are represented at nucleotide resolution on individual mRNA transcripts from a given genomic locus, transcriptome-wide. Libraries produced by this method sample the transcriptome messages at a rate consistent with traditional RNA sequencing, given a pool of total RNA as input as expected (Fig. 2a). Additionally, individual reads can be unique at tested sequencing levels, showing that library preparation remains highly productive with novel information at high sequencing depth and free from PCR artifacts that may otherwise skew abundance and length measurements (Fig. 2b). Additionally, resulting length distributions appear free from length

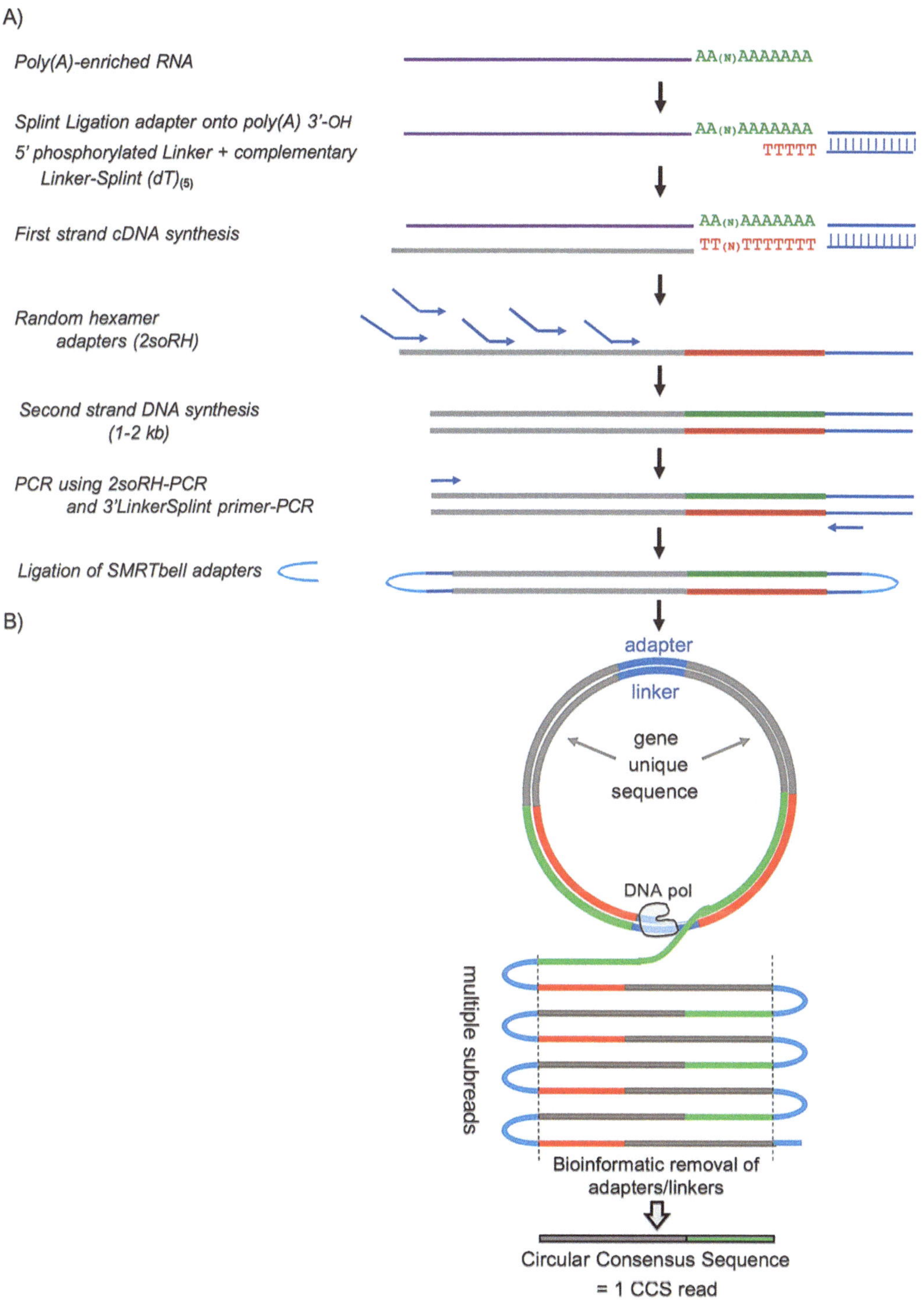

Fig. 1 Schematic representation of the library construction method for single molecule poly(A) tail sequencing (SM-PATseq). (**a**) Poly(A) enriched intact RNA is ligated to a splint adapter at the terminus of the poly(A) tail. Reverse transcription first strand synthesis is followed by a randomly primed second strand within the mRNA gene sequence. PCR amplification of the double-stranded DNA produces the final insert for sequencing, which creates sequence-able libraries by ligating optionally barcoded SMRTbell adapters. (**b**) Priming and loading of DNA polymerase occur within the SMRTbell adapter. The complex is sequenced on the Pacific Biosciences

biases and in agreement with previously published distributions measured by other methods (Figs. 3 and 4) [9–12]. The accuracy of the SM-PATseq method and ability to read homopolymer tails directly on the long-read platform also provide an opportunity to examine additional information, such as non-A bases appearing within the tail, and it can be modified or adapted to examine terminal uridylation/guanylation. The SM-PATseq protocol is portable across organisms and has been successfully applied to preps originating from human, mouse, and yeast samples. As described, this protocol will function for any sample for which sufficiently abundant high-quality total RNA can be obtained and an annotated transcriptome exists. The initial steps in processing and interpreting the data from the platform are also described, allowing for cross-comparison and statistical testing of conditions, and additionally providing an automated Snakemake workflow for primary data for direct use or to serve as an example for data handling (Fig. 5).

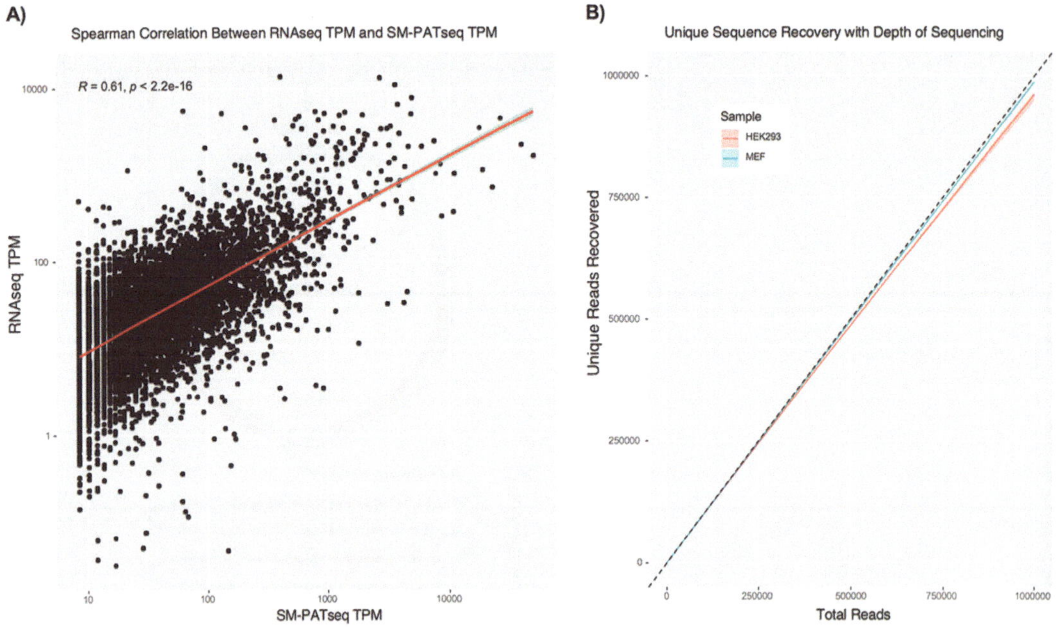

Fig. 2 Library recovery composition of SM-PATseq. (**a**) Spearman correlation of mRNA abundance in HEK293 cells measured by transcripts per million (TPM) of either ribosome-depleted RNAseq (GEO: GSM4567171) versus TPM measured by SM-PATseq [15]. Gene comparisons were limited to those observed in more than one read by SM-PATseq. (**b**) Unique fragment recovery vs. total read recovery for separate SM-PATseq libraries prepared from total RNA originating from HEK293 or MEF cells. Solid lines represent the mean trend line, while the shaded area represents a 95% confidence level across 6 or more individual library preps

Fig. 1 (continued) Sequel platform, producing a long continuous read that reads across the insert and poly(A) tail multiple times. These individual subread passes polish a singular insert circular consensus sequence (CCS) read and enable highly accurate direct sequencing of the poly(A) tail originating from a single mRNA molecule

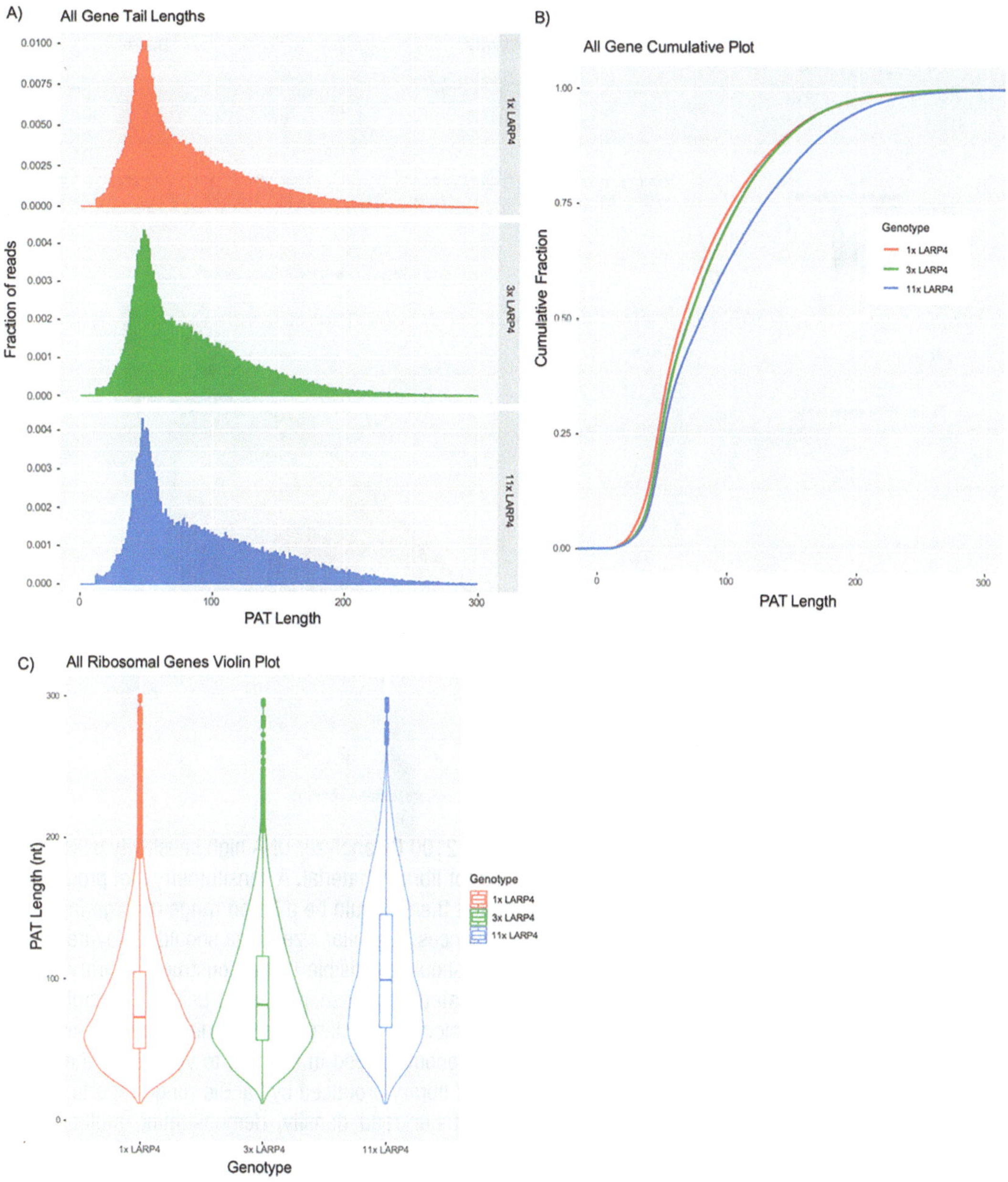

Fig. 3 Example plotting and presentation of data. Example visualization of analysis of a subset of previously published data obtained following this protocol [15]. (**a**) A histogram plot of the distribution of tail lengths across an experiment in human HEK293 cells in which a poly(A) tail-binding protein LARP4 was expressed at normal endogenous levels (1×) as well as over expression at 3× and 11×. A shifting of the density toward longer poly(A) tail (PAT) lengths generally across all mRNAs is observed concordant with overexpression. (**b**) A cumulative plot of PAT lengths observed in A showing the shift across message length population quartiles. (**c**) A violin plot of the total PAT lengths of ribosomal genes

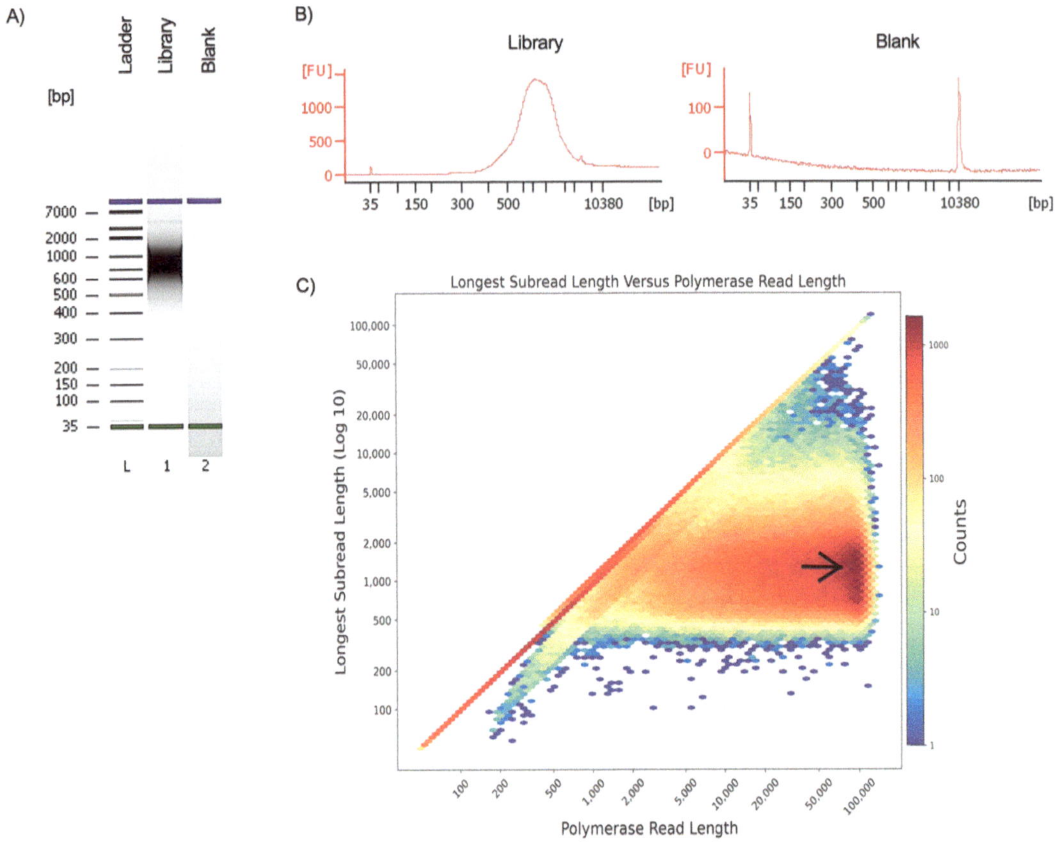

Fig. 4 Quality control of the constructed library. (**a**) Agilent 2100 Bioanalyzer DNA high sensitivity assay trace of a constructed library serving as a comparative control for library material. A densitometry plot producing a gel-like image of the trace visualized in panel b. Note that there should be a broad range of large inserts to facilitate the reading of both tails and gene-specific sequences. A similar size range should be favored when following instructions as described here. No small insert should be visible in the constructed library. (**b**) An electropherogram trace of covalently labeled DNA translated into base pair sizes using a control ladder sample. Due to both high yield and the possible occlusion of the large-size loading controls used for quantitation, measuring sample concentration by Qbit is recommended in addition to visualizing the library content here. (**c**) The Binned hex plot of Sequel sequenced library produced by PacBio runqc-reports. On the right side (black arrow), we aim to see the majority of our read density, demonstrating multiple read transcripts off the single-pass diagonal and corresponding to our expected insert size from the traces in (**a** and **b**) (the *y*-axis showing the within-adapter insert size, the *x*-axis representing the continuous polymerase read length)

2 Materials

Total RNA can be extracted by any method that produces high-quality, intact RNA without degradation. Here, we suggest and describe the use of the Maxwell instrument from Promega.

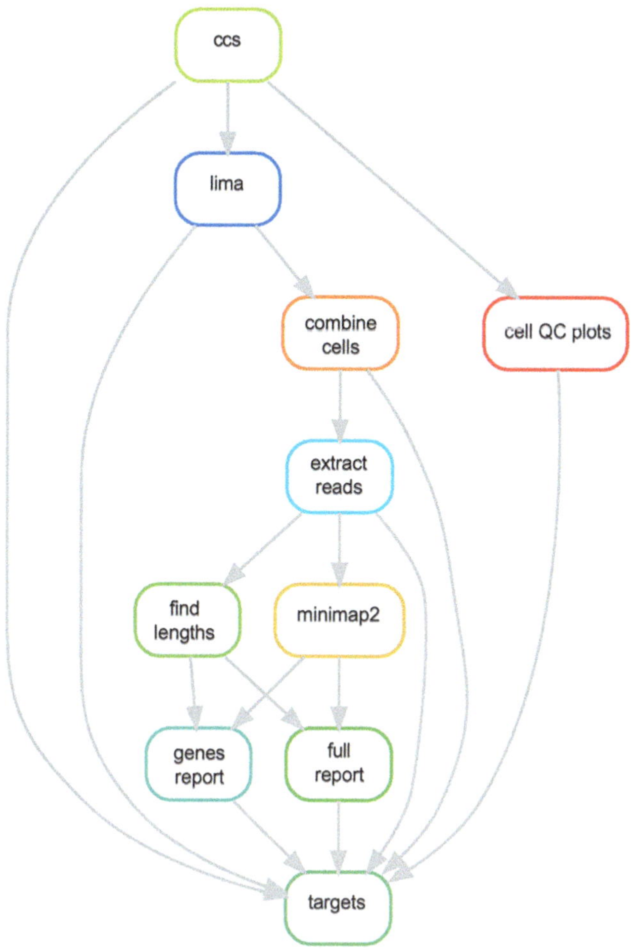

Fig. 5 Schematic representation of the initial processing workflow for sample subread data. CCS read generation is performed followed by sample barcode splitting with lima. Combined sample reads are mapped with minimap2 to a transcriptome for gene assignment of resolved poly(A) tails. Individual tails are measured by custom script using described criteria. Final analysis ready report files are compiled at gene-level statistics as well as in an exhaustive single per-read report file. Code is available on GitHub [16]

2.1 RNA Extraction

1. Maxwell 16 Instrument (Promega) configured with the Maxwell 16 High Strength LEV Magnetic Rod and Plunger Bar Adaptor.
2. Maxwell 16 LEV simplyRNA Cells kit.
3. PBS, phosphate buffered saline pH (7.4) (137 mM NaCl, 2.7 mM KCl, 8 mM Na_2HPO_4, 2 mM KH_2PO_4).

2.2 cDNA Preparation

1. Thermocycler.
2. Qubit Fluorometer.
3. Bioanalyzer 2100 (Agilent).
4. Tabletop vortex mixer.
5. Standard tabletop centrifuge for 1.5 mL tubes.
6. Custom oligos dissolved to 100 μM in 10 mM Tris-HCl pH (7.5). All oligos except for PCR oligos are HPLC purified.
   (a) *5′ Linker*: /5Phos/CTCTGCGTTGATACCACTGCTT GCGGCCGCATTA
   (b) *3′ LinkerSplint*: /5BiotinTEG/TAATGCGGCCGCAAG CAGTGGTATCAACGCAGAGTTTTTT
   (c) *2soRH*: GCCTGAGCTCTGCAGAGATCNNNNNN
   (d) *2soRH-PCR*: GCCTGAGCTCTGCAGAGATC
   (e) *3′-Linker-PCR*: AAGCAGTGGTATCAACGCAGA.
7. T4 RNA ligase 2.
8. RNase H.
9. Klenow Fragment 3′–5′ *exo-*.
10. Superscript II Reverse transcriptase (ThermoFisher).
11. RNAClean XP beads (Beckman-Coulter).
12. 10 mM Tris-HCl pH (7.5).
13. 10 mM Tris-HCl pH (8.0).
14. 5 M NaCl.
15. Nuclease-free water.
16. 70% (v/v) ethanol.
17. RiboLock RNase inhibitor.
18. Magnetic bead rack for 1.5 mL tubes.
19. 10 mM dNTPs.
20. Elution Buffer (Qiagen).
21. NEB Buffer 2.
22. KAPA HiFi HotStart Ready Mix (2×) (Roche).
23. TruSeq Stranded mRNA Prep Kit (Illumina).
24. dsDNA HS Assay kit (ThermoFisher).
25. Pacific Biosciences reagents and kits.
   (a) SMRTbell Template Prep Kit 1.0-SPv3.
   (b) Sequencing primer v3.
   (c) SMRTbell Barcoded Adapter Complete Prep Kit—96.
   (d) AMPure PB beads (*see* **Note 1**).

(e) Sequel Binding and Internal Control Kit 3.0.
(f) Sequel Sequencing Kit 3.0, 4 rxn.
(g) SMRT Cells 1M v3 Tray.
(h) Sequel Internal Control Complex 3.0.

3 Methods

3.1 RNA Preparation

1. Seed enough cells to yield at least 4 μg of total RNA, depending on the cell type. For HEK293 cells, this is two wells of a six-well plate, 5×10^5 cells/well. Ensure the cells are not more confluent than 80% at harvest so they still have room to grow.
2. Isolate total RNA using a Maxwell 16 LEV simplyRNA purification kit following kit instructions summarized here:
   1. Wash cells with 2 mL PBS per well.
   2. Lyse cells with 100 μL homogenization buffer per well containing thioglycerol from the Maxwell kit.
   3. Transfer lysate to a 1.5 mL Eppendorf tube.
   4. Continue to follow the Maxwell 16 LEV simplyRNA purification kit protocol adding 200 μL lysis buffer.
   5. Load the 400 μL samples onto the Maxwell 16 instrument.
   6. Use one simplyRNA cartridge per two wells of a six-well plate.
   7. Elute total DNase-treated RNA in 50 μL nuclease-free water into 0.5 mL tubes supplied in the Maxwell kit.
3. Proceed immediately, or store at −80 °C.

3.2 Library Preparation

1. Anneal 3′ Linker to 3′ LinkerSplint:
   (a) Combine equimolar amounts of each oligo by mixing:
   1 μL of 100 μM 3′ Linker oligo stock (100 pmol 3′ Linker)
   1 μL of 100 μM 3′ LinkerSplint oligo stock (100 pmol 3′ LinkerSplint)
   1 μL of 5 M NaCl stock (500 mM NaCl final)
   2 μL of 1 M Tris-HCl (pH 7.5) [20 mM Tris-HCl final]
   95 μL of nuclease-free water to a final volume of 100 μL
   (b) Heat to 80 °C for 2 min in a thermocycler.
   (c) Slowly cool to 25 °C (in thermocycler ramp down at 1 °C/min)

2. Enrich total RNA for polyadenylated transcripts using reagents and protocol from the TruSeq stranded mRNA kit (Illumina).
   (a) Dilute 4 μg of total RNA to a volume of 50 μL using nuclease-free water.
   (b) Mix RPB by vortexing until well dispersed (these are magnetic poly(A) annealing beads)
   (c) Add 50 μL RPB to each sample and mix aspirating six times.
   (d) Place in a thermocycler under a denaturing program as follows: preheat lid to 100 °C, 65 °C for 5 min, hold at 4 °C
   (e) Remove denatured samples to room temperature for 5 min.
   (f) Centrifuge at 280× *g* for 1 min.
   (g) Place on magnetic rack stand and wait for beads to settle and solution to clear (about 5 min).
   (h) Remove and discard the supernatant.
   (i) Remove tubes from the magnetic stand.
   (j) Add 200 μL BWB to the beads and mix aspirating six times.
   (k) Place on a magnetic stand and allow clearing (about 5 min)
   (l) Remove and discard the supernatant.
   (m) Remove samples from the magnetic stand.
   (n) Add 41 μL of 10 mM Tris-HCl (pH 8.0) to the beads, mixing by aspirating six times.
   (o) Heat to 75 °C for 2 min.
   (p) Centrifuge at 280 × *g* for 1 min.
   (q) Place on a magnetic rack and allow clearing.
   (r) Remove and retain poly(A) enriched RNA in the supernatant in a new tube.
3. Ligate the pre-annealed linker to the poly(A)-enriched RNA.
   (a) Mix a solution containing the following:

       40 μL of poly(A) enriched RNA

       2.5 μL of annealed 3′Linker solution (2.5 pmol of annealed linker)

       5 μL of T4 RNA ligase 2 10× buffer

       2 μL of T4 RNA ligase 2

       1 μL of RiboLock RNase inhibitor
   (b) Incubate the reaction mixture at 25 °C for 1 h.

(c) Perform cleanup of ligated RNA using 100 μL of RNA-Clean XP beads following manufacturer instructions briefly summarized here:

Add 100 μL volume bead suspension, mixing by pipette or vortex.

Incubate for 5 min on ice.

Move the tube to a magnetic strip and incubate for 5 min or until fully pelleted/clear.

Slowly remove the supernatant without disturbing the magnetic beads and discard.

Perform a wash of 200 μL 70% (v/v) ethanol, incubating for 30 s at room temperature before removing and discarding.

Repeat the ethanol wash three times.

Allow the pellet to air dry for 10 min without over-drying.

(d) Final elution from beads should be into 27.5 μL of Elution Buffer (Qiagen).

4. Synthesize first strand cDNA by reverse transcription of ligated RNA.

(a) Mix a reaction solution of the following:

26.5 μL of the final elution of ligated RNA from the prior step

5 μL of 0.1 M DTT (Superscript II kit)

5 μL of 10 mM dNTPs

10 μL of 5× Superscript II RT buffer

1 μL of RiboLock RNase Inhibitor

2.5 μL of Superscript II RT enzyme

(b) Incubate the reaction for 30 min at 42 °C.

(c) Incubate for 15 min at 70 °C.

(d) Cool to 37 °C.

(e) Add 1 μL of RNase H.

(f) Incubate for 15 min at 37 °C

(g) Cleanup the cDNA using 100 μL of RNAClean XP beads following manufacturer instructions (*see* **Note 2**)

(h) The final elution of cDNA should be in 25 μL of nuclease-free water.

5. Synthesis of second strand cDNA.

(a) (*see* **Note 3**)

(b) Mix a reaction solution of the following:

23.5 μL of purified first-strand cDNA from the prior step

1 μL of 100 μM oligo 2soRH

3 μL of NEB Buffer 2

1 μL of 10 mM dNTPs

(c) Heat in a thermocycler to 80 °C for 1 min.

(d) Cool to 37 °C and hold for 10 min.

(e) Add 1.5 μL of Klenow exo- and gently mix using a pipette.

(f) Incubate for 30 min at 37 °C.

(g) Incubate for 20 min at 75 °C.

(h) Purify PCR product by using 0.6× AMPure PB beads (20 μL) and follow Pacific Biosciences product protocol.

(i) Final elution from beads should be in 25 μL of Elution Buffer.

6. Amplification of cDNA fragments containing 3′ UTR and poly (A) tail.

(a) Mix a PCR reaction containing the following:

23 μL of purified cDNA from the prior step

25 μL of 2× Kapa HiFi HotStart Ready Mix

1 μL of 3′ Linker-PCR oligo stock 100 μM

1 μL of 2soRH-PCR oligo stock 100 μM

(b) Amplify in a thermocycler with the following program:

Step 1: 95 °C for 45 s

Step 2: 95 °C for 15 s

Step 3: 60 °C for 30 s

Step 4: 72 °C for 1 min

Step 5: Cycle steps 2–4 for 25 total times

Step 6: 72 °C for 2 min

Step 7: Hold at 4 °C

(c) Purify the PCR product using 0.6× (30 μL) of AMPure PB beads as described in the manufacturer's instructions.

(d) Final elution from beads should be in 20 μL of Elution Buffer.

7. Quantitate the resultant product using a Qubit dsDNA HS Assay kit.

(a) The expected yield from the initial total 4 μg input should be somewhere between 1 and 5 μg in the 20 μL of the final volume.

8. Check the size of the product using a Bioanalyzer HS DNA chip noting that dilution based on the Qubit quantitation should be required to avoid overloading.

   (a) There should be an observed range of fragment sizes over 500 bp, with the broad peak going from roughly 500 bp to 3000 bp and peaking around 1700 bp
   (b) Take note of the molar concentration reported by the bioanalyzer, correcting for any dilution factor needed for chip loading.

9. Final PCR product libraries are converted for sequencing on the PacBio Sequel platform using the protocol documented in "Preparing Amplicon Libraries using PacBio Barcoded Adapters for Multiplex SMRT Sequencing" and only briefly outlined here. Please refer to the original official documentation and carefully note the libraries' molar concentrations and size ranges and which barcodes are utilized for each sample, as these are important for pooling and performing the run.
10. End repair and barcoded adapter (SMRTbell) ligation.
    (a) Prepare a pre-mix of required reagents from the PacBio SMRTbell Barcoded Adapter Complete Prep Kit and aliquot the appropriate amount for each sample.
    (b) Add sample and barcoded adapter to each library aliquot in amounts based on previously measured concentrations.
    (c) Incubate at indicated temperatures and times, then hold at 4 °C.
11. The barcoded samples may now be mixed or multiplexed for further processing if desired.
12. Perform AMPure PB bead purification of the barcoded or multiplexed libraries as described in the product protocol (*see* **Note 4**).
13. Measure concentrations using a Qubit dsDNA HS Assay kit.
14. Perform DNA damage repair.
    (a) Mix DNA with reagents provided in the PacBio SMRTbell prep kit.
    (b) Incubate at 37 °C for 20 min, then hold at 4 °C
15. Perform exonuclease digestion to remove failed ligation products.
    (a) Add exonucleases ExoIII and ExoVII from the kit.
    (b) Incubate at 37 °C for 60 min, then hold at 4 °C.
16. Perform a double AMPure PB bead cleanup as described in the product protocol (*see* **Note 4**).
17. Measure the concentration of the product using the Qubit dsDNA HS Assay kit.
18. It is strongly suggested but optional to verify the prepared SMRTbell library using a Bioanalyzer High Sensitivity DNA Analysis kit while taking note of Qubit concentrations to avoid overloading (Fig. 4).

3.3 Sequencing

1. Use the guided SMRTlink software to set up the PacBio Sequel run following instructions which goes step-by-step through preparing primed complexes and loading the instrument (select "Sample Setup" from the Sequel Instrument Control Software on your instrument server webserver).
2. Follow the software guide to prepare the sample or pooled library for loading based on earlier size and concentration measurements. Steps are also outlined in the SRMT_Link_Users_Guide document. You must enter both concentration (from Qubit) and average library size (from Bioanalyzer). For loading concentration on the plate, we typically use a 6–8 pM target loading concentration.
3. Follow guided instructions to anneal the sequencing primer to the SMRTbell Template library. A proper ratio of primer to template library is critical at this stage. Use the software Binding Calculator to aid in the process.
4. Heat the primer to 80 °C for 2 min, then cool on ice.
5. Mix the multiplexed library with the primer and buffer from the kit.
6. Incubate at 20 °C for 30 min, then hold at 4 °C.
7. Load the Sequel instrument sample and consumables strictly following SMRTlink generated step-by-step instructions.

3.4 Processing of Sequence Data

1. Sequencing reads from the Sequel instrument come in the form of continuous reads around a "rolling circle." The insert is read over multiple passes in each direction. These insert passes are referred to as subreads. Typically to form high-accuracy circular consensus sequences (CCS; also referred to as HiFi), these multiple passes are used to polish both base identity and resolve low complexity regions such as the poly(A) tail itself. Thus, we can accurately determine the length at high resolution and confidence. From here on, CCS refers to CCS/HiFi polished reads interchangeably (*see* **Note 5**).
2. Processing steps are carried out on the NIH high-performance computing cluster using programs freely available from Pacific Biosciences and open-source software.
    1. An example workflow and associated scripts using Snakemake is available from https://github.com/ibenjame/SM-PATseq [16], starting from an entry point of subreads (*see* Fig. 5)
3. If CCS reads are not the default output from your Sequel experiment, subreads may be converted to CCS reads using the program "ccs" with default settings. Optionally, the "`--min-passes`" argument may be increased from 3 to improve per-read accuracy at the cost of some lower sequenced messages (`ccs {subreads.bam} {output.bam}`).

4. To further improve the accuracy of the experiment, subreads plus CCS reads may be polished using the program DeepConsensus [17] or similar; however, note that this does require both subread and CCS inputs and a large amount of processing power that can be multithreaded to more manageable runtimes. Here we do not perform this as the workflow was developed before this became widely available, but it would be recommended to improve eventual outputs if possible. Output from this step should be an equivalent further polished CCS bam file.
5. CCS or DeepConcensus polished CCS reads are demultiplexed to individual samples using the program "lima" (v2.6 or above) using the appropriate sample barcodes utilized during library construction. Here we use the symmetric barcode argument for our libraries to eliminate any potential artifact reads (`lima {input.bam} {barcodes.fa} {output.bam} --hifi-preset SYMMETRIC`).
6. Use samtools to convert demultiplexed sample CCS reads to fastq format for further processing and alignment (`samtools fastq {input.bam} > {output.fastq}`).
7. CCS read IDs are assigned to transcripts through the use of PacBio minimap2 to align reads to a transcriptome fasta, such as those obtained from GENCODE or Ensembl (`pbmm2 align --sort --preset CCS {transcripts.fa} {input.fastq} {output.bam}`).
8. To determine and assign a poly(A) tail length to a CCS ID, reads are evaluated in either direction for a poly(A) stretch abutting the adapter utilized in library construction (CTCTGCGTTGATACCACTGCTT). A one base mismatch is tolerated for the adapter sequence pattern matching.
   1. Due to the random nature of the circularized fragment priming, CCS sequences may be found in either the forward or reverse direction with a roughly 50/50 split between the two expected.
   2. Single base interruptions to the poly(A) tail are tolerated if bracketed by a minimum of two A bases on either side of the alternate base (*see* **Note 6**).
   3. Optionally these non-A base occurrences may be recorded and evaluated critically; however, these may be due to downstream library construction and amplification errors rather than true biological origin. We find rates that are consistent with published literature on both fronts.
9. Following the assignment of both gene identity (via mapping) and poly(A) tail length (via custom script) for each mRNA fragment, we tabulate the two key pieces of information to a single tab-delimited file assigned by CCS read ID, alongside

any other stats we want to include for optional filtering or statistics such as mapping score, non-A bases, read length, and similar. Data in this form is conducive to import into R or similar suites for analyses, such as compiling and plotting overall or by-gene length distributions or deriving specific statistical tests between samples. Example plots derived from this workflow and associated scripts are presented above in Fig. 3.

4 Notes

1. AMPure PB beads are specifically formulated for SMRTbell libraries for the PacBio Sequel system and should not be substituted with non-PB beads where PB beads are called for in the protocol. Take careful note as differing AMPure stocks are used at different points in the workflow.
2. Though the material is cDNA, we use RNAClean rather than AMPure XP (Beckman-Coulter) because we already have them at room temperature from before, and they are slightly less expensive to use. Either bead type could be used here.
3. The size range of final products can be attenuated at this step by adjusting the concentration of oligo added. As described here, final product yields should have a median size of roughly 1700 bp.
4. The quantity of beads needed will vary based on the product size range observed in the bioanalyzer trace according to product specifications.
5. Explicitly, HiFi refers to CCS reads with $Q \geq 20$. This is generally reached for any CCS constructed over a default minimum pass of 3. If the default minimum pass of 3 is reduced for CCS creation, the two are no longer equivalent.
6. In a regular expression, this appears as a capture class ((AA+.)*AA+) followed immediately by an adapter sequence in most programming languages. Where parentheses are grouping characters, "+" denotes greedy matching of the preceding character, "." denotes a wildcard, and "*" denotes 0 to any number of matches of the preceding.

Acknowledgments

We thank Steven Coon (NICHD) for valuable methodology development, Ryan Dale (NICHD) for his help with refining a portable Bioinformatic workflow, as well as the NICHD Division of Developmental Biology labs and Bioinformatics and Scientific

Programming Core for very helpful discussions. This work utilized computational resources of the NIH High-Performance Computing Biowulf cluster (http://hpc.nih.gov) and was supported by Intramural Research Programs of the National Human Genome Research Institute (NHGRI), the *Eunice Kennedy Shriver* National Institute of Child Health and Human Development, National Institutes of Health; and the NIH Office of the Director's Common Fund.

Conflict of Interest The authors have no conflicts of interest to declare.

References

1. Curinha A, Oliveira Braz S, Pereira-Castro I et al (2014) Implications of polyadenylation in health and disease. Nucleus 5(6):508–519
2. Elkon R, Ugalde AP, Agami R (2013) Alternative cleavage and polyadenylation: extent, regulation and function. Nat Rev Genet 14(7): 496–506
3. Tian B, Manley JL (2013) Alternative cleavage and polyadenylation: the long and short of it. Trends Biochem Sci 38(6):312–320
4. Nicholson AL, Pasquinelli AE (2019) Tales of detailed poly(A) tails. Trends Cell Biol 29(3): 191–200
5. Eisen TJ, Eichhorn SW, Subtelny AO et al (2020) The dynamics of cytoplasmic mRNA metabolism. Mol Cell 77(4):786–799
6. Rabani M, Levin JZ, Fan L et al (2011) Metabolic labeling of RNA uncovers principles of RNA production and degradation dynamics in mammalian cells. Nat Biotechnol 29:436–442
7. Batra R, Manchanda M, Swanson MS (2015) Global insights into alternative polyadenylation regulation. RNA Biol 12(6):597–602
8. Decker CJ, Parker R (1993) A turnover pathway for both stable and unstable mRNAs in yeast: evidence for a requirement for deadenylation. Genes Dev 7:1632–1643
9. Subtelny AO, Eichhorn SW, Chen GR et al (2014) Poly(A)-tail profiling reveals an embryonic switch in translational control. Nature 508:66–71
10. Chang H, Lim J, Ha M et al (2014) TAIL-seq: genome-wide determination of poly(A) tail length and 3′ end modifications. Mol Cell 53: 1044–1052
11. Lim J, Lee M, Son A et al (2016) mTAIL-seq reveals dynamic poly(A) tail regulation in oocyte-to-embryo development. Genes Dev 30:1671–1682
12. Yu F, Zhang Y, Cheng C et al (2020) Poly(A)-seq: a method for direct sequencing and analysis of the transcriptomic poly(A)-tails. PLoS One 15(6):e0234696
13. Workman RE, Tang AD, Tang PS et al (2019) Nanopore native RNA sequencing of a human poly(A) transcriptome. Nat Methods 16(12): 1297–1305
14. Mangin A, de Pontual L, Tsai Y-C et al (2021) Robust detection of somatic mosaicism and repeat interruptions by long-read targeted sequencing in myotonic dystrophy type 1. Int J Mol Sci 22(5):2616
15. Mattijssen S, Iben JR, Li T et al (2020) Single molecule poly(A) tail-seq shows LARP4 opposes deadenylation throughout mRNA lifespan with most impact on short tails. elife 9: e59186
16. Iben J (2022) Snakemake workflow for SM-PATseq. https://github.com/ibenjame/SM-PATseq. Accessed 20 Dec 2022
17. Baid G, Cook DE, Shafin K et al (2022) DeepConsensus improves the accuracy of sequences with a gap-aware sequence transformer. Nat Biotechnol. https://doi.org/10.1038/s41587-022-01435-7

Chapter 18

Mathematical Modeling of mRNA Poly(A) Tail Shortening Process

Benedict Fabia, Minju Kim, Jongmin Lim, and Young-suk Lee

Abstract

A sequence of deadenylation events, or the shortening of the poly(A) tail, is a highly regulated process during the life cycle of mRNAs. Advances in biochemistry have enabled the study of deadenylation events at single-nucleotide resolution. Here we describe mathematical models and their applications to estimate the kinetics of a single deadenylation event in vitro. We demonstrate how this quantitative approach is used for assessing reactions with synthetic RNA with poly(A) tails and the CCR4-NOT complex. This method is also applicable to investigating the catalytic activities of other exonucleases and RNA substrates. All example data and custom software are available on GitHub: https://github.com/2yngsklab/deadenylation-kinetics.

Key words Deadenylation, RNA modification, RNA stability, Poly(A) tail, Exonuclease, CCR4-NOT, CCR4, CAF1

1 Introduction

Most eukaryotic mRNAs have a long stretch of adenosines at their 3′ end [1, 2]. This stretch of adenosines, also known as the poly (A) tail, plays an important role in gene expression by regulating mRNA translation [3–5], export [6–8], and stability [9, 10]. For example, the poly(A) tail acts as a physical barrier against promiscuous exonuclease activity [11] and is also a sequence-specific platform for RNA-binding proteins [12] and post-transcriptional gene regulation [13]. Therefore, closely examining how this poly(A) tail is regulated is key to understanding the mechanism of gene expression.

The shortening of the poly(A) tail is a sequence of deadenylation events. This removal of single adenosine is accomplished by specific exonucleases such as CCR4 (CCR4a/CNOT6 and CCR4b/CNOT6L) [14–16] and CAF1 (CAF1/CNOT7 and POP2/CNOT8/CALIF) [17, 18] which are thought to recognize

Eugene Valkov and Aaron C. Goldstrohm (eds.), *Deadenylation: Methods and Protocols*, Methods in Molecular Biology, vol. 2723, https://doi.org/10.1007/978-1-0716-3481-3_18,

the first few nucleotides of the 3′ end of poly(A). It is worth noting that CCR4 and CAF1 are the two distinct catalytic subunits of the CCR4-NOT (CNOT) multi-protein complex [19]. The CNOT complex consists of six additional non-enzymatic subunits that tightly regulate the length of the poly(A) tail [20].

In vitro reconstitution experiments provide the means to control the biochemical reaction and investigate the molecular mechanism of deadenylation. For example, the full reconstitution of the human CNOT complex enabled the systematic characterization of the role of each subunit in the context of deadenylation [20]. Such advances in biochemistry complement cell-based and animal-based findings. This allows careful analyses of specific trans-factors and reveals insights for targeted mRNA deadenylation [21].

Here we describe a mathematical approach to analyze in vitro deadenylation experiments quantitatively. Unlike existing methods, we estimate the kinetics of a single deadenylation event which reveals the change in deadenylation kinetics during the poly (A) tail-shortening process. The utility of our approach is two-fold: (1) it leverages all information instead of tracking the modal poly(A) tail length or the time of completion, for example, and (2) it does not enforce strong assumptions regarding the biochemical reaction such as that the kinetics is constant for each deadenylation event. Owing to the model's simplicity, this approach is general and applicable to any sequence of catalytic events for DNA/RNA metabolism.

2 Materials

Mathematical modeling and parameter estimation involve proprietary software for image processing, the R programming language and its associated packages, and custom R functions which are described below in detail (Table 1). This computational approach is accomplished with a combination of manual and automated steps dependent on successful in vitro deadenylation experiments with denaturing gels at single-nucleotide resolution. Details of the in vitro experiment are covered extensively in the literature and only briefly described here, emphasizing important aspects utilized during the computational steps. All example data and related custom code are available on GitHub: https://github.com/2yngsklab/deadenylation-kinetics.

2.1 Example In Vitro Deadenylation Experiment

We utilize a published in vitro deadenylation experiment `gel.tif` to help describe each step of the computational analysis (Fig. 1). Briefly, a synthetic RNA sequence of 7 nucleotides (5′--UCUACAU-3′) and a 20 adenosine stretch (A20) was used as a substrate to assess the deadenylation reaction of the full human CNOT complex. Other synthetic RNAs such as A1, A0, and UCU

Table 1
List of existing and custom software

Software	Version	Description
Multi Gauge (Fujifilm)	3.0	Image processing software
R	4.2.1	Programming language
tidyverse	1.3.1	A collection of R packages
RStudio Desktop	2022.07.2.	Integrated development environment (IDE)
deSolve	1.32	Parameter estimation
minpack.lm	1.2–2	Parameter estimation
rxnrate	Custom	Mathematical model for deadenylation
simulation	Custom	Computer simulation for deadenylation
ssq	Custom	Objective function for optimization
GetMetaPlot	Custom	Data visualization for cumulative intensity values
GetLanePlot	Custom	Data visualization for per-lane intensity values
GetHeatMap	Custom	Data visualization for data digitization
GetStepPlot	Custom	Data visualization for parameter estimates

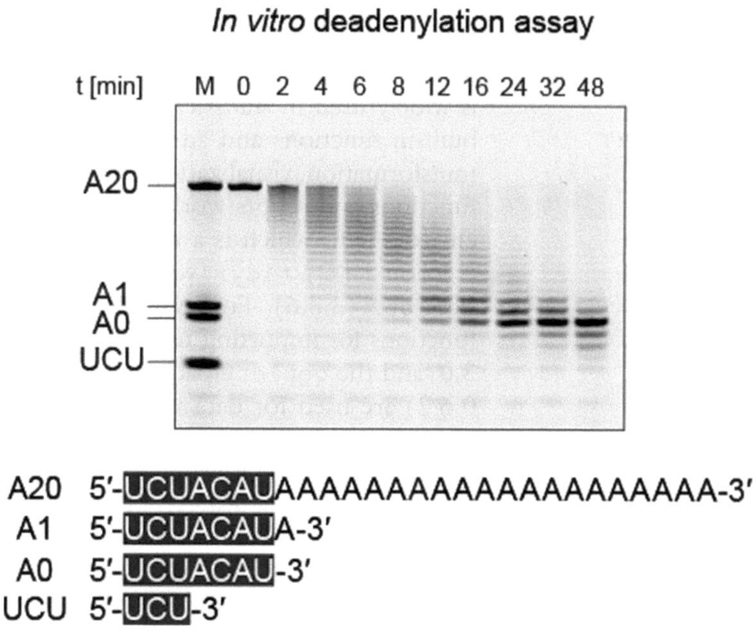

Fig. 1 Anatomy of in vitro deadenylation experiments. Sequences of the synthetic 7-mer-A20 (A20), 7-mer-A (A1), 7-mer (A0), UCU RNA substrates. In vitro deadenylation experiment with A20 RNA substrates (50 nM) and human CCR4-NOT complex (25 nM)

were used as size markers. The first two lanes are the marker and control lanes. The neighboring gel lanes correspond to deadenylation reactions after predefined time points such as 2 min, 4 min, 6 min, and 8 min. Each band corresponds to an RNA species of poly(A) tail length, and adjacent bands indicate RNA species with a single-adenosine difference. Band intensity roughly correlates with the amount of RNA and is used to measure the deadenylation reactions for the particular RNA species quantitatively. For example, high band intensity is observed at the A0 position after 48 min, indicating the completion time for removing 20 adenosines. We recommend counting the number of bands via visual inspection and confirming that it corresponds to the length of the RNA substrate. As expected, this example data with A20 exhibits 20 distinct bands between A20 and A1.

2.2 Image Processing Software

Multi Gauge V3.0 (Fujifilm) is proprietary software used here to process the gel image of an in vitro deadenylation experiment. Multi Gauge provides a user-friendly graphical user interface (GUI) to help calibrate and quantify the densitometric profiles for each lane in the gel image. The computational analysis is not limited by a particular image processing software, and any software with similar features, such as ImageJ, Image Studio, and QuantityOne, should also be applicable.

2.3 R Programming Language and System Environment

R (version 4.2.1) is a programming language for mathematical modeling and parameter estimation. The R programming language is widely used in statistics and data science due to its convenient built-in functions and advanced packages for data preprocessing, transformation, visualization, and mathematical modeling. Custom functions and scripts work under the `tidyverse` (version 1.3.1) environment, which is a collection of R packages such as `readr` (version 2.1.2), `tidyr` (version 1.2.0), `dplyr` (version 1.0.9), and `ggplot2` (3.3.6). For example, the R package `readr` contains functions for importing text files (.txt) generated by Multi Gauge 3.0, and the `ggplot2` package and the `viridis` package (version 0.6.2) are used for data visualization. Lastly, we recommend the integrated development environment (IDE) RStudio for those unfamiliar with the R programming language. The following computation steps were tested on RStudio Desktop version 2022.07.2.

2.4 R Packages for Mathematical Modeling

Mathematical modeling and parameter estimation utilize the R functions available in the `deSolve` (version 1.32) and `minpack.lm` (version 1.2–2) packages. The `deSolve` package contains functions that solve initial value problems (IVP) of systems of ordinary differential equations (ODE). In terms of the poly(A) tail-shortening process, we assume that the sequence of deadenylation events is independent and model it as a system of ODEs. In particular, the `ode` function is used for computer simulation of the mathematical model, such as computing the concentration of each RNA species at certain time points. The `minpack.lm` R package computes the Levenberg–Marquardt (LM) routine for solving nonlinear least squares (NLS) problems, such as a system of ODEs. It is used here to estimate the parameters of the mathematical model of deadenylation.

2.5 Custom R Functions

All custom R functions are available in the R script `deadenylation.R`.

`rxnrate`: The *`rxnrate`* (Reaction Rate) function defines the mathematical model for a sequence of deadenylation events and is used as input for the ODE solver. As input, this function takes three parameters: (1) a list of the time points of interest, (2) the concentration of each RNA species, and (3) the rate parameters for each deadenylation event.

`simulation`: The `simulation` function computes the concentration of each RNA species for a given time point and solves the ODE defined by the `rxnrate` function. As input, this function takes three parameters: (1) the rate parameters for each deadenylation event, (2) the length of synthetic RNA (i.e., the number of RNA species), and (3) the list of time points of interest.

`ssq`: The `ssq` (Sum of Squared Residuals) function calculates the difference between the predicted and observed concentrations of each RNA species and is key for the LM routine in the `minpack.lm` R package. As input, this function takes the preprocessed image data and the rate parameters for each deadenylation event. The function generates the predicted concentration by the `simulation` function.

`GetMetaPlot`: The `GetMetaPlot` function returns a single line plot that represents the cumulative intensity value for each pixel row of the text file (.txt) converted from the image processing software (Fig. 2a). The vertical lines in red indicate the local minima computed based on the meta line plot which may include false positives. This metaplot and the following lane plots are used to curate the local minima manually and include false negatives.

`GetLanePlot`: The `GetLanePlot` function returns a multipanel plot that explores the intensity values for each lane (Fig. 2b). The vertical lines in red indicate the manually curated local minima used to bin individual bands and digitize the preprocessed text file (.txt).

`GetHeatMap`: The `GetHeatMap` function returns a heatmap that represents the digital version of the original gel image to improve data visualization and interpretation (Fig. 3). Analogous to the in vitro experiment, the reaction time points are in the *x*-axis and the individual RNA species are shown in the *y*-axis. This digitization corrects the nonlinear effects of electrophoresis and enables linear interpretation of the deadenylation reaction. The heatmap values are the normalized intensity values.

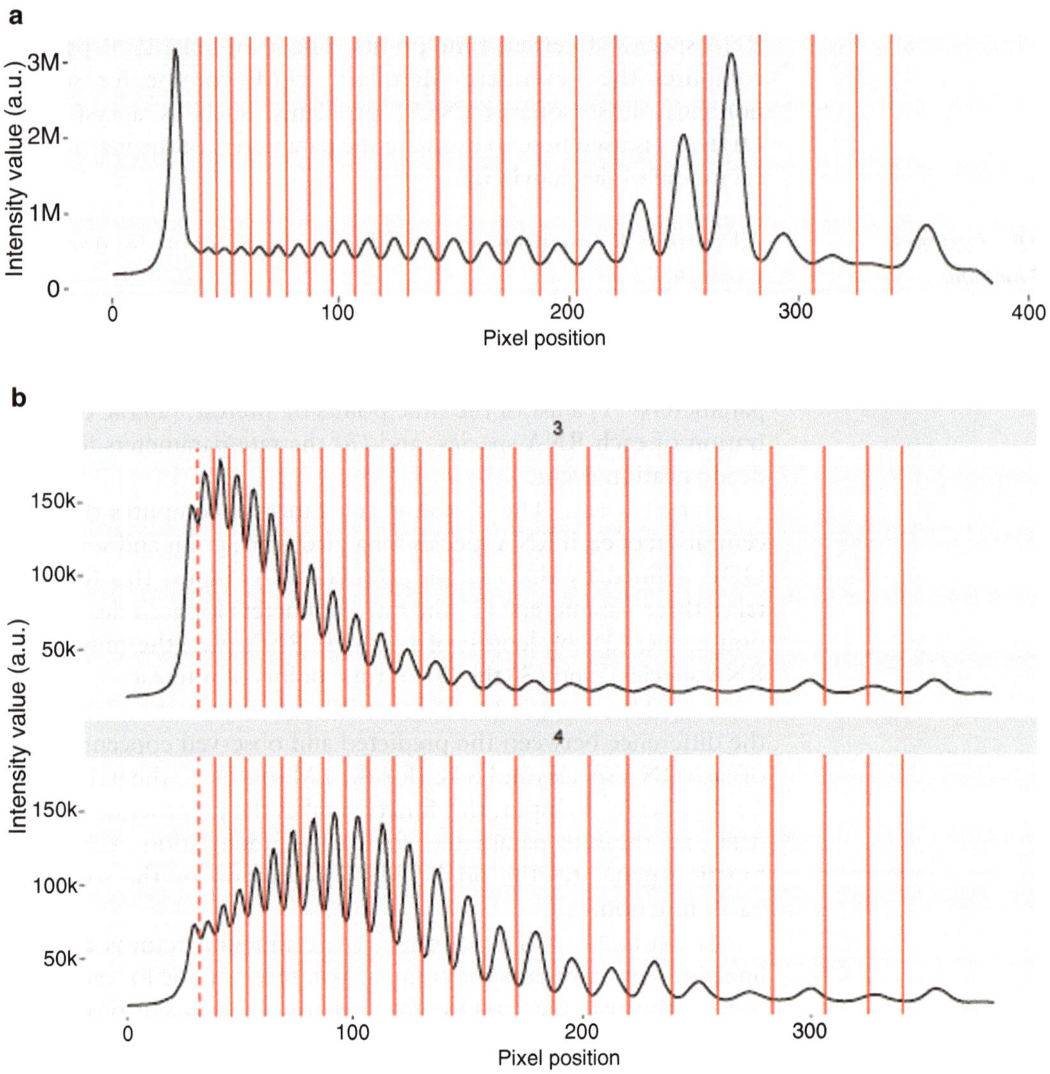

Fig. 2 Manual inspection of local maxima and minima. (**a**) The aggregated intensity values for each pixel row of the provided imaged processed text file (`converted.txt`) are shown as a line plot and generated by the `GetMetaPlot` function. Vertical red lines mark the predicted local minima. a.u.: arbitrary units; M: million. (**b**) The intensity values for lanes 3 and 4 are shown as line plots and generated by the `GetLanePlot` function. Vertical red lines mark the predicted local minima, and a vertical dashed line marks the manual addition. a.u.: arbitrary units; k: thousand

`GetStepPlot`: The `GetStepPlot` returns a step plot that represents the parameter estimate of deadenylation at each nucleotide position (Fig. 5d). The *x*-axis of the step plot indicates the nucleotide position, and the y-axis indicates the estimated rate of deadenylation. The error bars represent the standard error of the parameter estimation.

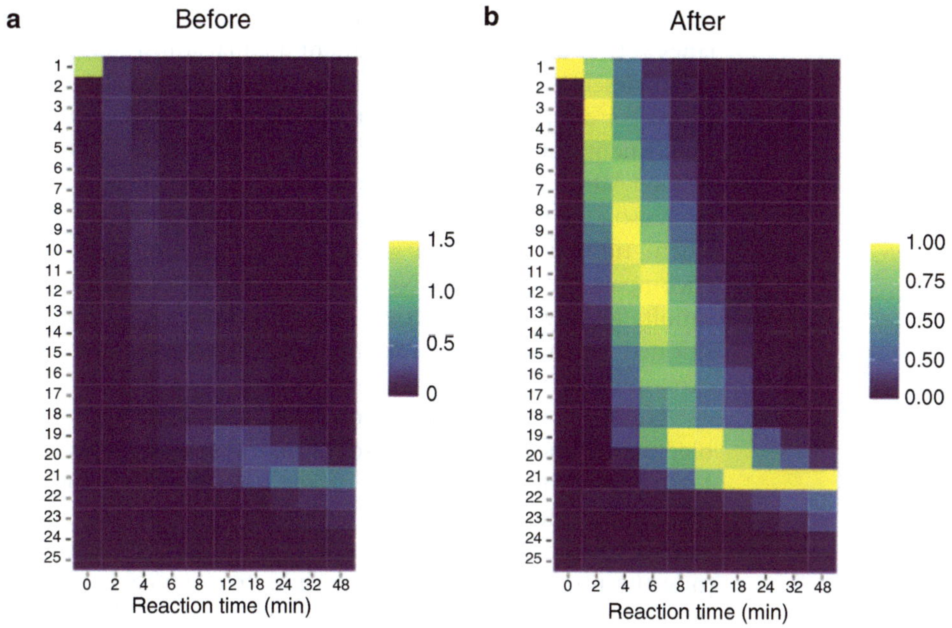

Fig. 3 Data digitization, normalization, and visualization. The heatmap of the band-level intensity values (**a**) before normalization in unit millions and (**b**) after lane-wise normalization, which is used for data visualization. Note that lane-wise normalization highlights the most abundant RNA species for a given reaction time point

3 Methods

The following computational approach involves two distinct procedures: (1) data digitization and (2) mathematical modeling. The data digitization step consists of data preprocessing (Fig. 2), data normalization, and data visualization (Fig. 3). The mathematical modeling step includes computer simulation and parameter estimation, as well as the mathematical assumptions and limitations. Example data analysis is available in the R script `script.R`.

3.1 Image Processing

Pixelated gel images (.tif) are converted to text files (.txt) using the image processing software Multi Gauge V3.0 (Fujifilm). Image inversion and contrast adjustments are done via the image software (*see* **Note 1**). Here, the image background is of lighter color and the gel bands are of darker color. RNA quantification is accomplished by densitometric analysis of the gel image such that each pixel row is given an intensity value of an arbitrary unit. The pixel rows near the gel bands are given high-intensity values according to the relative concentration of the RNA species. Each lane (i.e., column) is then processed with some manual adjustments and exported as a text file of intensity values for each pixel row and column.

To convert the two-dimensional image to a set of one-dimensional profiles, we utilize the *profile* option of the *measure* mode, which requests the user to enter the number of lanes (i.e., profiles) of interest. Count and include the marker and control lanes as they are used for subsequent manual adjustments. This step creates a selection frame that consists of the main quadrilateral, four circles for each corner of the quadrilateral, and a rectangle for each experimental lane. Drag each circle to fit the corners of the in vitro deadenylation experiments (*see* **Note 2**). Further, adjust the four circles so each rectangle covers its corresponding experimental lane. Then, tighten the width of these rectangles so that they only encompass the gel band. Note that all rectangles will be tightened simultaneously. The position of individual rectangles may be repositioned manually to capture the intensity values of the experimental lane.

Next, to ensure that each one-dimensional profile is comparable, further fine tuning of the selection frame is necessary. To do so, first, display the densitometric profile based on the adjusted selection frame. The selection of a set of lanes plots a line plot of the intensity values for each lane. Adjust the selection frame by dragging the top right circle to align the intensity peaks across profile lanes. The local peaks and troughs of the line plots should be positioned in parallel. We recommend that you start by selecting the marker lane and the last experimental lane that is near completion of the poly(A)-shortening process. For the provided example data, ensure that the intensity profiles of the third and fourth lanes are aligned to fine tune the intensity peaks at early deadenylation events. Finally, select all lanes and export the intensity values as a text (.txt) file containing intensity values for each pixel row and lane.

3.2 Data Preprocessing

Data preprocessing is applied to the `converted.txt` file for later data visualization and mathematical interpretation. The main objective of this step is to transform the pixel-level intensity values into interpretable values that represent the RNA abundance for each gel band. To accomplish this, this step involves automatic and manual trough (i.e., local minima) detection, data binning, and other manual optimizations (Fig. 2). Intensity metaplots and multi-panel line plots, which we describe below in detail, are useful data visualization tools that are made available to help with the manual curation process. Data preprocessing for the example data is made available in supplementary R script `script.R`.

Each column of the `converted.txt` file is the intensity profile of the corresponding experimental lane. As such, the local maximum values represent the RNA abundance for that gel band, and the local minima demarcate each gel band. We use these local minima to convert each gel band's pixel-level values to quantitative values. Following data import, set the number of expected minima (*num.troughs*) and the reaction time points (*timepoints*). For instance, the example data is of an A20 RNA substrate, so expect 24 troughs based on the image processing step (*see* **Note 3**). Ensure that the number of reaction time points matches the number of columns of the converted text file. The exact values of the reaction time points are used for parameter estimation. You may assign arbitrary negative integers for marker lanes filtered out in subsequent steps.

Execute the `GetMetaPlot` function to examine the automatic troughs (Fig. 2a). We recommend examining the metaplot before adding or removing pixel positions via manual inspection. The local minima detection algorithm aggregates the intensity values for each pixel row. Then it computes the discrete analog of the second derivative to find the local minima of the aggregated intensity values. The metaplot shows the pixel rows on the *x*-axis and the aggregated intensity values on the *y*-axis, where the vertical lines indicate the automatic troughs. The highest intensity peaks should indicate the control RNAs in the marker lane.

Manual inspection is needed to remove false and add missing *local minima* (*see* **Note 4**). Execute the `GetLanePlot` function to examine the intensity values for each experimental lane (Fig. 2b). It is critical to manually inspect the automatic troughs and confirm that they successfully demarcate the gel bands for each experimental lane. If the number of automatic and manual troughs is not equal to the number of expected minima (*num.troughs*), then the program returns a warning with the total number of automatic and manual troughs. For the example data where we expect the local minima detection algorithm to detect 24 local minima, 23 troughs, and all confirmed as true positives via a manual inspection of the metaplot (Fig. 2a). When examining the lane plots for lanes 3 and 4, we find that the algorithm missed an early local minimum that represents the RNA products of the first deadenylation reaction (Fig. 2b). For the example data, added a local minimum at the pixel position 33. Applying the final set of *local minima*, converts the pixel-level to band-level intensity values.

3.3 Data Normalization and Visualization

Data normalization rescales values to a common scale and enables quantitative comparison across experiments, such as when estimating mathematical model parameters (Fig. 3a). Here, we apply unity-based normalization (also known as min-max normalization) across experimental lanes but excluding the marker lane. As a result, the gel band with the highest intensity (e.g., the single band at 0 min reaction time) is assigned a normalized value of 1 and the gel band with the lowest intensity is set a normalized value of 0.

Heatmap is used to visualize the digitalized gel image where the *x*-axis represents the reaction time points (mins), the *y*-axis represents the 3′ end position from the initial RNA substrate, and the heatmap values represent the intensity values. For data visualization, lane-wise normalization is applied to highlight the most abundant RNA species for a given reaction time point (Fig. 3b). This approach highlights the mode of the poly(A) tail length at each

$$\frac{dA_{max}}{dt} = -\gamma_{max} A_{max}$$

$$\frac{dA_{max-1}}{dt} = -\gamma_{max} A_{max} - \gamma_{max-1} A_{max-1}$$

$$\vdots$$

$$\frac{dA_2}{dt} = \gamma_3 A_3 - \gamma_2 A_2$$

$$\frac{dA_1}{dt} = \gamma_2 A_2 - \gamma_1 A_1$$

$$\frac{dA_0}{dt} = \gamma_1 A_1$$

Fig. 4 Mathematical model of the poly(A) tail-shortening process. The system of ordinary differential equations (ODEs) and parameters used to model the deadenylation process. γ_i: deadenylation kinetics; A_i: RNA abundance

reaction time point and serves as a visual queue of the average deadenylation rate of the experiment. Expect a distribution of intensity values centered at the modal poly(A) tail length. That is because poly(A) tail shortening is an irreversible, stochastic process. Only accessing the modal poly(A) tail length overlooks the stochasticity from a sequence of deadenylation events. This completes the digitization step.

3.4 Mathematical Assumptions

We model the poly(A) tail shortening process as a sequence of independent deadenylation reactions (Fig. 4). Mathematically, this is equivalent to a first-order Markov process where each state variable (e.g., RNA abundance) only depends on the immediately preceding state. In the context of deadenylation, this model assumes that the amount of a particular RNA species of poly (A) tail length n is only determined by that of poly(A) tail length $n + 1$. And so, the mathematical model is described as a system of ordinary differential equations (ODE) such that each parameter γ_i represents the deadenylation kinetics at nucleotide position i.

3.5 Parameter Estimation

The mathematical model is dynamic, and its parameter estimation is a nonlinear least squares (NLS) problem. We use the Levenberg–Marquardt (LM) algorithm to estimate the parameters of the mathematical model of the poly(A) tail-shortening process. The LM algorithm computes an iterative sequence of optimization steps to minimize the residual or the difference between the observed and predicted values (Fig. 5a–c). Parameter estimation is accomplished by the `nls.lm` function. The `nls.lm` returns a list of (1) optimized parameters and the standard error of estimation, (2) the residual sum of squares, (3) the reason for termination, and other information related to the optimization procedure.

As for all NLS problems, parameter initialization is critical in reaching a local optimum. Poor initialization may lead to divergent optimization paths or may converge to non-feasible solutions. In practice, we find that using the average deadenylation rate as the initial value results in feasible solutions, which we confirm by computer simulation (*see* Subheading 3.7). For the example data, we initialize the parameters to 0.8 to reach a feasible solution. We recommend that users confirm whether this value is close to average deadenylation kinetics by visually inspecting the heatmap described in Subheading 3.3.

A critical step in optimization is defining the residual function, also known as the loss function, that computes the difference between the observed and predicted values. We accomplish this by utilizing the custom `ssq` function. The observed values are the normalized intensity values from the digitization step. As expected, these values are not updated during the optimization procedure. Unlike observed values, computer simulation generates predicted values, which requires a well-defined mathematical model and its associated parameters (*see* Subheading 3.7). We create an R construct that represents the mathematical model using the custom `rxnrate` function. Then, given the parameter values (e.g., initialized values), we compute the predicted values with the R construct described above and the `ode` function from the `deSolve` R package. Finally, the loss function returns the residual sum of squares between the observed and predicted values.

3.6 Data Visualization

Data visualization of the parameter estimates is accomplished with step plots representing the discrete nature of the poly(A) tail-shortening process (Fig. 5d). We use error bars to plot the standard error of each parameter estimate. The standard errors are computed based on the Hessian at the parameter estimates and represent the range in which the true parameter values reside. Non-overlapping error bars indicate a statistically significant change in deadenylation kinetics.

3.7 Computer Simulation

Computer simulation plays a crucial role in parameter optimization (*see* Subheading 3.5) and model validation (*see* Subheading 3.3). Model specification is accomplished by the custom `rxnrate` function that creates an R construct of the mathematical model. First, initial values for each state variable (i.e., RNA abundance) are set to represent reaction time point 0, where only the initial RNA substrate is present. Of note, this initial value is used to solve the initial value problem (IVP) of differential equations. For the example

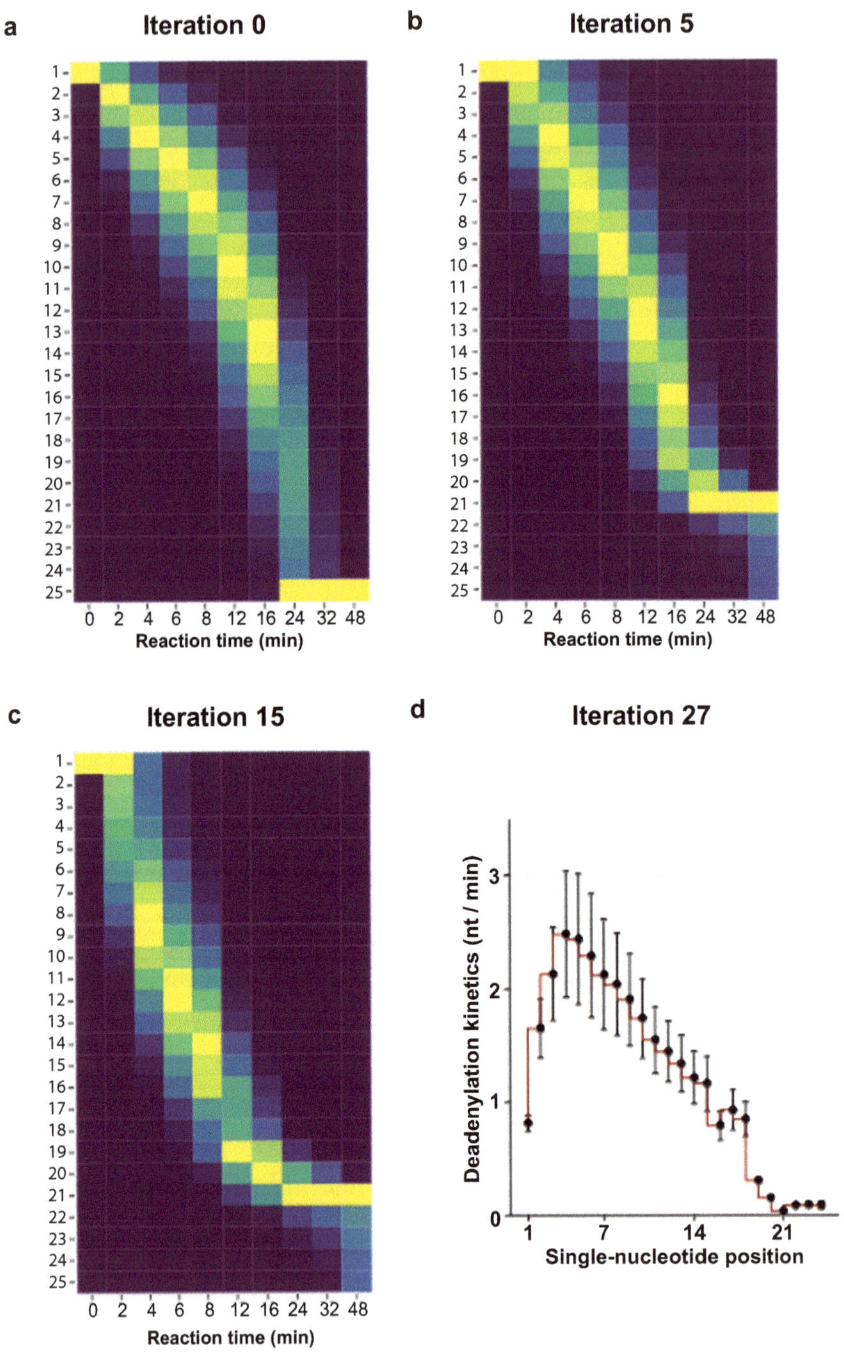

Fig. 5 Parameter estimation by the LM algorithm. A series of heatmaps representing the iterative steps of parameter optimization. Computer simulation based on parameter estimates at (**a**) iteration 0, (**b**) iteration 5, and (**c**) iteration 15. (**d**) The parameter estimates at convergence (iteration 27) are shown as a step plot and generated by the `GetStepPlot` function. Error bars represent the standard error of parameter estimation

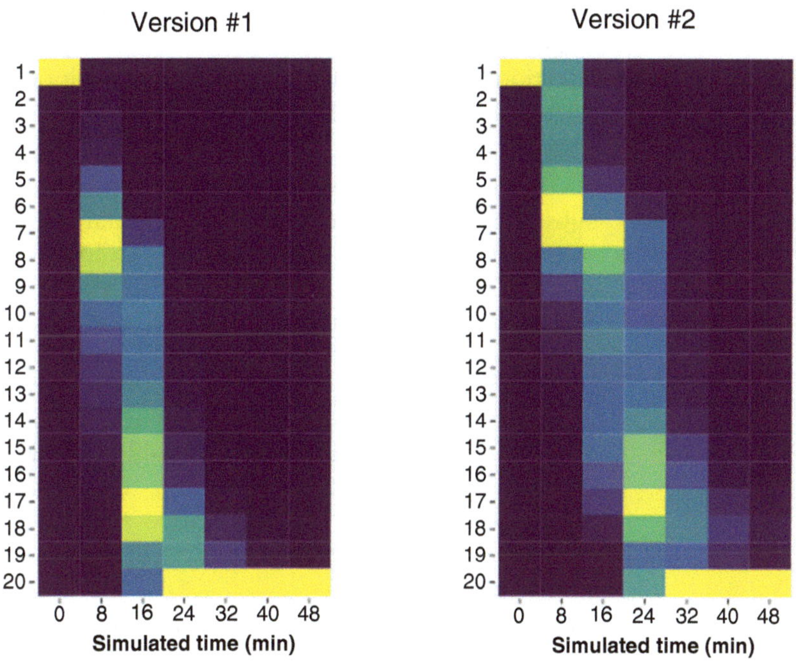

Fig. 6 Computer simulation of the poly(A) tail-shortening process. Heatmaps represent the computer simulation with two different set of arbitrary parameter values. See the provided R script `simulate.R` for details

data, the initial values are set to 1 for only the A20 RNA substrate and 0 for all other substrates. Then, given a set of parameters and a reaction time point of interest, the poly(A) tail-shorting process is simulated using the `ode` function. In all, the `ode` function takes as inputs (1) the R construct of the mathematical model, (2) the predefined or estimated parameters, (3) initial state values, and (4) reaction time point(s) and returns as output an R object of class `deSolve` which contains a matrix containing the values of each state at the given time points. An example computer simulation (Fig. 6) is available in the R script `simulate.R`.

4 Notes

1. Handling raw image data

   Depending on the original gel image data, you may have to invert and adjust the image to facilitate the image processing step. Use image processing software that supports these basic features for handling raw image data. For Multi Gauge V3.0 (Fujifilm), this feature is available in the *display* mode.

2. Adjusting the selection frame for lane profiling

   Take advantage of the marker lane when adjusting the selection frame. We recommend including a marker substrate that indicates the completion of the deadenylation process.

These markers are the A0, A1, and UCU substrates in the example data. There are four circles for each corner of the selection frame, and this substrate helps position the top and bottom left circles of the selection frame.

3. Counting the number of intermediate substrates

   It is essential to note the expected number of intermediate substrates before executing the computational steps. Each intermediate substrate corresponds to an individual gel band in the in vitro deadenylation experiment, which is then quantified as peaks of the intensity profiles. For the example data which use A20 substrates, we expect 25 peaks: one represents the initial substrate (i.e., A20), 20 peaks correspond to the subsequent removal of single adenosines, which includes A1 and A0, and four peaks indicate the additional degradation of the synthetic 3′UTR. Conversely, this corresponds to 24 expected minima which are used to guide the manual inspection step. For A30 substrates, there are 35 peaks and 34 troughs.

4. Manual inspection and curation of local minima

   Two variables add missing and remove false local minima in the main Rscript file `script.R`. The *add.list* variable is a list of pixel positions that are manually curated as local minima. The *remove.list* variable is a list of indexes of predicted minima from the local minima detection algorithm that are considered false troughs via manual inspection. For example, including the number 10 on this list removes the tenth predicted minimum.

Acknowledgements

We thank Yevgen Levdansky and Eugene Valkov for sharing the in vitro deadenylation data. This research was also supported by grant no. NRF-2021R1C1C1009282 and no. NRF-2021R1A4A303278912 from the Basic Science Research Program through the National Research Foundation of Korea, funded by the Ministry of Science and ICT (Y.-s.L.).

References

1. Darnell JE, Wall R, Tushinski RJ (1971) An adenylic acid-rich sequence in messenger RNA of HeLa cells and its possible relationship to reiterated sites in DNA. Proc Natl Acad Sci U S A 68:1321–1325
2. Lee SY, Mendecki J, Brawerman G (1971) A polynucleotide segment rich in adenylic acid in the rapidly-labeled polyribosomal RNA component of mouse sarcoma 180 ascites cells. Proc Natl Acad Sci U S A 68:1331–1335
3. Jacobson A, Favreau M (1983) Possible involvement of poly(A) in protein synthesis. Nucleic Acids Res 11:6353–6368
4. Carrazana EJ, Pasieka KB, Majzoub JA (1988) The vasopressin mRNA poly(A) tract is unusually long and increases during stimulation of vasopressin gene expression *in vivo*. Mol Cell Biol 8:2267–2274

5. Imataka H, Gradi A, Sonenberg N (1998) A newly identified N-terminal amino acid sequence of human eIF4G binds poly(A)-binding protein and functions in poly(A)-dependent translation. EMBO J 17:7480–7489
6. Wickens MP, Gurdon JB (1983) Post-transcriptional processing of simian virus 40 late transcripts in injected frog oocytes. J Mol Biol 163:1–26
7. Eckner R, Ellmeier W, Birnstiel ML (1991) Mature mRNA 3′ end formation stimulates RNA export from the nucleus. EMBO J 10: 3513–3522
8. Afonina E, Stauber R, Pavlakis GN (1998) The human poly(A)-binding protein 1 shuttles between the nucleus and the cytoplasm. J Biol Chem 273:13015–13021
9. Huez G, Marbaix G, Hubert E et al (1974) Role of the polyadenylate segment in the translation of globin messenger RNA in Xenopus oocytes. Proc Natl Acad Sci U S A 71:3143–3146
10. Berger SL, Cooper HL (1975) Very short-lived and stable mRNAs from resting human lymphocytes. Proc Natl Acad Sci U S A 72:3873–3877
11. Brewer G, Ross J (1988) Poly(A) shortening and degradation of the 3′ A+U-rich sequences of human c-myc mRNA in a cell-free system. Mol Cell Biol 8:1697–1708
12. Mangus DA, Evans MC, Jacobson A (2003) Poly(A)-binding proteins: multifunctional scaffolds for the post-transcriptional control of gene expression. Genome Biol 4:223
13. Passmore LA, Coller J (2022) Roles of mRNA poly(A) tails in regulation of eukaryotic gene expression. Nat Rev Mol Cell Biol 23:93–106
14. Dupressoir A, Morel AP, Barbot W et al (2001) Identification of four families of yCCR4- and Mg2+-dependent endonuclease-related proteins in higher eukaryotes, and characterization of orthologs of yCCR4 with a conserved leucine-rich repeat essential for hCAF1/hPOP2 binding. BMC Genomics 2:9
15. Piao X, Zhang X, Wu L, Belasco JG (2010) CCR4-NOT deadenylates mRNA associated with RNA-induced silencing complexes in human cells. Mol Cell Biol 30:1486–1494
16. Morita M, Suzuki T, Nakamura T et al (2007) Depletion of mammalian CCR4b deadenylase triggers elevation of the p27Kip1 mRNA level and impairs cell growth. Mol Cell Biol 27: 4980–4990
17. Bianchin C, Mauxion F, Sentis S et al (2005) Conservation of the deadenylase activity of proteins of the Caf1 family in human. RNA 11:487–494
18. Yamashita A, Chang T-C, Yamashita Y et al (2005) Concerted action of poly(A) nucleases and decapping enzyme in mammalian mRNA turnover. Nat Struct Mol Biol 12:1054–1063
19. Webster MW, Chen Y-H, Stowell JAW et al (2018) mRNA deadenylation is coupled to translation rates by the differential activities of Ccr4-not nucleases. Mol Cell 70:1089–1100.e8
20. Raisch T, Chang C-T, Levdansky Y et al (2019) Reconstitution of recombinant human CCR4-NOT reveals molecular insights into regulated deadenylation. Nat Commun 10:3173
21. Wiederhold K, Passmore LA (2010) Cytoplasmic deadenylation: regulation of mRNA fate. Biochem Soc Trans 38:1531–1536

Index

A

B

C

Eugene Valkov and Aaron C. Goldstrohm (eds.), *Deadenylation: Methods and Protocols*, Methods in Molecular Biology, vol. 2723,
https://doi.org/10.1007/978-1-0716-3481-3,

D

S

MIX
Papier aus verantwortungsvollen Quellen
Paper from responsible sources
FSC® C105338

If you have any concerns about our products,
you can contact us on
ProductSafety@springernature.com

In case Publisher is established outside the EU,
the EU authorized representative is:
Springer Nature Customer Service Center GmbH
Europaplatz 3, 69115 Heidelberg, Germany

Printed by Libri Plureos GmbH
in Hamburg, Germany